# Introduction to the Modelling of Marine Ecosystems

# Introduction to the Modelling of Marine Ecosystems

Second edition

## Wolfgang Fennel[1]
[1]Leibniz Institute for Baltic Sea Reseach,
Seestraße 15, D - 18119 Rostock, Germany,
Tel. +49-381-5197-110, e-mail: wolfgang.fennel@io-warnemuende.de

## Thomas Neumann[2,3]
[2]Leibniz Institute for Baltic Sea Reseach,
Seestraße 15, D - 18119 Rostock, Germany,
Tel. +49-381-5197-113, e-mail: thomas.neumann@io-warnemuende.de
[3]Marine Science and Technology Center, Klaipeda University
H. Manto 84, LT-92294 Klaipeda, Lithuania

**ELSEVIER**   AMSTERDAM · BOSTON · HEIDELBERG · LONDON · NEW YORK · OXFORD
PARIS · SAN DIEGO · SAN FRANCISCO · SINGAPORE · SYDNEY · TOKYO

Elsevier
Radarweg 29, PO Box 211, 1000 AE Amsterdam, Netherlands
The Boulevard, Langford Lane, Kidlington, Oxford OX5 1GB, UK
225 Wyman Street, Waltham, MA 02451, USA

Second edition 2015

**Notices**
Knowledge and best practice in this field are constantly changing. As new research and experience broaden our understanding, changes in research methods, professional practices, or medical treatment may become necessary.

Practitioners and researchers must always rely on their own experience and knowledge in evaluating and using any information, methods, compounds, or experiments described herein. In using such information or methods they should be mindful of their own safety and the safety of others, including parties for whom they have a professional responsibility.

To the fullest extent of the law, neither the Publisher nor the authors, contributors, or editors, assume any liability for any injury and/or damage to persons or property as a matter of products liability, negligence or otherwise, or from any use or operation of any methods, products, instructions, or ideas contained in the material herein.

ISBN: 978-0-444-63363-7

**British Library Cataloguing in Publication Data**
A catalogue record for this book is available from the British Library

**Library of Congress Cataloging-in-Publication Data**
A catalog record for this book is available from the Library of Congress

For information on all Elsevier publications visit
our web site at store.elsevier.com

This book has been manufactured using Print On Demand technology.

Working together
to grow libraries in
developing countries

www.elsevier.com • www.bookaid.org

# Contents

Please find the companion website at http://booksite.elsevier.com/9780444633637/.

# Preface

Ten years after the publication of the first version of this book, we present a revised second edition. The general aims and structure of the textbook outlined in the preface of the first edition are practically unchanged. We used the opportunity to edit this version, with several corrections and a few expansions. The examples of models are complemented by pelagic ecosystem models of the subarctic Atlantic. The scope of the modelling approaches is expanded by the inclusion of a mass class-dependent fish model. As in the original version, the tutorial aspect of the book was supported by the booksite with MATLAB codes, which allow the reader to reproduce most of the findings in Chapters 2–4.

**Warnemünde, February 2014**
**Wolfgang Fennel and Thomas Neumann**

## Preface of the First Edition

During the last few decades, the theoretical research on marine systems, particularly in numerical modelling, has developed rapidly. A number of biogeochemical models, population models and coupled physical chemical and biological models have been developed and are used for research. Although this is a rapidly growing field, as documented by the large number of publications in the scientific journals, there are very few textbooks dealing with modelling of marine ecosystems. We found in particular that a textbook giving a systematic introduction to the modelling of marine ecosystem is not available and, therefore, it is timely to write a book that focuses on model building, and that helps interested scientists to familiarize themselves with the technical aspects of modelling and start building their own models.

The book begins with very simple first steps of modelling and develops more and more complex models. It describes how to couple biological model components with three-dimensional circulation models. In principle, one can continue to include more processes into models, but this would lead to overly complex models as difficult to understand as nature itself. The step-by-step approach to increasing the complexity of the models is intended to allow students of biological oceanography and interested scientists with only limited experience in mathematical modelling to explore the theoretical framework. The book may also serve as an introduction to coupled models for physical oceanographers and marine chemists.

We, the authors, are physicists with some background in theory and modelling. However, we had to learn ecological aspects of marine biology from the scientific literature and discussions with marine biologists. Nevertheless, when physicists are dealing with biology, there is a danger that many aspects of biology or biogeochemistry are not as well represented as experts in the field may expect. On the other hand, the most important aim of this text is to show how model development can be done. Therefore, the textbook concentrates on the approach of model development, illustrating the mathematical aspects and giving examples. This tutorial aspect is supported by a set of MATLAB programmes on the booksite, which can be used to reproduce many of the results described in Chapters 2–4.

For many discussions, in particular for the coupling of circulation and biological models, we have to choose example systems. We used the Baltic Sea, which can serve as a testbed. Hence, the models are not applicable to all systems because there are always site-specific aspects. On the other hand, the models have also some universal aspects and the understanding of the approach can be a useful guide for model development in other marine areas. The book was not intended to give a comprehensive overview of all existing models, and only a subset of papers on modelling was quoted.

We owe thanks to many colleagues—in particular from our institute, the Baltic Sea Research Institute (IOW)—for helpful discussions and invaluable support. We are indebted to Prof. Oscar Schofield and Dr. Katja Fennel for valuable comments on a draft version of the manuscript, and we thank Leon Tovey for his careful proofreading of the English. We further thank Thomas Fennel for helping with the production of several of the figures.

**Warnemünde, January 2004**
**Wolfgang Fennel and Thomas Neumann**

# Chapter 1

# Introduction

## 1.1 MODELS OF MARINE ECOSYSTEMS

Understanding and quantitatively describing of marine ecosystems requires an integration of physics, chemistry and biology. Coupling biology and physical oceanography in models has many attractive features: We can do experiments with models of marine systems while the real system can only be observed in the state at the time of the observation. We can also employ the predictive potential of models for applications such as environmental management or, on a larger scale, we can study past and future developments with the aid of experimental simulations. Moreover, a global synthesis of sparse observations can be achieved by using coupled three-dimensional models to extrapolate data in a coherent manner.

However, running complex coupled models requires substantial knowledge and skill. To approach the level of skill needed to work with coupled models, it is reasonable to proceed step-by-step from simple to complex problems.

What are biological models? We use the term 'model' synonymously for theoretical descriptions in terms of sets of differential equations that describe the food web dynamics of marine systems. The food webs are relatively complex systems, which can sketched simply as a flux of material from nutrients to phytoplankton to zooplankton to fish and recycling paths back to nutrients. Phytoplankton communities consist of a spectrum of microscopic single-cell plants, microalgae. Many microalgae in marine or freshwater systems are primary producers, which build up organic compounds directly from carbon dioxide and various nutrients dissolved in the water. The captured energy is passed along to components of the aquatic food chain through the consumption of microalgae by secondary producers, the zooplankton. The zooplankton in turn is eaten by fish, which is caught by man. Moreover, there are pathways from the different trophic levels to nutrients through respiration, excretion and dead organic material, detritus, which is mineralized by microbial activity. The regenerated nutrients can then again fuel primary production.

It is obvious that the complex network of the full marine food web cannot practically be covered by one generic model. There are many models that were developed for selected, isolated parts of the food web. For these models the links to the upper or lower trophic levels must be prescribed or parameterized

Introduction to the Modelling of Marine Ecosystems. http://dx.doi.org/10.1016/B978-0-444-63363-7.00001-5

effectively in an appropriate manner. In order to construct such models, we may consider a number of individuals or we can introduce aggregated state variables to characterize a system. State variables must be well defined and measurable quantities, such as concentration of nutrients and biomass or abundance (i.e. number of animals per unit of volume). The dynamics of the state variables (i.e. their change in time and space), is driven by processes, such as nutrient uptake, respiration or grazing, as well as physical processes such as light variations, turbulence and advection.

Ecosystem models can be characterized roughly by their complexity, i.e. by the number of state variables and the degree of process resolution. The resolution of processes can be scaled up or down by aggregation of variables into a few integrated ones or by increasing the number of variables, respectively. For example, zooplankton can be considered as a bulk biomass or can be described in a stage resolving manner. Models with very many state variables are not automatically better than those with only a few variables. The higher the number of variables, the larger the requirements of process understanding and quantification. If the process rates are more or less guesswork, there is no advantage to increase the number of poorly known rates and parameters. Moreover, not every problem requires a high process resolution, and the usage of a subset of aggregated state variables may be sufficient to answer specific questions. Models should be only as complex as required and justified by the problem at hand. Model development needs to be focused, with clear objectives and a methodological concept that ensures that the goals can be reached.

Alternatively, models can be characterized also by their spatial dimension, ranging from zero-dimensional box models to advanced three-dimensional models. In box models, the physical processes are largely simplified while the resolution of chemical biological processes can be very complex. Such models are easy to run and may serve as workbenches for model development. The next step is one-dimensional water column models, which allow a detailed description of important physical controls of biological processes by, for example, vertical mixing and light profiles. Such models may be useful for systems with weak horizontal advection. In order to couple the biological models to full circulation models, it is advisable to reduce the complexity of the biological representations as far as reasonable. If, for example, advection plays an important part for the life cycles of a species, the biological aspects may be largely ignored. Extreme cases of reduced biology coupled to circulation models are simulations of trajectories of individuals, cells or animals, which are treated as passively drifting particles.

The process of constructing the equations, i.e. building a model, will be described in the following chapters, starting with simple cases followed by increasingly complex models. Ecologists often are uncomfortable with the numerous simplifying assumptions that underlie most models. However, modelling marine ecosystems can benefit from looking at theoretical physics, which demonstrates how deliberately simplified formulations of cause–effect relationships help to reproduce predominant characteristics of some distinctly identifiable, observable phenomena. It is illuminating to read the viewpoint of G.S. Riley, one of the pioneers in modelling marine plankton, to this problem, as mentioned in his famous paper (Riley, 1946). He wrote: '...physical oceanography, one of the youngest branches in the actual years, is more mature than the much older study of marine biology. This is perhaps partly due to the complexities of the material. More important, however, is the fact that physical oceanography has aroused the interest of a number of men of considerable mathematical ability, while on the other hand marine biologists have been largely unaware of the growing field of bio-mathematics, or at least they have felt that the synthetic approach will be unprofitable until it is more firmly backed by experimental data'.

There is another important issue that deserves some consideration. The biological model equations are basically nonlinear and, in general, not analytically solvable. Thus, early attempts to model marine ecosystems were retarded or even stopped by mathematical problems. These difficulties could only be resolved by numerical methods that require computers, which were not available in an easy-to-use way until the 1980s. This frustrating situation may be one of the reasons that many marine biologists or biological oceanographers were not very much interested in mathematical models in the early years.

With the advent of easily accessible computers, the technical problems were removed. However, it seems the interdisciplinary discussion on modelling develops slowly. Some biologist are doubtful whether anything can be learned from models, but a growing community sees a beneficial potential in modelling. Why do we need models? The reasons include,

- to develop and enhance understanding,
- to quantify descriptions of processes,
- to synthesize and consolidate our knowledge,
- to establish interaction of theory and observation,
- to develop predictive potential,
- to simulate scenarios of past and future developments.

Models are mathematical tools by which we analyse, synthesize and test our understanding of the dynamics of the system through retrospective and predictive calculations. Comparison with data provides the process of model validation. Owing to problems of observational undersampling of marine systems, data are often insufficient when used for model validation. Validation of models often amounts to fitting the data by adjusting parameters, i.e. calibrating the model. It is important to limit the number of adjustable parameters, because a tuned model with too many fitted parameters can lose any predictive potential. It might work well for one situation but could break down when applied to another case. Riley stated this more than 50 years ago, when he wrote '...(analysis based on a model, expressed by a differential equation,)... is a useful tool in putting ecological theories to a quantitative test. The disadvantage is that it requires detailed quantitative information about many processes, some of which are only poorly understood. Therefore, until more adequate knowledge is obtained, any application of the method must contain some arbitrary assumptions and many errors due to over-simplification' (Riley, 1947).

Because many modelers have backgrounds in physics and mathematics, ecosystem modelling is an interdisciplinary task that requires a well-balanced dialogue of biologists and modelers to address the right questions and to develop theoretical descriptions of the processes to be modelled. The development of the interdisciplinary dialogue is a process which should start at the universities where students of marine biology should acquire some theoretical and mathematical background to be able to model marine ecosystems or to cooperate closely with modelers. The main goal of this book is to help to facilitate this process.

## 1.2 MODELS FROM NUTRIENTS TO FISH

If, by assumption, all relevant processes and transformation rates of a marine ecosystem are known and formulated by a set of equations, we might expect to be able to predict the state of an ecosystem by solving the equations for given external forces and initial conditions. However, this amounts to an interesting philosophical question, which was considered by Laplace in the context of the physics of many-particle systems. Assuming that there is a superhuman ghost (Laplace's demon) who knows all initial conditions for every single particle of a many particle system, such as an ideal classical gas, then the future state of the many-particle system could be predicted by integration of the equations of motion. The analysis of this problem had shown that many-particle systems can be treated reasonably only by introducing statistical methods. In chemical-biological systems, the problems are even much more complex than in a nonliving system of many particles. The governing equations of the chemical-biological dynamics cannot be derived explicitly from conservation laws as in Newton's mechanics or in geophysical fluid dynamics. The biological equations must be derived from observations. Experimental findings must be translated into mathematical formulations, which describe the processes. Parameterizations of relationships are necessary to formulate process rates and interactions between state variables in the framework of mathematical models.

### 1.2.1 Models of Individuals, Populations and Biomass

In nature, the aquatic ecosystems consist of individuals (unicellular organisms, copepods, fish), which have biomass and form populations. Different cuts through the complex network of the food web with its many facets are motivated by striving for answers to specific questions with the help of models. These cuts may amount to different types of models, which look on individuals, populations or biomass. Therefore, it is not only reasonable, but the only tractable way, to reduce the complexity of the nature with models. There are several models types:

- models of individuals,
- population models,
- biomass models,
- combined individual and population models,
- models combining aspects of populations, biomass and individual dynamics.

Individual-based models consider individuals as basic units of the biological system, while state-variable models look at numbers or mass of very many individuals per unity of volume. Individual-based models can include genetic variations among individuals while population and biomass models use mean rates and discard individual variations within ensembles of very many individuals of one or several groups. Although individuals are basic units of ecosystems, it is not always necessary to resolve individual properties. State-variable models represent consistent theories, provided the state variables are well-defined, observable quantities. The rates used in the model equations should be, within a certain range of accuracy, observable and independently reproducible quantities.

There is no general rule for how far a system can be simplified. An obvious rule of thumb is that models should be as simple as possible and as complex as necessary to answer specific questions. Simple models with only one state variable, which refers to a species of interest, may need only one equation but are, to a larger extent, data driven compared to models with several state variables. For example, in a one-species model with one state variable, the variation of process rates in relation to nutrients requires externally prescribed nutrient data, such as maps in terms of monthly means derived from observations. Similarly, particle-tracking models for copepods or fish larvae that include some biology, such as individual growth, require prescribed data of food or prey distributions.

Models with several state variables require more equations and more parameters, which must be specified. However, such models can describe dynamical interactions between the state variables, e.g. food resources and grazers or predator–prey interaction, in such a way that the variations of the rates can be calculated consistently within the model system. Models connecting elements of biomass and/or population models with aspects of individual based models can be internally more consistent, with less need for externally prescribed parameter fields than one-equation models. Owing to the rapid developments in computer technology, the higher CPU demands of more complex models are no longer a serious obstacle.

Models are being used for ecological analysis, quantification of biogeochemical fluxes and fisheries management. One class of models looks at the lower part of the food web, i.e. from nutrients to the zooplankton. The model food web is truncated at a certain level, e.g. zooplankton, by parameterization of the top-down-acting higher trophic levels. In particular, predation by fish is implicitly included in zooplankton mortality terms.

There is a long list of biological models that describe the principal features of the plankton dynamics in marine ecosystems. We quote some early papers that can be considered as pioneering work in modelling: (Aksnes and Lie, 1990; Cushing, 1959; Evans and Parslow, 1985; Jansson, 1976; Radach and Moll, 1990; Riley, 1946; Sjöberg and Willmot, 1977; Sjöberg, 1980; Steele, 1974; Wroblewski and O'Brien, 1976). A widely used model, in particular in the context of JGOFS projects (Joint Global Flux Studies), is the chemical-biological model of Fasham et al. (1990). One of the most complex chemical-biological models with very many components is the so-called ERSEM (European Regional Seas Ecosystem Model), which was developed for the North Sea; see, e.g. Baretta-Bekker et al., 1997; Ebenhöh et al., 1997. Extensive lists of references can be found in Fransz et al. (1991), Totterdell (1993), Moll and Radach (2003) and Travers et al. (2007). Individual-based models may focus on phytoplankton cells or copepods (e.g. Batchelder et al., 2002; Miller et al., 1998; Woods and Barkmann, 1994) or larvae and fish (e.g. Hinrichsen et al., 2001); see also the books of DeAngelis and Gross (1992), Grimm and Railsback (2005) and references therein. Another class of models describes the fish stock dynamics, which largely ignores the bottom-up effect of the lower part of the food web (see, e.g. Gulland, 1974; Rothschild, 1986).

## 1.2.2  Fisheries Models

Fishery management depends on scientific advice for management decisions. One question that is not easily answered is: How many fish can we take out of the sea without jeopardizing the resources? But there are also the economic conditions of fisherman and fishery industries that make management decisions complex and sometimes difficult. Making predictions implies the use of models of fish population.

Fish stock assessment models may be grouped into analytical models and production models (e.g. Gulland, 1974; Rothschild, 1986). In analytical models, a fish population is considered as sum of individuals, whose dynamics is controlled by growth, mortality and recruitment. Some knowledge of the life cycle of the fish is usually taken into account. A basic relation describes the decline of fish due to total mortality, which is the sum of fishery mortality and the natural death rate. The estimated abundance can be combined with the individual mass (weight) at age classes to assess the total biomass.

In analytical models, a fish population of abundance, $n$, is considered as the sum of individuals, whose dynamics is controlled by growth, mortality and recruitment. A basic relation used is

$$\frac{dn}{dt} = -\phi n - \mu n = -\zeta n,$$

where $\zeta = \phi + \mu$ is the total mortality consisting of fishery mortality, $\phi$, and natural mortality, $\mu$. The integration is easy and gives, for an initial number of individuals, $n_0$,

$$n(t) = n_0 \exp(-\zeta t).$$

This can be combined with the individual mass (weight), $w$, to estimate the corresponding biomass. The individual mass obeys an equation of the type $\frac{d}{dt} w = \gamma w$ with the solution $w = w_0 \exp(\gamma t)$, where $w_0$ is the initial mass and the effective growth rate, $\gamma$, depends on stages or year classes. The growth equation for the individual mass has a different form when the Bertalanffy approach is used, $\frac{d}{dt} w = \alpha w^{2/3} - \beta w$. These types of models are population models combined with models of individual growth (weight at age).

Production, or logistic, models treat fish populations as a whole, considering the changes in total biomass as a function of biomass and fishing effort without explicit reference to its structure, such as age composition. The equations are written for the biomass $B$,

$$\frac{dB}{dt} = f(B),$$

where the function, $f(B)$, follows after some scaling arguments as $f(B) = \alpha B(B_0 - B)$, where $\alpha$ is a constant. It is clear that the change, i.e. the derivative of the biomass, tends to zero for both zero biomass and a certain equilibrium value, $B_0$, where the population stabilizes. This model type is basically a biomass model.

A further separation of models can be made into single species and or multi-species models. Multispecies models taken trophic interactions between different species into account, which are ignored in single-species population models. Fishery models largely ignore the linkages to lower trophic levels. In particular, environmental data and other bottom up information are widely disregarded.

Usually, fishery models depends on data to characterize the stock. The external information comes from surveys, e.g. acoustic survey of pelagic species, and catch per unit of effort from commercial fishery. However, these data are often poorly constrained and may involve uncertainties due to undersampling and other well-known other reasons. The consistency of the data may be improved by statistical methods and incorporating biological parameters that are independent of catch data. While biomass-based production models require integrated data, such as catch and fishery effort, the age-structured multispecies models need much more detailed information (e.g. Horbowy, 1996). Though fishery models truncate more or less the lower part of the food chain, they assimilate fishery data, which carry a lot of implicit information included in the used observed data.

While life cycles of fish may span several years, the time scales of the environmental variations are set by seasonal cycles and annual and interannual variations. Modelling efforts, which include bottom-up control, have to deal with different processes at different time scales and have to take the memory effects of the ecosystem into account. The integration of 'bottom-up' modelling into models used in fishery management and stock assessment is a difficult task and poses a modelling challenge that needs further research. Today, it seems to be feasible to involve those aspects of environmental control into fisheries models, which act directly on the trophic level of fish. Examples are perturbations of the recruitment by oxygen depletion or by unusual dispersion of fish eggs by changed wind patterns in areas that are normally retention regions.

Another field of modelling addresses fish migration. This is a typical example of individual-based modelling, where the trajectories can be computed from archived data or directly with a linked circulation model. Aspects of active motion of the individuals' swimming, swarming and ontogenetic migration can be combined with the motion of water. It is not easy to find model formulations that govern the behavioural response to environmental signals, e.g. reaction to gradients of light, temperature, salinity, prey, avoidance of predation, etc. The individual-based models consist of two sets of equations, one set to prescribe the development of the individual in given environments and the second to determine the trajectories by integrating velocity field, which can be provided by a circulation model or as archived data.

The main focus of the following chapters of this book will be the lower part of the food web, truncated at the level of zooplankton. Fish, then, play only an indirect role by contributing to the mortality of zooplankton. However, in several parts of this book, we will also show how biogeochemical and fish models can be connected in a two-way interaction mode.

### 1.2.3 Unifying Theoretical Concept

As mentioned, there are several classes of models, including individual-based models, population models, biomass models and combinations thereof. The question of whether there is unifying theoretical concept that connects the different classes of models was addressed; in, e.g., DeAngelis and Gross, (1992) and Grimm, (1999). In the food web, there are individuals (cells, copepods, fish) at different trophic levels that mediate the flux of material. The individuals have biomass and form populations that interact up, down and across the food web. The state variables' biomass concentration or abundance represent averages over very many individuals. Hence, a unifying approach should start with individuals and indicate by which operations the individuals are integrated to form state variables.

The state of an individual, e.g. phytoplankton cell, copepod, fish, can be characterized by the location in the physical space, $\mathbf{r}_{ind}(t)$, and by the mass, $m_{ind}(t)$. More parameters may be added to describe the biological state, when required. However, in the following, we restrict our considerations to the

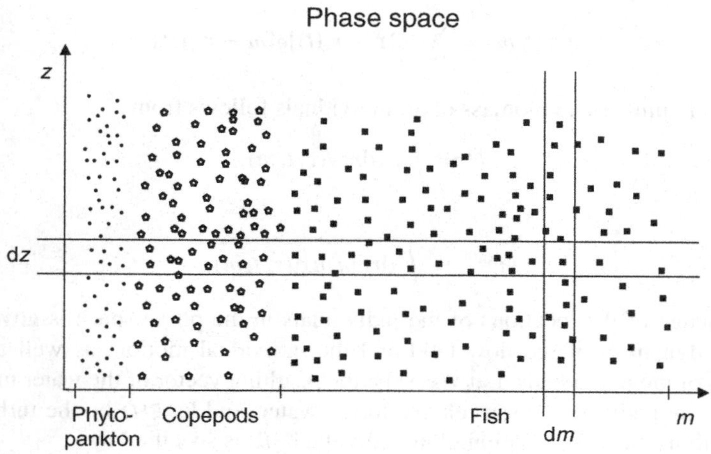

## Phase space

**FIGURE 1.1** Illustration of the concept of the phase space for the two-dimensional $(z, m)$ example. Each point refers to one individual. The size classes are reflected in terms of mass.

parameter 'individual mass'. We define an abstract 'phase space' that has the four dimensions, $(x, y, z, m)$. Each individual occupies a point in the phase space for every moment, $t$. Different slices through the phase space can look at a set of individuals, a population in terms of abundance, or represent functional groups by their biomass. An example of the phase space for the two dimensions, $z$ and $m$, is sketched in Fig. 1.1. The points change their locations in the phase space with time due to physical motion and growth and will move along a trajectory. Underneath the cloud of points, which characterize the individuals, lies a set of continuous fields representing physical quantities such as temperature, currents and light, as well as chemical variables, such as nutrients. These fields are independent of $m$, i.e. they do not vary parallel to the $m$-axis in the phase space. The state density of one individual can be defined as

$$\varrho(\mathbf{r}, t, m) = \delta(\mathbf{r} - \mathbf{r}_{ind}(t))\delta(m - m_{ind}(t)). \tag{1.1}$$

Here we use Dirac's delta function, which is defined by

$$\delta(x) = \begin{cases} \infty & x = 0 \\ 0 & x \neq 0, \end{cases}$$

in such a way that

$$\int dx \delta(x) = 1.$$

A system of $N$ individuals of a species represents a cloud of points in the phase space for every moment, $t$. The cloud will change its shape with time. This approach uses concepts known in the statistical theory of gases and plasmas; (see Klimontovich, 1982). We can write the state density for a many-individual system as sum of the state densities of the individuals

$$\varrho(\mathbf{r}, t, m) = \sum_{i=1}^{N} \delta[\mathbf{r} - \mathbf{r}_i(t)]\delta[m - m_i(t)]. \tag{1.2}$$

The total number and biomass of all individuals follows from

$$N = \int \mathrm{d}\mathbf{r}\mathrm{d}m \varrho(\mathbf{r}, t, m),$$

and

$$m^{\text{total}} = \int \mathrm{d}\mathbf{r}\mathrm{d}m\, m\, \varrho(\mathbf{r}, t, m).$$

The change of the locations of the individuals in the phase space is governed by the dynamics of the flow field and the individual motion, as well as the growth of the individuals. Let $\mathbf{v}(\mathbf{r}, t)$ be the resulting vector of the water motion and the individual's motion relative to the water, and let $\xi(t)$ be the turbulent fluctuations, then the location of an individual, '$i$', is specified by

$$\mathbf{r}_i(t) = \int_0^t \mathrm{d}t'(\mathbf{v}(\mathbf{r}_i, t') + \xi(t')). \tag{1.3}$$

To describe the growth of individuals, we may consider the example of the development of a copepod from the egg to the adult stage, as governed by an equation of the type

$$\frac{\mathrm{d}}{\mathrm{d}t}m_i = (g - l)m_i^p,$$

where the rate $g$ stands for ingestion and $l$ for losses and $p$ is an allometric exponent less than unity.

Changes in abundance are obtained by integrating a set of model equations for many individuals and taking averages after certain time intervals. Another way is to introduce the population density, $\sigma$, for a certain parcel of water, $\Delta V$. The population density is related to the state density by

$$\sigma(\mathbf{r}_V, m, t) = \frac{1}{\Delta V}\int_{\Delta V} \mathrm{d}\mathbf{r}\varrho(\mathbf{r}, m, t), \tag{1.4}$$

where $\mathbf{r}_V$ is the location of the parcel, $\Delta V$, in the sense of hydrodynamics. We drop the index, '$V$', in the following discussion. Then $\sigma(m, t)\mathrm{d}m$ is the number of individuals in the mass interval $(m, m + \mathrm{d}m)$ in a considered parcel. For a Lagrangian parcel of fluid, moving with the flow, the number of individuals is controlled by birth and death rates as well as by active motion, swimming or sinking, of the individuals, which enables them to leave or enter the parcel.

The state variable 'abundance', i.e. the number of individuals per unit volume, $\Delta V$, follows from Equation (1.4) as

$$n(\mathbf{r}, t) = \int_{m_{\min}}^{m_{\max}} \mathrm{d}m\, \sigma(m, t), \tag{1.5}$$

and the state variable 'biomass concentration' follows from

$$B(\mathbf{r}, t) = \int_{m_{min}}^{m_{max}} dm\, m\, \sigma(m, t). \tag{1.6}$$

Thus, by different choices of integrations along the mass axis, different state variables are defined, including bulk state variables, which lump together several groups. Such state variables also represent continuous fields that are, after the integration, independent of the mass variable.

Dynamic changes of the number and biomass of individuals is driven through cell division (primary production) and reproduction (egg-laying female adults), as well as mortality (natural death, hunger, predator–prey interaction, fishery, etc.). The processes can be formulated for each individual. However, if we are not interested in the question of which particular individuals die, divide or lay eggs, the process can be prescribed by rates that define which percentage of a population dies, divides or lays eggs.

In population models, the state variables are abundance, i.e. number of individuals per unity of volume, and the change of the abundance, $n(\mathbf{r}, t)$, is driven by births and death rates as well as immigration and emigration,

$$\frac{d}{dt}n = (r_{birth} - r_{deaths})n + \text{migration-flux}, \tag{1.7}$$

while for biomass models the change of concentration, i.e. the mass per unity volume, is considered as driven by processes such as nutrient uptake, grazing, mortality. This type of model was used by Riley in his pioneering study (Riley, 1946).

The introduction of state densities, which take into account the individual nature of the species at different levels of the food web, allows a unified approach where population and biomass model emerge from individuals. The link to hydrodynamics is established by introducing the population density as integral over a volume element. This volume element can be considered either as a Lagrangian parcel of water, moving with the flow, or a fixed Eulerian cell. The dynamics of water motion in the volume element is governed by the equations of the fluid dynamics.

## 1.2.4 The Plan of the Book

In the following chapters we will describe the development of models of marine ecosystems in a step by step approach. Chapters 2, 3, and 4 deal with box models, where the role of physics is largely reduced and only indispensable physical control mechanisms are involved by suitable parameterizations, e.g. the role of light is parameterized by the length of the day and vertical stratification by two or three vertically stacked boxes with prescribed exchange rates. State variables will be introduced and their mutual chemical-biological interaction, and later, predator–prey interaction processes will be outlined and quantified.

We start with truncated model food webs and then introduce higher parts of the food chain, i.e. life-cycle resolving model of copepods and fish. The models are available as MATLAB codes so that most of the shown results can be reproduced by running the model codes.

A more detailed description and discussion of physical biological interaction follows in Chapter 5. Remaining Chapters 6 and 7 are devoted to the coupling of the chemical-biological model with a full three-dimensional circulation model, starting with embedding a biogeochemical model, truncated at the level of zooplankton. The coupling of a stage-resolving zooplankton model and a mass-class structured fish model is addressed in Chapter 7. The remainder of the book deals with technical issues, such as an short introduction to MATLAB as far as it is required to reproduce the results in the first four chapters.

Chapter 2

# Chemical-Biological Models

Biogeochemical models provide tools to describe, understand and quantify fluxes of matter through the food web or parts of it, and interactions with the atmosphere and sediments. Such models involve very many individuals of different functional groups or species and ignore differences among the individuals. The role of phytoplankton or zooplankton is reduced to state variables, which carry the summary effects for any chemical and biological transformations. Ideally, the state variables and the process rates should be well defined, observable and, within a certain range of accuracy, independently reproducible quantities. However, there are rates such as mortality, that are hard to quantify. Thus, models may contain some poorly constrained rates that are established *ad hoc* by reasonable assumptions. This and the next two chapters deal basically with box models, which look at spatially averaged state variables. We describe important functional relationships for the chemical-biological dynamics and consider approaches to construct biomass models. Starting with simple models, it is demonstrated how the complexity of the model dynamics can be developed step-by-step.

## 2.1 CHEMICAL-BIOLOGICAL PROCESSES

In order to illustrate the process of model development we start with the construction of simple models that describe parts of the ecosystem. As opposed to physics, where the model equations are mathematical formulations of basic principles, the model formulations for ecosystems have to be derived from observations. Ecological principles such as the Redfield ratio (Redfield et al., 1963), i.e. a fixed molar ratio of the main chemical elements in living cells, Liebig's law (de Baar, 1994), and size considerations (the larger ones eat the smaller ones), provide some guidance to constructing the equations. However, they do not define the mathematical formulations needed for modelling.

First, we have to identify a clear goal in order to define what we wish to describe with the model. The example we are going to use in this chapter aims at the description of the seasonal cycle of phytoplankton in mid-latitudes. We wish to formulate models that help to quantify the transfer of inorganic nutrients through parts of the lower food web and to estimate changes of biomass

Introduction to the Modelling of Marine Ecosystems. http://dx.doi.org/10.1016/B978-0-444-63363-7.00002-7
**13**

in response to changes in external forcing. Second, we have to determine what state variables and processes must be included in the model and which mathematical formulation describes the processes.

A model is characterized by the choice of state variables. The state variables are concentrations or abundance, which can be quantified by measurements. They depend usually on time and space coordinates, and their dynamics are governed by processes that are, in general, functions of space, time and other state variables specific to the system being studied. State variables consist of a numerical value and a dimension, e.g. mass per volume or number of individuals per volume. The dimension is expressed in corresponding units, such as 'mmol m$^{-3}$' or number of cells per litre. The processes are usually expressed by rates with the dimension of inverse time, e.g. one per day, d$^{-1}$, or one per second, s$^{-1}$.

### 2.1.1 Biomass Models

We start with a sketch of a simple model of a pelagic ecosystem, which is reasonably described by the dynamics of the nutrients. Primary production driven by light generates phytoplankton biomass, which is dependent on the uptake of dissolved nutrients. A portion of the phytoplankton biomass dies and is transformed into detritus. The detritus in turn will be recycled into dissolved nutrients by mineralization processes and becomes available again for uptake by phytoplankton. This closes the cycling of material through the model food web.

One of the fundamental laws which the biogeochemical models have to obey is the conservation of mass. The total mass, $M$, may be expressed by the mass of one of the chemical elements needed by the plankton cells and most often the Redfield ratio is used to estimate the other constituents. Typical choices are, for example, carbon or nitrate used as a 'model currency' to quantify the amount of plankton biomass and detritus (i.e. the state variables that refer to living and nonliving elements of the ecosystem). The fact that changes of mass in a model ecosystem are controlled entirely by sources and sinks can be expressed by

$$\frac{\mathrm{d}}{\mathrm{d}t}M = \text{sources} - \text{sinks}. \tag{2.1}$$

The use of differential equations implies that the state variables can be considered as continuous functions in time and space. Dissolved nutrients are represented as *in situ* or averaged concentrations. For organic and inorganic particulate matter, such as plankton or detritus, we assume that the number of particles is high enough that the concentrations (biomass per unit volume) behave like continuous functions. Let $C_n$ be a concentration representing a state variable indexed by $n$ in a box of the volume $V$. Then the total mass in the box is

$$M = \sum_n C_n V, \tag{2.2}$$

and hence

$$V\frac{d}{dt}C_n = \text{sources}_n - \text{sinks}_n \pm \text{transfers}_{n-1,n+1}. \tag{2.3}$$

Examples for sources and sinks in natural systems are external nutrient inputs by river discharge and burial of material in sediments, respectively. The transfer term reflects the propagation of nutrients through different state variables, i.e. matter bound in the state variable with index $n$ flows into state variables with index $n\pm1$, and vice versa. The transformation is driven by biological processes such as nutrient uptake during primary production or by microbial conversion, i.e. mineralization.

The key problem in model building is to find adequate mathematical formulations that describe these transformations. Processes such as primary production or mineralization are complex, and their mathematical descriptions may involve substantial simplification. To illustrate the procedure, we start with the simple example of transformation processes characterized by a first-order chemical reaction, where a substrate, $S$, will be converted into a product, $P$, at a constant rate, $k$, as sketched in Fig. 2.1. Initially, at $t=0$, there is only the substrate $S=S_0$ while $P_0=0$. Assuming that the substrate concentration decreases at the same rate as the product increases, we find the equations,

$$\frac{d}{dt}S = -kS, \quad \frac{d}{dt}P = kS. \tag{2.4}$$

Adding these equations gives

$$\frac{d}{dt}(S+P) = 0.$$

This expresses the conservation of mass in the absence of external sources or sinks. The integration gives $S + P = C$, where the constant of integration, $C$, is determined by the initial condition, $S(0) = S_0 = C$ and hence,

$$P = S_0 - S.$$

The integration of Equation (2.4) gives

$$S = S_0 e^{-kt}, \quad P = S_0(1 - e^{-kt}). \tag{2.5}$$

This solution describes a chemical process, controlled by a constant process rate, where a substrate is completely converted into a product. The initial concentration of the product is zero and the conversion of the substrate is independent of the amount of the product. The solution is shown in Fig. 2.2 (top panel), where the parameter are chosen as $S_0 = 5\,\text{mmol/m}^3$, $P_0 = 0$ and $k = 1/d$.

**FIGURE 2.1** Conceptual diagram for a first-order chemical reaction.

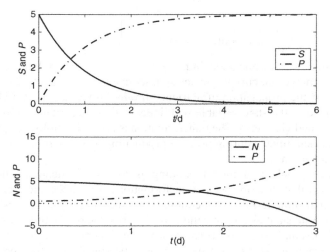

**FIGURE 2.2**   Upper panel: First-order chemical reaction where the substrate is transferred in the product $P$. Lower panel: An attempt to model the uptake of nutrients, $N$, by phytoplankton, $P$, in an analog manner as a chemical reaction. Note that the model nutrient becomes negative, i.e. this approach failed.

Can we use such an approach to describe the phytoplankton development? Apart from the role of light, it is known that primary production depends on the availability of nutrients. We may assume that the substrate corresponds to the dissolved nutrient, $N$, and the product is the phytoplankton, $P$. However, there is an obvious difference to the considered chemical reaction, which is driven by the intrinsic chemical properties of the substrate, such as the decay of radioactive material. We have to take into account that primary production can commence only if there are already plankton cells, a seed population, that can conduct photosynthesis and cell division. Plankton growth depends on the concentration of the phytoplankton, $P$, and the concentration of nutrients, $N$, which are taken up as the plankton grows. Thus, we may try the approach

$$\frac{\mathrm{d}}{\mathrm{d}t}P = kP, \quad \frac{\mathrm{d}}{\mathrm{d}t}N = -kP. \tag{2.6}$$

The initial conditions are $P(0) = P_0$ and $N(0) = N_0$. Conservation of mass requires

$$P + N = P_0 + N_0, \tag{2.7}$$

and the solutions are

$$P = P_0 e^{kt}, \quad N = N_0 + P_0(1 - e^{kt}).$$

The result seems to be not very different from Equation (2.5) for the initial phase of the development. However, as time proceeds, the solution becomes unstable because after a certain time, $\tau$, the nutrients are exhausted and the growth must stop. The time period, $\tau$, after which the nutrient concentration, $N$, reaches a zero crossing is determined by $N_0 + P_0 = P_0 e^{k\tau}$, which gives $\tau = \frac{1}{k} \ln(1 + \frac{N_0}{P_0})$. This is illustrated in Fig. 2.2 (bottom), where the parameter are chosen as $N_0 = 5\,\text{mmol}\,\text{N}\,\text{m}^{-3}$, $P_0 = 0.5\,\text{mmol}\,\text{N}\,\text{m}^{-3}$ and $k = 1\,\text{d}^{-1}$.

In order to circumvent this difficulty, we may try an approach where the growth rate depends explicitly on the nutrient concentration and becomes small when the nutrients are depleted. We may choose

$$\frac{\mathrm{d}}{\mathrm{d}t}N = -k'NP, \quad \frac{\mathrm{d}}{\mathrm{d}t}P = k'NP. \tag{2.8}$$

Here $k'$ is the rate $k$ divided by a reference value of $N$, which may be chosen as the initial value $N_0$. With Equation (2.7), we first solve the equation

$$\frac{\mathrm{d}}{\mathrm{d}t}P = k'(P_0 + N_0 - P)P, \tag{2.9}$$

and then use Equation (2.7) to calculate $N$. The solution can be found in a straightforward manner by separation of the variables,

$$\frac{\mathrm{d}P}{(N_0 + P_0 - P)P} = k'\mathrm{d}t.$$

Estimating the involved standard integrals we obtain

$$\ln \frac{P}{N_0 + P_0 - P} = (N_0 + P_0)(k't + C),$$

where the constant of integration, $C$, follows from the initial condition as $C = \frac{1}{N_0 + P_0} \ln \frac{P_0}{N_0}$. Thus, the result is

$$P = P_0 \frac{P_0 + N_0}{P_0 + N_0 \exp[-k'(N_0 + P_0)t]}.$$

With Equation (2.7), we find

$$N = N_0 \frac{(N_0 + P_0) \exp[-k'(N_0 + P_0)t]}{P_0 + N_0 \exp[-k'(N_0 + P_0)t]}.$$

The solution is shown in Fig. 2.3, where the parameters are chosen as $N_0 = 5\,\text{mmol}\,\text{N/m}^3$, $P_0 = 0.5\,\text{mmol}\,\text{N/m}^3$ and $k' = k/N_0 = 0.2/\text{d/mmol}\,\text{N/m}^3$. For large times $t \rightarrow \infty$, all nutrients are assimilated by the plankton, $P = N_0 + P_0$ and $N = 0$. This solution is stable. Mathematically, this is due to the quadratic term of $P$ in Equation (2.9). However, a weak point of the model equations (2.8) is the underlying assumption that the growth rate increases with the nutrient concentrations even at high values, while the experimental experience says that above a certain nutrient level, the growth does not respond

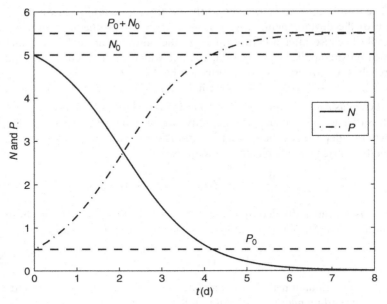

**FIGURE 2.3**  Nutrient and plankton dynamics for a growth rate controlled by the nutrient concentration.

to further addition of nutrients. A more sophisticated approach will be discussed in the next section. Unfortunately, it turns out that more elaborated approaches amount to equations that cannot be solved analytically, and hence, numerical methods must be employed.

### 2.1.2  Nutrient Limitation

It is known that the rates that control primary production are not constants but depend on various factors such as light and nutrients. Thus, process rates, such as $k$ in the chemical reaction in Equation (2.4), cannot in general be assumed as constant but need a more detailed consideration. An important concept for model development is Justus von Liebig's famous law of the minimum (de Baar, 1994). The law of the minimum states that if only one of the essential nutrients becomes rare, then the growing of plants is no longer possible. This discovery was the scientific basis for fertilization of plants. A second important fact is the existence of a robust quantitative molar ratio of carbon to nitrogen to phosphorous, $C:N:P = 106:16:1$, in living cells of marine phytoplankton, the so-called Redfield ratio. The important implication for modelling is that we can focus on only those one or two nutrients that are exhausted first and hence are limiting the further biomass development. Knowledge of one element allows the calculation of the other elements.

We commence with a consideration of how nutrients control the rate of phytoplankton development. Let $N$ be a nutrient concentration and $P$ the phytoplankton biomass concentration that increases proportional to $P$, then

$$\frac{\mathrm{d}}{\mathrm{d}t}P = r_{\max}f(N)P,$$

where $r_{\max}$ refers to the maximum rate, which constitutes intrinsic cell properties at given light and temperature. If the nutrient is plentiful, then the increase of $P$ is practically not affected by the nutrient concentration and we can choose $f(N) = 1$. If the nutrient concentration becomes small, then we may expect that $f(N) \sim N^{\nu}$. Based on experimental studies, Monod (1949) found an empirical relationship

$$f(N) = \frac{N}{k_N + N}, \tag{2.10}$$

with $k_N$ being a half-saturation constant, i.e. $f(k_N) = 0.5$. For $N \gg k_N$ we have $f(N) = 1$, while for $N \ll k_N$ it follows $f(N) \approx N/k_N$. Thus, these formulations comprise the cases (2.6) and (2.8).

The expression (2.10) describes the 'velocity' of a reaction and is also known as Michaelis and Menten kinetics (Michaelis and Menten, 1913). It was derived in the context of enzyme-catalysed reactions. A substrate, $S$, and enzyme, $E$, form an intermediate complex, $C = [SE]$, at rate $k_1$. The complex breaks into a product, $P$, and the enzyme, $E$, operating at rate $k_2$. There is also a backward reaction where the complex breaks into substrate, $S$, and enzyme, $E$, at rate $k_{-1}$.

$$S + E \underset{k_{-1}}{\overset{k_1}{\rightleftharpoons}} C \overset{k_2}{\to} P + E \tag{2.11}$$

These reactions can be expressed by three-dynamical equations, which describe the changes of the concentrations,

$$\frac{\mathrm{d}}{\mathrm{d}t}S = -k_1 ES + k_{-1}C, \tag{2.12}$$

$$\frac{\mathrm{d}}{\mathrm{d}t}C = k_1 ES - (k_{-1} + k_2)C, \tag{2.13}$$

$$\frac{\mathrm{d}}{\mathrm{d}t}P = k_2 C, \tag{2.14}$$

and by the conservation of enzyme,

$$E = E_0 - C. \tag{2.15}$$

For an efficient enzyme the forward reaction will be faster than the backward process, $k_{-1} < k_1, k_2$, and a steady state will be reached where $C$ is practically constant, $\frac{\mathrm{d}}{\mathrm{d}t}C \approx 0$. With Equation (2.13), this implies $k_1(E_0 - C)S = (k_{-1} + k_2)C$ or

$$C = \frac{E_0 S}{k_1 \left( \frac{k_{-1} + k_2}{k_1} + S \right)}.$$

After substitution of $C$ in Equation (2.14), we obtain

$$\frac{d}{dt}P = \frac{k_2 E_0 S}{\frac{k_{-1}+k_2}{k_1} + S}.$$  (2.16)

Comparing this with Equation (2.10), we may set $k_2 E_0 = r_{max}$ and $\frac{k_{-1}+k_2}{k_1} = k_N$. Thus, the rate of reaction depends on the concentration of the substrate. However, in this case, $P$ does not occur on the right-hand side of Equation (2.16).

Because it is not clear whether the nutrient uptake of phytoplankton cells corresponds to such an enzyme reaction, we may prefer to assume that the growth of cells stops if the nutrients are depleted, while the growth rate is independent of the nutrient concentration at high nutrient supply. Then, we can use any reasonable mathematical expression, $f(N)$, with a limiting property, i.e. $f \to 1$ for high concentration and $f \to N^v$ where $v$ is a positive integer. Examples are

$$f(N) = \frac{N^2}{k_N^2 + N^2},$$  (2.17)

or

$$f(N) = \frac{2}{\pi}\arctan\left(\frac{N}{k_N}\right), f(N) = 1 - \exp\left(-\frac{N}{k_N}\right).$$

These function are shown in Fig. 2.4 for different choices of the half-saturation constant. Note that for the exponential formula, the limiting function at $N = k_N$ assumes the value $f(k_N) = 0.6321$ rather than 0.5 as in the other three cases. Thus, in this case, $k_N$ is not a half-saturation constant in a strict sense.

Now we can rewrite the equations in Equation (2.8) with one of the limiting functions, e.g.

$$\frac{d}{dt}N = -r_{max}\frac{N}{k_N + N}P, \frac{d}{dt}P = r_{max}\frac{N}{k_N + N}P.$$  (2.18)

Then we find the behaviour shown in Fig. 2.5, where the parameters are, somewhat arbitrarily, chosen as $r_{max} = 0.8\,d^{-1}$ and $k_N = 2\,mmol\,m^{-3}$. The resulting uptake rate is somewhat smaller than $k'$ for $N > k_N$ but exceeds $k'$ for $N < k_N$. Comparing Figs. 2.3 and 2.5 shows that the gross results are similar. The advantage of the choice of a limiting function in Equation (2.18) is that no arbitrary normalization concentration is required, and the rate has the dimension one over time.

### 2.1.3 Recycling

So far we have considered nutrient uptake by phytoplankton in Equations (2.8) and (2.18), which describe a process where the nutrients are transferred

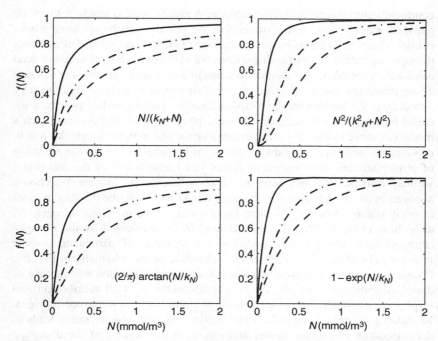

**FIGURE 2.4** Four different choices of functions with limiting properties plotted for half-saturation constant $k_N = 0.1 \, \text{mmol/m}^3$ (solid), $k_N = 0.3 \, \text{mmol/m}^3$ (dash dot), $k_N = 0.5 \, \text{mmol/m}^3$ (dash).

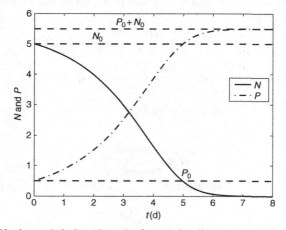

**FIGURE 2.5** Nutrient and plankton dynamics for a nutrient limiting the growth rate.

completely into the model phytoplankton. A steady state is reached when the nutrients are depleted. In this subsection, we introduce pathways from phytoplankton back to nutrients by two processes: (1) fast direct release of nutrients through respiration and extracellular release and (2) slow mineralization of dead cells, which are taken into account in terms of a new state variable, the detritus. These processes can be included to construct simple models of algal primary production. We begin with the simplest case to describe such a system, a so-called NPD model that includes nutrients, phytoplankton and detritus. Such a model can serve to describe processes in a chemostat with no zooplankton in it, or we can assume that zooplankton grazing is included in an effective mortality of phytoplankton. With regard to Justus von Liebig's law of the minimum, we may consider only one nutrient, which is limiting the system. We choose nitrogen as the model 'currency' because it is known to be the limiting nutrient in many marine systems. The state variables dissolved inorganic nitrogen, $N$, as well as phytoplankton, $P$, and detritus, $D$, are expressed by the nitrogen captured in living and dead particles, i.e. in terms of particulate organic nitrogen. The structure of the model is sketched in the schematic in Fig. 2.6: Primary production is driven by solar radiation in conjunction with uptake of dissolved inorganic nitrogen, $N$, by phytoplankton, $P$. Cell metabolism and respiration establish a direct path from phytoplankton to dissolved nitrogen. Mortality converts phytoplankton into detritus, $D$, and mineralization leads to decomposition of organic matter and results in an increase of the dissolved nitrogen pool.

We describe the processes by transfer terms written as $L_{XY}$ or $l_{XY}$, which can be read as 'loss of $X$ to $Y$', where the lowercase $l$ refers to constant rates while the uppercase $L$ refers to rates that are functions of other parameters or state variables. In this section, we commence with constant rates, i.e. we use the lowercase '$l$'.

Next we 'translate' the model structure into a set of three equations for the three state variables:

$$\frac{d}{dt}N = -r_{\max}\frac{NP}{k_N + N} + l_{PN}P + l_{DN}D,$$

$$\frac{d}{dt}P = r_{\max}\frac{NP}{k_N + N} - l_{PN}P - l_{PD}P, \qquad (2.19)$$

$$\frac{d}{dt}D = l_{PD}P - l_{DN}D.$$

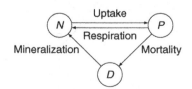

**FIGURE 2.6** Conceptual diagram of a simple NPD model.

The involved parameters are the maximum uptake rate for phytoplankton, $r_{max}$, which depends on light and temperature, losses by respiration, $l_{PN}$, and mortality $l_{PD}$. Mineralization is simulated by the rate, $l_{DN}$. If we add the three equation we obtain

$$\frac{d}{dt}(N + P + D) = 0,$$

i.e. the conservation of mass is fulfilled. We look first at the steady-state solution of the set of equations, which describe the asymptotic behaviour of the model for large times. We set the right-hand side of Equation (2.19) equal to zero and find

$$0 = -r_{max}\frac{N}{k_N + N}P + l_{PN}P + l_{DN}D,$$

$$0 = r_{max}\frac{N}{k_N + N}P - l_{PN}P - l_{PD}P, \qquad (2.20)$$

$$0 = l_{PD}P - l_{DN}D.$$

There is a trivial solution, $P = 0$, which is of no interest. Using the third equation to eliminate $D$ in the first one, we find that the first two equations give

$$r_{max}\frac{N}{k_N + N} - l_{PN} - l_{PD} = 0,$$

which implies that an equilibrium among the state variables can be reached for a certain nutrient level, $N^*$, where the growth and loss terms of the phytoplankton are balanced. This nutrient level is given by

$$N^* = k_N\frac{l_{PN} + l_{PD}}{r_{max} - l_{PN} - l_{PD}}.$$

Assuming that in the course of a bloom, $N$ has decreased from the initial value $N_0$ to $N^*$, then the amount of $N_0 - N^*$ is accumulated in the $P$ and $D$ pool. Their ratio is set by

$$\frac{P}{D} = \frac{l_{DN}}{l_{PD}}.$$

The dynamics of this still very simple model are illustrated in Fig. 2.7 for different choices of the rate parameters, $l_{PN}$, $l_{PD}$ and $l_{DN}$, where the uptake parameters are $r_{max} = 1\,d^{-1}$ and $k_N = 0.3\,mmol/m^3$ and the initial values are $N_0 = 5\,mmol/m^3$, $P_0 = 0.05\,mmol/m^3$ and $D_0 = 0$. The plot illustrates that the steady-state ratios of $P/D$ are set by the ratio of the rates $l_{DN}/l_{PD}$. The time scale at which the steady state is reached depends on the mineralization rate, $l_{DN}$, which corresponds to the longest pathway in the model cycle. In contrast to the uptake description without recycling where the steady state was reached for zero nutrients and maximum phytoplankton, we find that with recycling the steady state is reached at finite levels of the state variables. The ratios of the state

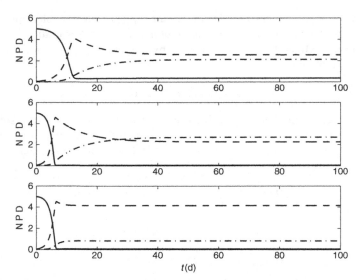

**FIGURE 2.7** Dynamics of nutrients, $N$ (solid), phytoplankton, $P$ (dashed), and detritus, $D$ (dash dotted) for different choices of the loss parameters. Top panel: $l_{PN} = 0.50\,\mathrm{d}^{-1}$, $l_{PD} = 0.05\,\mathrm{d}^{-1}$ and $l_{DN} = 0.06\,\mathrm{d}^{-1}$; middle: $l_{PN} = 0.10\,\mathrm{d}^{-1}$, $l_{PD} = 0.06\,\mathrm{d}^{-1}$ and $l_{DN} = 0.05\,\mathrm{d}^{-1}$; bottom: $l_{PN} = 0.10\,\mathrm{d}^{-1}$, $l_{PD} = 0.10\,\mathrm{d}^{-1}$ and $l_{DN} = 0.5\,\mathrm{d}^{-1}$. Note that the steady state nutrient level is small but nonzero.

variables are set by the mortality and mineralization rates. It can easily be seen that these findings apply also for the modified uptake function, $N^2/(k_N^2 + N^2)$. Then the critical nutrient level is given by

$$N^* = \sqrt{k_N \frac{l_{PN} + l_{PD}}{r_{\max} - l_{PN} - l_{PD}}}.$$

### 2.1.4 Zooplankton Grazing

The next trophic level in the marine food web is the herbivore zooplankton which consumes phytoplankton. Thus, grazing by zooplankton is an important loss term for phytoplankton. The simplest way to include this top-down control on phytoplankton is the inclusion of the grazing effects in the phytoplankton mortality. However, the dynamical variations of zooplankton, and hence their effect on phytoplankton, can better be accounted for by means of a new state variable referring to zooplankton biomass. Phytoplankton is a limiting resource for zooplankton growth. Let $Z$ be the zooplankton biomass per unit of volume. Then the zooplankton growth obeys a relation described by

$$\frac{\mathrm{d}}{\mathrm{d}t}Z \sim g(P)Z, \qquad (2.21)$$

where $g(P)$ is a grazing rate that quantifies the ingestion of phytoplankton. Similar to the uptake of nutrients by phytoplankton, we expect that a relationship exists which becomes independent of the food if the food is abundant but is proportional to the available food concentration at scarce resources. An often-used limiting formulation for $g(P)$ is the so-called Ivlev function (Ivlev, 1945), which was originally derived from a variety of fish-feeding experiments. Ivlev found that the ration of food eaten by fish increased with the amount of food offered only up to a certain limit. This observation led to the expression

$$g(P) = g_{max}(1 - e^{-I_v P}), \qquad (2.22)$$

where $I_v$ is the Ivlev parameter. There is no *a priori* principle that defines the analytical form of the limiting functions. Only the fact that growth and grazing are limited by nutrients and food is a fundamental relationship. Other choices are, for example,

$$g(P) = g_{max}[1 - e^{-I_v^2 P^2}], \qquad (2.23)$$

or Monod's limitation function,

$$g(P) = g_{max} \frac{P}{I_v^{-1} + P}.$$

To complete the formulation of the differential equation (2.21) for the state variable zooplankton, we need to include loss terms, which are controlled by respiration rates, $l_{ZN}$, and mortality rates, $l_{ZD}$. The mortality comprises natural deaths of, e.g. copepods as well as the removal of copepods eaten by fish. In principle, we could try to include a further state variable for fish. However, this is a difficult issue, and for many applications it is advisable to truncate the model food web at the level of zooplankton by an effective mortality rate. However, in Chapter 4, an approach to develop consistent nutrient-to-fish models will be shown.

## 2.2 SIMPLE MODELS

### 2.2.1 Construction of a Simple NPZD Model

In the previous section, we discussed a few aspects and process descriptions that are relevant to start ecosystem modelling. In this section, we use the findings to construct a model of a marine system. As mentioned earlier, first we have to define what we wish to describe with the model. The following discussion aims at modelling of the yearly cycle of nutrients and plankton in a simplified boxlike system that may serve as a crude first-order description of a part of a marine system, such as a sub-basin of the Baltic Sea. The model is required to simulate both phytoplankton spring blooms as well as yearly cycles of the state variables.

A representation of the yearly course of chemical and biological parameters in the Baltic Sea is shown in Fig. 2.8 and can be summarized as follows: After the onset of the spring bloom, the phytoplankton biomass reaches maximum levels in a few days and decreases with a time scale of 30 to 40 days to low levels (Schulz et al., 1978). A secondary bloom occurs in the autumn. The zooplankton biomass develops slowly and reaches its maximum in summer, about 90 to 100 days after the spring bloom peak. The nutrients are depleted after the spring bloom but increase to high levels during winter until the next spring bloom starts. The system in the upper layer switches back and forth from eutrophic to oligotrophic conditions during the year. Despite the low values of phytoplankton biomass, the production rates are relatively high during summer. It was also noted in Schulz et al. (1978) that the maximum values of the zooplankton biomass shown in Fig. 2.8 (lower panel) were overestimated. The net hauls that were used to collect zooplankton included often aggregates of cyanobacteria that were not removed from the samples.

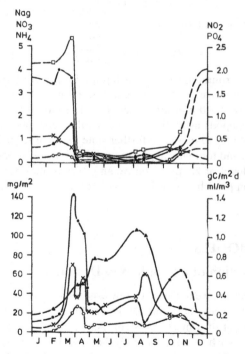

**FIGURE 2.8**    Annual cycle of nutrients (top) in the upper 10 m and plankton (bottom) in the upper 25 m measured at a central station in the Arkona basin of the Baltic Sea. Upper panel: squares – total dissolved inorganic nitrogen, dots - $NO_3$, triangles - $NH_4$, and crosses - $PO_4$, all in mmol m$^{-3}$. Lower panel: dots – chlorophyll $a$, in mg m$^{-2}$, circles – phaeopigment in mg m$^{-2}$, crosses – primary production in g C m$^{-2}$ d$^{-1}$, and triangles – zooplankton biomass in ml m$^{-3}$ (Redrawn after Schulz et al. (1978)).

Although the yearly cycle shown in Fig. 2.8 gives a good qualitative picture of the dynamical behaviour of nutrients and plankton, it also indicates the general problem of undersampling. The number of measurements is too small to provide adequate temporal resolution for the processes involved. Moreover, the data were recorded at a fixed location, and spatial variations due to the advection of different water masses through the sampling station cannot be distinguished from the local changes of the observed quantities. This is a general problem in marine sciences, where processes are often oversampled in time and undersampled in space. In systems like the Baltic Sea, the vertical stratification shows a pronounced yearly cycle with a relatively shallow thermocline in summer and a deep vertical mixing in the winter. The physical–biological interaction, which controls the onset of the spring bloom is determined by the ratio of mixed layer depth and the depth of the euphotic zone. The general concept was outlined by Sverdrup (1953). A simpler concept was proposed by Kaiser and Schulz (1978), where it was shown on the basis of observations in the Baltic Sea that the spring bloom starts if the thickness of the mixed layer is smaller than the euphotic zone depth. This problem will be discussed in more detail in Chapter 5.

A further important observation is that phytoplankton shows significant sinking rates, especially for diatoms. This implies that a substantial amount of the phytoplankton biomass leaves the euphotic zone and can no longer contribute to the primary production. The spring bloom is apparently more controlled by sinking than by zooplankton grazing; (see, e.g., Bodungen et al., 1981; Smetacek et al., 1978).

Marginal and semi-enclosed seas, such as the Baltic, are subject to a substantial external nutrient supply from river loads and atmospheric deposition. Hence, such systems are not closed, but directly influenced by material fluxes from the corresponding drainage areas and can be subject to significant eutrophication (e.g. Larsson et al., 1985; Nehring and Matthäus, 1991; Matthäus, 1995).

It is obviously reasonable to start with a simple model, which comprises integrated knowledge of the processes from observations but which also isolates the most important biology and physics to describe essential aspects of a marine ecosystem. The comparison of results obtained with simplified models and field data needs some consideration because the measurements may also reflect processes, which are neglected in the model. Thus, we cannot expect a one-to-one correspondence of model results and observations, but we can check the plausibility of the model by considering integrated quantities, such as response times and the range of variations of the state variables. Such system properties are more or less well established from many observations.

In order to construct a simple model, which can describe, at least qualitatively, the observational findings in the Arkona basin of the Baltic Sea shown in Fig. 2.8, we take four state variables into account: a limiting nutrient, $N$, bulk phytoplankton, $P$, bulk zooplankton, $Z$, and bulk detritus, $D$. We consider

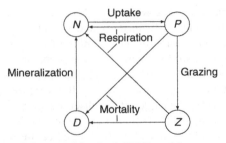

**FIGURE 2.9**   A conceptual diagram of the simple NPZD model.

a horizontally averaged basin with two vertical layers, which are represented by two vertically stacked boxes. As the limiting nutrient, we choose nitrate. The temporal changes of the state variables are completely described by the dynamics of the biological and chemical sources and sinks and by the fluxes through the boundaries of the model boxes. In order to obtain results that are not too site specific, we use nondimensional variables; i.e. the state variables are given as concentrations relative to the winter concentration of the limiting nutrient, $N_{ref}$. The conceptional diagram of the model is sketched in Fig. 2.9. As minimum requirements for a description of the physical control, we introduce the formation of a thermocline in spring and the destruction of the stratification in late autumn. An illustration of the annual cycle of formation and destruction of the thermocline is shown in Fig. 2.10. There is a simple switch between the biologically active seasons and the wintertime, when the whole water column will be mixed; i.e. the two model boxes are vertically separated by a thermocline from spring to autumn. From late autumn to early spring, the whole water column is well mixed, and it is assumed that the boxes have the same values of the state variables. It is clear that the development of the thermal stratification is related to the incoming solar radiation, which is also necessary for photosynthesis and primary production. We assume for the moment that the

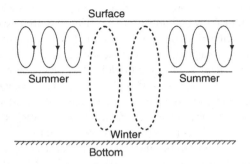

**FIGURE 2.10**   Annual cycle of stratification and mixing in the simple, first approach box model.

depth of the thermocline is smaller than the euphotic zone, and hence the spring bloom starts after the thermocline is formed (Kaiser and Schulz, 1978). Simply spoken, the biology is 'switched on and off' by formation and destruction of the stratification, as sketched in Fig. 2.10. The switch is implemented formally by a step function, which is equal to 1 if the length of the day exceeds a certain threshold value. The analytical expression for the length of the day will be derived in Chapter 5.

The phytoplankton growth with rate $R = r_{max}f(N)$ is a source for phytoplankton, $P$, but a sink for nutrients, $N$. The zooplankton, $Z$, consumes $P$ with a grazing rate, $G$. The grazing is a source for zooplankton but a sink for phytoplankton. As in the previous section, we describe several transfer processes by terms such as $L_{XY}$ or $l_{XY}$, which can be read as 'loss of $X$ to $Y$', where the lowercase $l$ refers to constant rates while the uppercase $L$ refers to functions of parameters or state variables. At several places we use 'switches' in terms of the step function $\theta(x)$. We define the step function as

$$\theta(x) = 1 \text{ for } x > 0 \quad \text{and} \quad \theta(x) = 0 \text{ for } x \leq 0. \quad (2.24)$$

The state variables and the relevant process parameters are listed in Table 2.1. The losses of $P$ and $Z$ are sinks for the plankton but sources for $N$ and $D$. The total losses are controlled by the rates $L_P$ and $L_Z$, which are sums of the loss rates, e.g. respiration, $l_{PN}$, $l_{ZN}$, and the mortality rates, $l_{PD}$, $l_{ZD}$, respectively.

$$L_P = l_{PN} + l_{PD} \quad \text{and} \quad L_Z = l_{ZN} + l_{ZD}.$$

In this simple approach, the mortality rates, $l_{PD}$, $l_{ZD}$, also include sinking, which prescribes how fast dead organic material sinks from the upper into the lower box. During the biologically active season, the state variable, $D$, refers to the fraction of dead organic matter that sinks down into the lower layer and accumulates there. It is assumed that the detritus will be recycled in the lower box and will be available as potential nutrient. This implies that during wintertime, when the two layers are connected and mixed, the variables $D$ and $N$ assume the same values and virtually cannot be distinguished. Thus, $D$ refers to the portion of nutrients that is potentially available in the lower box. Although bacterial biomass is not explicitly considered in the model, we note that the mineralization rate, $l_{DN}$, reflects microbial activity implicitly. For the sake of simplicity, we ignore the processes of accumulation and transformation of material at the bottom in the first model version.

The uptake and grazing rates, $R$ and $G$, depend on a switch $\theta^+ = \theta(\Delta d - d_0)$; i.e. the rates are nonzero only for that part of the year where the normalized length of the day, $\Delta d$, exceeds the threshold $d_0$. The definition of $\Delta d$ is given in Chapter 5, and examples of the annual cycle of $\Delta d$ for different latitudes are shown in Fig. 5.1. Because the formation of a seasonal thermocline

**TABLE 2.1** State Variables, Rates and Parameters

| Symbol | Meaning |
|---|---|
| $N$ | Bulk nitrogen |
| $P$ | Bulk phytoplankton |
| $Z$ | Bulk zooplankton |
| $D$ | Bulk detritus |
| $P_0$ | Background concentration |
| $Z_0$ | Background concentration |
| $l_P$ | Total loss of phytoplankton |
| $l_{PN}$ | Fraction of $L_P$ recycled in the upper box |
| $l_{PD}$ | Fraction of $L_P$ sinking down into the lower box |
| $l_Z$ | Loss of zooplankton |
| $l_{ZN}$ | Fraction of $L_Z$ recycled in the upper box |
| $l_{ZD}$ | Fraction of $L_Z$ sinking to the lower box |
| $l_D$ | Loss of detritus by injections into the upper layer |
| $S_N^{ext}$ | External sources and sinks of $N$ |
| $S_D^{ext}$ | External sources and sinks of $D$ |
| $A_{mix}$ | Winter mixing of the boxes |
| $N_{ref}$ | Reference nutrient level |

depends on the heat flux through the sea surface and the mixing in the upper layer, we can use the normalized length of the day as a proxy for the annual variation of the irradiation. Thus, $\theta^+$ defines the onset of the biologically active season in our model. The mixing rate, $A_{mix}$, which controls the mixing of the upper and lower layers from late autumn to early spring, is proportional to $\theta^- = \theta(d_0 - \Delta d)$. During that time, phyto- and zooplankton assume certain small background concentrations, $P_0$ and $Z_0$, which mimic the overwintering. With these assumptions, the set of differential equations that defines the model is explicitly:

$$\frac{d}{dt}N = -R(N, t)P + l_{PN}(P - P_0) + l_{ZN}(Z - Z_0) \tag{2.25}$$

$$+ L_{DN}D + A_{mix}(D - N) + S_N^{ext},$$

$$\frac{d}{dt}P = +R(N, t)P - G(P, t)Z - (l_{PN}P + l_{PD})(P - P_0), \tag{2.26}$$

$$\frac{d}{dt}Z = G(P,t)Z - (l_{ZD} + l_{ZN})(Z - Z_0), \tag{2.27}$$

$$\frac{d}{dt}D = l_{ZD}(Z - Z_0) + l_{PD}(P - P_0) - L_{DN}D \tag{2.28}$$

$$-A_{mix}(D - N) + S_D^{ext}.$$

The uptake rate $R$ of phytoplankton is chosen as

$$R(N,t) = \theta^+ r_{max} f(N) = \theta(\Delta d - d_0) r_{max} \frac{N^2}{k_N^2 + N^2}. \tag{2.29}$$

The maximum uptake rate, $r_{max}$, is a function of light and temperature. In this discussion, both daily variations in available light as well as random changes by varying cloud coverage remain unresolved. Our principal time scale is 1 day. The grazing rate is defined as

$$G(P,t) = \theta^+ g_{max} g(P) = \theta(\Delta d - d_0) g_{max}[1 - \exp(-I_v^2 P^2)]. \tag{2.30}$$

Both $R$ and $G$ are 'switched-off' from late autumn to early spring. These formulas are modified Michaelis–Menten and the Ivlev approaches, Equations (2.17) and (2.23), respectively. The maximum grazing or ingestion rate, $g_{max}$, is in general a function of the temperature. This will be taken into account later. The set of equations that describes the model food web is truncated at the trophic level of zooplankton. In particular, the loss rate of zooplankton, $l_{ZD}$ involves both mortality due to natural death, e.g. starvation, and predation by higher trophic levels, without a detailed description. The term $A_{mix}(D - N)$ in Equations (2.25) and (2.28) describes the vertical mixing of the two model layers during winter at the rate, $a_{mix}$; i.e.

$$A_{mix} = a_{mix}\theta(d_0 - \Delta d). \tag{2.31}$$

Due to this mixing, the whole water column is equally filled with nutrients during winter, i.e. $D = N$. That the mixing term provides this adjustment can be shown explicitly by solving the equations

$$\frac{d}{dt}D = -a_{mix}(D - N),$$

$$\frac{d}{dt}N = a_{mix}(D - N).$$

We choose the initial conditions $D = D_0$ and $N = N_0$ for the time $t_0$, when the mixing starts. For simplicity, we set $t_0 = 0$. Using the conservation of mass, which amounts to $N + D = N_0 + D_0$, we have

$$\frac{dt}{D} = -2a_{mix}D + a_{mix}(N_0 + D_0).$$

The solution for $D$ is obviously

$$D = \frac{1}{2}\left[D_0 + N_0 + (D_0 - N_0)e^{-2a_{mix}t}\right].$$

For $N$ it follows

$$N = \frac{1}{2}[D_0 + N_0 - (D_0 - N_0)e^{-2a_{mix}t}].$$

For large times, i.e. $t \ll (2a_{mix})^{-1}$, we find

$$N = D = \frac{1}{2}(D_0 + N_0).$$

The term $L_{DN}$ in Equations (2.25) and (2.28) was included to simulate upward nutrients fluxes, such as a nutrient injection that can be forced by strong winds eroding the thermocline and releasing nutrients into the upper layer. Finally, we have added the terms $S_N^{ext}$ and $S_D^{ext}$ to Equations (2.25) and (2.28) in order to include sources and sinks that prescribe external fluxes of matter in or out of the system.

## 2.2.2 First Model Runs

Our simple model is defined by the set of nonlinear ordinary differential equations (2.25)–(2.28), with the rates described in the previous section. We can now consider the preparations required to run the model on a computer. This involves three steps: (1) writing a computer code to solve the set of differential equations, (2) testing and running the code, and (3) visualizing the results. We can choose one of many existing programming languages to code the set of differential equations. A well-known and appropriate mathematical standard software is the MATLAB package, which we will use in this and the following chapters at several places. The programming aspects involve more technical problems, which will be considered in Chapter 8, where a brief introduction to MATLAB, with several example codes, is given. In particular, this model program is described as an illustrative example. Here we focus on the structure of the model equations and use the visualized products of the simulation to discuss the results.

In this section, we study the properties of the model for constant rates. The numerical values of the parameters are listed in Table 2.2. We start the investigation of the model with the following choice of parameters:

The model is initialized with the winter values of the nutrients and the detritus concentrations, i.e. $N = D = 0.99$, and phyto- and zooplankton concentrations are set to the background values, $P(0) = P_0$ and $Z(0) = Z_0$. We note that the total sum of the initial values of all normalized state variables is and that all state variables are normalized by $N_{ref}$.

First we simulate the dynamics of the state variables for the period of 1 year. The results are shown in Fig. 2.11. The spring bloom starts after day 75 and reaches its maximum after 5–7 days when about 80% of the nutrients have been consumed by the model phytoplankton (Fig. 2.11, P-dashed). About 25 to 30 days after the onset of the bloom (day 100 to 105 of the simulated year), the phytoplankton concentration is about half the peak value. The spring bloom is

**TABLE 2.2** Model Parameters and Rates

| Parameter | Symbol | Unit | Numerical Value |
|---|---|---|---|
| Maximum growth rate | $r^0_{max}$ | $d^{-1}$ | 1 |
| Half-saturation | $k_N$ | – | 0.1 |
| Plant metabolic losses | $l_{PN}$ | $d^{-1}$ | 0.1 |
| Plant death rate | $l_{PD}$ | $d^{-1}$ | 0.02 |
| Mineralization rate | $l_{DN}$ | $d^{-1}$ | 0.001 |
| Maximum grazing rate | $g_{max}$ | $d^{-1}$ | 0.5 |
| Ivlev constant | $I_v^{02}$ | – | 1.2 |
| Zooplankton metabolic losses | $l_{ZN}$ | $d^{-1}$ | 0.01 |
| Zooplankton death rate | $l^0_{ZD}$ | $d^{-1}$ | 0.02 |
| Nutrient initial value | $N(0)$ | – | 0.99 |
| Phytoplankton initial value | $P_0$ | – | 0.01 |
| Phytoplankton background level | $P_0$ | – | 0.01 |
| Zooplankton initial value | $Z(0)$ | – | 0.01 |
| Zooplankton background level | $Z_0$ | – | 0.01 |
| Detritus initial value | $D(0)$ | – | 0.99 |
| Winter mixing rate | $a_{mix}$ | $d^{-1}$ | 0.5 |
| Nutrient import rate | $S_N^{ext}$ | $d^{-1}$ | 0 |
| Detritus export rate | $S_D^{ext}$ | $d^{-1}$ | 0 |
| Detritus injections | $l_D$ | $d^{-1}$ | 0 |
| Reference nutrient level | $N_{ref}$ | $mmol/m^3$ | 4.5 |

Numerical values of the parameters of the constant process rates and initial value of the state variables for the first version of the NPZD model.

limited by high sinking, as indicated by the detritus response in the lower layer (Fig. 2.11, bottom panel) and by the grazing. The response of the zooplankton follows the spring bloom with a delay of about 15 days, and the maximum value of $Z$, which is about 40% of the peak value of the phytoplankton bloom, is reached after 20 to 30 days.

The nutrient concentration decreases rapidly after the onset of the bloom by about one order of magnitude; i.e. the nutrient level is about 10% of the winter value, while the detritus concentration approaches twice the winter value

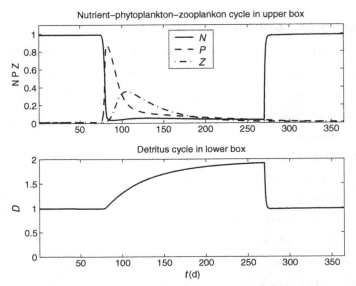

**FIGURE 2.11**   Annual cycle of the state variables nutrient, $N$, phytoplankton, $P$, zooplankton, $Z$ (upper panel), and detritus, $D$ (lower panel).

in late summer. At day 270, the stratification is assumed to be destroyed by wind mixing and winter convection. Then the detritus and the nutrients reach equal levels in the whole water column, while $P$ and $Z$ decrease towards their background levels. The model is able to simulate the spring bloom but fails to describe the observed summertime development. The observation showed a much slower response of the zooplankton to the spring bloom signal, and the strongest zooplankton signal was detected 90–100 days after the spring bloom peak; see Fig. 2.8. Obviously, the model is too simple.

### 2.2.3   A Simple NPZD Model with Variable Rates

The constant rates used in the first model version are not sufficient to describe the seasonal cycle after the spring bloom. To improve the model performance, we make several plausible assumptions for the process rates, while we keep the number of state variables; i.e. we use the same bulk variables as in the previous section. As a first step in refining the model, we try to improve the sinking rates. In particular, we take the different magnitude of sedimentation of organic material in the course of the year by means of time-dependent rates into account. In the first model version, the sinking rate was $0.02 \, \text{d}^{-1}$. For an upper layer thickness of $20 \, \text{m}$ this implies a sinking velocity of $0.4 \, \text{m} \, \text{d}^{-1}$. Slightly higher values have been adopted by Wolff and Strass (1993). In their model, the sinking velocity switches between $1\,\mathrm{m\,d^{-1}}$ during spring bloom to $0.3\,\mathrm{m\,d^{-1}}$ during oligotrophic conditions. Fasham et al. (1990) studied two cases of sinking rates, $10\,\mathrm{m\,d^{-1}}$ and $1\,\mathrm{m\,d^{-1}}$, and found that high sinking rates lead to limited summer production. Stigebrandt and Wulff (1987) used maximum rates of $4\,\mathrm{m\,d^{-1}}$ and minimum rates of $0.1\,\mathrm{m\,d^{-1}}$, where the sinking rates were assumed to be a quadratic function of the biomass in the uppermost layer. This approach has the advantage that the variation of the sinking speed is controlled by model variables rather than switches. Their model has no state variable for zooplankton, and hence, the sinking is more critical to limiting the phytoplankton blooms.

Time-dependent sinking rates reflect different sinking properties of the main functional groups, which are diatoms, flagellates and cyanobacteria (blue-green algae). These functional groups are lumped together in the state variable, $P$. The diatoms develop first and are known to sink relatively fast. The flagellates are able to swim and do not sink, while the blue-green algae are slightly buoyant and can accumulate at the sea surface. Thus, we may try a time-dependent sinking rate with high values in spring, as was suggested from observations (Bodungen et al., 1981), but very small values during the summer. A possible formulation of this rate could be

$$L_{\mathrm{PD}} = \theta(\Delta d - d_0) l_{\mathrm{PD}} \exp(-(t - t_{\mathrm{start}})/30) + \theta(d_0 - \Delta d) l_{\mathrm{PD}}^0, \qquad (2.32)$$

where $t_{\mathrm{start}}$ is the time of the onset of the spring bloom. This implies that the state variable, $P$, reflects different functional groups at different times.

In the second step of model improvement, we try to include the effects of temperature on the process rates. Many of the relevant processes depend on temperature, which has, in temperate latitudes, a clear seasonal cycle in the upper mixed layer. In order to prescribe the temperature in the upper layer, which corresponds to the upper box of our model, we use the mean annual cycle of the near surface temperature of the Arkona Basin in the Baltic Sea. The temperature cycle is shown in Fig. 2.12. For the sake of simplicity we use the results of Eppley (1972), to prescribe the temperature effects on several rates. Then the effect of temperature may be expressed by the 'Eppley factor' as

$$\exp(a_E T), \qquad (2.33)$$

where $T$ is the temperature in degrees Celsius. The numerical value of the Eppley constant is $a_E = 0.063(^\circ\mathrm{C})^{-1}$. A more thorough discussion on the influence of temperature on algal growth rates can be found in Parson et al. (1984). The empirical relation of the maximum growth rate of algae as function of temperature is (Eppley, 1972),

$$\log_{10} r_{\mathrm{max}} = 0.0275 \frac{T}{^\circ\mathrm{C}} - 0.070.$$

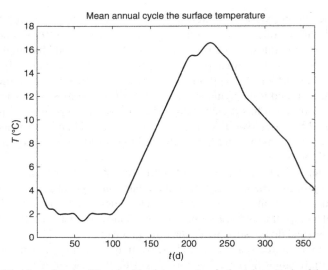

**FIGURE 2.12**    Annual cycle of the mixed-layer temperature in the Arkona Basin of the Baltic Sea.

This can be expressed as

$$r_{max} = \exp\left[\ln 10 \left(0.0275\frac{T}{°C} - 0.070\right)\right] = \exp\left(0.063\frac{T}{°C}\right)1.175.$$

A similar relation, with the constant $a_R = 0.069/°C$, was earlier proposed by Riley (1946) to prescribe the temperature effect on the respiration rate.

The growth rate depends on the temperature through the maximum uptake rate,

$$r_{max} = r_{max}^0 \exp(a_E T)\Delta d,$$

where $r_{max}^0$ is the rate at $T = 0\,°C$. Here we have also included the normalized length of the day $\Delta d$ as a measure of the seasonal variations of the available light, see formula (5.2) in Chapter 5. Moreover, because during summer the bulk phytoplankton variable refers to flagellates rather than diatoms, which dominate the spring bloom, we assume also that the half-saturation constant depends on temperature, $k_N = k_N^0 \exp(-a_E T)$. As a further modification, we link the respiration rate to the uptake,

$$L_{PN} = l_{PN}R(N,T).$$

Thus, not a fraction of the biomass but a fraction of the assimilated nutrient is lost through respiration.

Similar modifications can be introduced in the grazing or ingestion formula, i.e.

$$g_{max} = g_{max}^0 \exp(a_E T) \Delta d,$$

and

$$I_v^2 = I_v^{0^2} \exp(a_E T) \Delta d.$$

The egestion can be considered as a certain fraction of the ingested food,

$$L_{ZN} = l_{ZN} G(P, T).$$

We may also allow for a variation of the zooplankton mortality by assuming higher predation during longer days,

$$L_{ZD} = l_{ZD}^0 (1 - \Delta d).$$

Finally, we assume a small flux of nutrients from the lower box into the surface box at a rate $l_{DN} = 0.001 \, d^{-1}$. The parameters are summarized in Table 2.3. The simulation with these changes for 1 year is shown in Fig. 2.13. The only progress we have made compared to the first model version with constant rates, shown in Fig. 2.11, is that the spring bloom is now limited by the sinking, while the top-down control by zooplankton grazing is less important. However, the values are too small after the spring bloom, and the nutrient level cannot support enough production to generate the summer peak of zooplankton. What is missing? In systems like the Baltic Sea, there is a substantial nutrient input due to river discharges, which carry nutrient loads from the drainage area, as well as atmospheric deposition of nutrients. In order to include such external sources, we may prescribe a certain external input of nutrients at a given rate, say, e.g. $S_N^{ext} = 5 \times 10^{-3} \, d^{-1}$. This implies that the total mass in the system increases linearly with time. In order to keep a fixed mass balance, we assume for the moment that the same amount of matter leaves the system by burial in the sediments, i.e. $S_D^{ext} = -S_N^{ext}$. Then the simulation gives the results shown in Fig. 2.14. The results are qualitatively close to the observation (Fig. 2.8). The model zooplankton assumes maximum values in summer, and we even find a small secondary bloom in autumn. The external supply of nutrients obviously plays an important role. We repeat the simulation for a period extended three model years with an increased rate of the nutrient supply, $S_N^{ext} = 2 \times 10^{-3} \, d^{-1}$. We assume $S_N^{ext} = -S_D^{ext}$. The results are shown in Fig. 2.15 and indicate that the excess nutrients supplied to the surface layer are passed through the model food web and accumulate in the model zooplankton. Obviously, the annual cycles are reproduced for each year. Therefore, because all external controls of the model are the same for each year, the initial conditions at the onset of the spring bloom are exactly reproduced after the winter seasons. The model simulation show that the spring bloom depends basically on the wintertime levels of nutrients and a switch to start the primary production. However, in order to simulate secondary

**TABLE 2.3** Model Parameters and Rates

| Parameter | Symbol | Unit | Numerical Value |
|---|---|---|---|
| Maximum growth rate | $r_{max}^0$ | $d^{-1}$ | 1 |
| Half-saturation | $k_N^0$ | – | 0.05 |
| Plant metabolic losses | $l_{PN}$ | $d^{-1}$ | 0.1 |
| Plant death rate | $l_{PD}^0$ | $d^{-1}$ | 0.1 |
| Mineralization rate | $l_{DN}$ | $d^{-1}$ | 0.001 |
| Eppley factor | $a_E$ | $°C^{-1}$ | 0.063 |
| Onset of spring bloom | $t_{start}$ | d | 75 |
| Maximum grazing rate | $g_{max}^0$ | $d^{-1}$ | 0.5 |
| Ivlev constant | $I_v^{02}$ | – | 3 |
| Zooplankton metabolic losses | $l_{ZN}$ | $d^{-1}$ | 0.3 |
| Zooplankton death rate | $l_{ZD}^0$ | $d^{-1}$ | 0.08 |
| Nutrient initial value | $N(0)$ | – | 0.99 |
| Zooplankton initial value | $Z(0)$ | – | 0.01 |
| Phytoplankton initial value | $P_0$ | – | 0.01 |
| Detritus initial value | $D(0)$ | – | 0.99 |
| Winter mixing rate | $a_{mix}$ | $d^{-1}$ | 0.5 |
| Nutrient import rate | $S_N^{ext}$ | $d^{-1}$ | 0.005 |
| Detritus export rate | $S_D^{ext}$ | $d^{-1}$ | −0.005 |

Numerical values of the parameters and rates for the modified version of the NPZD model with variable process rates.

blooms and in particular, a reasonable annual cycle of zooplankton, the external addition of nutrients is required. As a general feature the system in the upper layer has switched from eutrophic to oligotrophic conditions after the spring bloom. Series of experiments with variations of the involve parameters (not shown) revealed that the model is stable and robust against variations of the parameters.

## 2.2.4 Eutrophication Experiments

After the first test runs with the model, we can now perform some model experiments. First, we may look at the effect of a strong mixing event due to wind forcing, which erodes the thermocline and provides a nutrient input

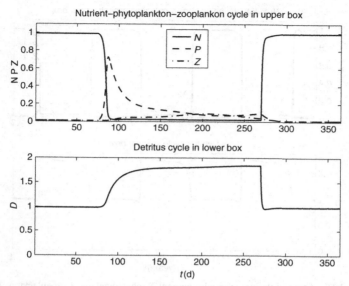

**FIGURE 2.13**  Annual cycle of the state variables nutrients, $N$, phytoplankton, $P$, zooplankton, $Z$ (upper panel), and detritus, $D$ (lower panel), for variable rates.

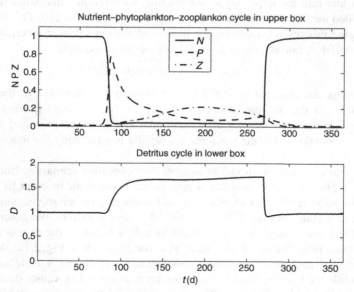

**FIGURE 2.14**  As Fig. 2.13, but with an external supply of nutrients into the upper box at a rate of $S_N^{ext} = 5 \times 10^{-4}$ d$^{-1}$ and a removal of detritus from the lower box at the same rate.

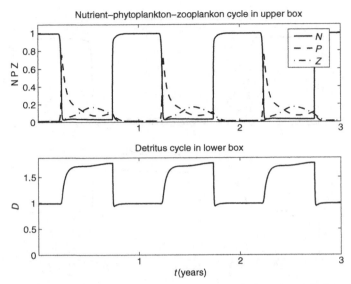

**FIGURE 2.15**   Cycle of the state variables nutrient $N$, phytoplankton, $P$, zooplankton, $Z$ (upper panel), and detritus, $D$ (lower panel), over 3 years for variable rates and external supply of nutrients into the upper box at a rate of $S_N^{ext} = 2 \times 10^{-4}\,d^{-1}$ and a removal of detritus from the lower box at the same rate.

from below into the upper layer. We assume, for example, that wind forcing has eroded the thermocline for 10 days, say, from day 210 to 220. This implies that during this period, a substantial amount of nutrients flow at a certain rate, $l_{DN} = 0.01 d^{-1}$, into the upper layer. We choose as an example

$$L_{DN} = l_{DN}[\theta(t - 210) - \theta(t - 220)].$$

The results are depicted in Fig. 2.16 and show a fast response of the phytoplankton to the nutrient injection. Owing to the enhanced grazing pressure in late summer, the model phytoplankton is quickly consumed by the model zooplankton; i.e. the nutrients are rapidly passed along the model food chain.

We can also use the model to explore eutrophication scenarios. Eutrophication implies excessive nutrient supply to the ecosystem. In order to create eutrophication in the model, we may add nutrients, i.e. dissolved inorganic matter, at a constant rate, $S_N^{ext} = 2 \times 10^{-3}\,d^{-1}$, to the system, but contrary to runs in the previous section, we do not include a burial of the excess matter at the same rate. The result of a three-year run is shown in Fig. 2.17. We find that the total amount of the winter concentrations of $D$ and $N$ increases like $S_N^{ext}t$, while the nutrient level in the upper layer assumes low values during the summer (Fig. 2.17, solid line). The peak values of the spring phytoplankton bloom, as well as the peaks of the zooplankton, increase, while the timing of

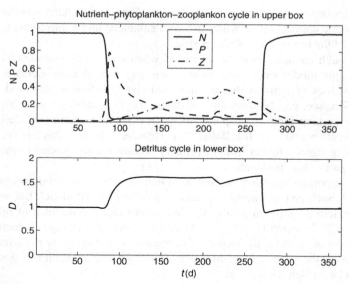

**FIGURE 2.16** Annual cycle of the state variables nutrient, $N$, phytoplankton, $P$, zooplankton, $Z$ (upper panel), and detritus, $D$ (lower panel), with a nutrient injection due to a mixing event for the days 210–220.

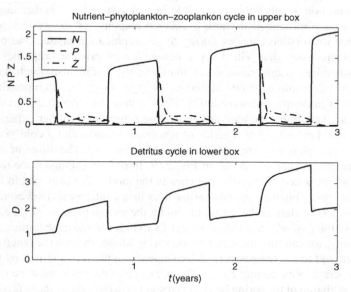

**FIGURE 2.17** Eutrophication effect on the annual cycles of the state variables nutrient $N$, phytoplankton, $P$, zooplankton, $Z$ (upper panel), and detritus, $D$ (lower panel), for variable process rates and with external nutrient supply at a rate of $S_N^{\text{ext}} = 2 \times 10^{-3}\,\text{d}^{-1}$.

the phytoplankton bloom remains unchanged. Substantial changes occur in the zooplankton concentration, which assumes enhanced levels during the first few months of the biologically active period.

Although caution has to be exercised when comparing simulations based on a simple model with a real system, we may note that an increase of the winter values of nutrient due to increased river loads was observed in the Baltic Sea; see, e.g. Nehring and Matthäus, 1991. A similarly clear signal of the zooplankton biomass response to the loads was not detected during the eutrophication phase of the Baltic Sea. However, the fish catches increased suggesting that the higher level of primary production was passed to the upper trophic part of the food web.

An important question is how a system reacts if the eutrophication is reversed. Such an experiment can easily be done with the model. Let us allow for an external supply at a rate, $S_N^{ext}$, which is twice the removal of nutrients through $S_D^{ext}$, for a period of 3 years. After three years, this changes in such a way that the export is twice the import of nutrients; i.e. nutrients are removed from the model at the, reversed rate during the following 3 years. With $\tau = 3 \times 365$ d, this can be written formally as

$$S_N^{ext} = 0.002(1 + \theta(\tau - t)) \quad \text{and} \quad S_D^{ext} = -0.002(1 + \theta(t - \tau)).$$

The results of this simulation are shown in Fig. 2.18. Obviously, we arrive at initial state of the model system; i.e. the model is reversible with regard to the eutrophication. This reversibility is an interesting feature, which leads to the question whether this is possible in real systems or whether important processes are missing that cause a delayed response of the system or even introduce irreversible changes during the eutrophication phase. One obvious class of processes that will delay a restoration of pristine properties is the biogeochemistry transformations of matter in the sediments, which have not yet been taken into account. Moreover, changes in species composition can delay or even destroy the reversibility of the system. In order to involve aspects of climate variations, we look at temperature effects in a further experimental simulation. Let us assume a series of a normal, a warm and a cold year. The normal year refers to the temperature shown in Fig. 2.12. The differences to the cold and warm year are shown in Fig. 2.19. In order to simulate the different years, we must first change the switches in the model. For example, in a warm year the spring bloom can start earlier than in a colder year. However, so far we have set the start of the spring bloom by the switch $\theta(\Delta d - d_0)$, i.e. by the length of the day, which is independent of variations in the temperature. As one possibility, we can link the onset of the spring bloom to both the length of the day, $\Delta d$, and to the temperature, $T$. Assume that a thermocline will be formed for temperatures exceeding $T_{start} = 2.5\,°C$, and that the daily radiation required for the initiation of the spring bloom corresponds to day 60, i.e. $d_0 = 60$ with the corresponding normalized length of the day, $\Delta d$. Then we can replace the switch $\theta(\Delta d - d_0)$ in Equations (2.29)–(2.31) by $\theta(\Delta d - d_0)\theta(T - T_{start})$. Because this

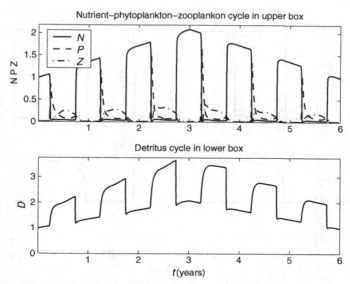

**FIGURE 2.18** Annual cycle of the state variables nutrient $N$, phytoplankton, $P$, zooplankton, $Z$ (upper panel), and detritus, $D$ (lower panel), for variable process rates and external sources and sinks. During the first three years, nutrient are supplied at a rate of $S_N^{ext} = 2 \times 10^{-3} \mathrm{d}^{-1}$, and during the last three years, nutrients are removed at the same rate.

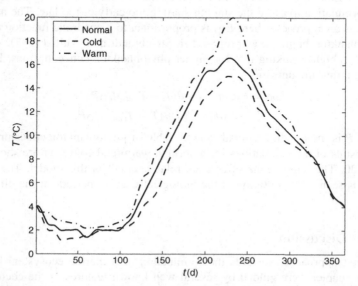

**FIGURE 2.19** Annual cycle of the mixed layer temperature (upper box) of three consecutive years (solid: normal year, dash: cold year, dash-dotted: warm year).

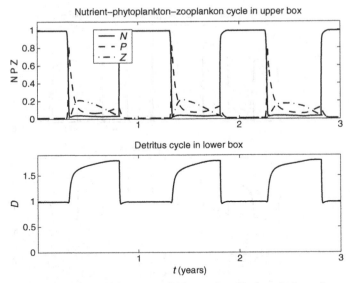

**FIGURE 2.20** Annual cycle of the state variables nutrient $N$, phytoplankton, $P$, zooplankton, $Z$ (upper panel), and detritus, $D$ (lower panel), for three consecutive years with different temperature variation as shown in Fig. 2.19.

change implies that the start time set in the sinking rate (2.32) is now different from the time of the onset of the model bloom, we have to modify the sinking rate formulation to avoid the sinking starting too early or too late. The natural choice is an approach where $L_{PD}$ is proportional to a quadratic function of the phytoplankton biomass, as proposed in Stigebrandt and Wulff (1987), which implies a higher sinking rate for higher phytoplankton concentrations. In our notation, this amounts to

$$L_{PD} = \theta(\Delta d - d_0)\theta(T - T_{start})l_{PD}P$$
$$+ \theta(d_0 - \Delta d)\theta(T - T_{start})l_{PD}P,$$

where $P$ is the nondimensional state variable of phytoplankton concentration. The results of the simulations for a normal, warm and cold year are shown in Fig. 2.20. The temperature effects are relative small in this model. The main effect is seen in the duration of the biologically active periods of the different years.

## 2.2.5 Discussion

The first approaches towards the formulation of a marine ecosystem model in this chapter were guided by several well-known features of the chemical–biological dynamics in the Baltic Sea. The processes controlling the dynamics of the state variables are the physical biological interaction, which provides the condition for the initiation of the spring bloom, nutrient uptake, grazing, respiration, mortality, sinking and recycling. The model simulations with time-dependent rates reproduced the gross features of the yearly cycles of the state variables as shown in Fig. 2.14, and the general response of the system to external nutrient loads during a eutrophication phase. The model results are qualitatively consistent with the observational findings in the Baltic Sea; (see Bodungen et al., 1981; Jansson, 1978; Kaiser et al., 1981; Nehring, 1981; Nehring and Matthäus, 1991; Schulz et al., 1978). The MATLAB programs for the models that were used to generate most of the figures are listed in Appendix and available on the booksite.

Clearly, these model versions are still very simple and, according to the plan of this book, only one step on our way from simple to more complex models. Apart from the very simple 'chemical biological dynamics', the physics is also simplified in a radical manner. Caution has to be exercised when comparing the results of a horizontally integrated model with observations that are not horizontally integrated. Averages over basins will generally deviate from local *in situ* observations. This problem can be reduced through a linkage of circulation models with biological and chemical model components. Such a coupled model can be constructed by embedding, for example, a model of the type defined by Equations (2.25)–(2.28) into a circulation model. Coupled model systems include a more complex physical control and can generate spatial variations of the state variables, often referred to as patchiness. For example, wind driven coastal upwelling which will be discussed in Chapter 5, can supply nutrients into the upper layer and stimulate the formation of plankton patches. Those patches will be transported by currents and dispersed by turbulent diffusion. The resulting uneven spatial distribution of the biological quantities illustrates the difficulty to validate biological models by means of time series at fixed positions. The interesting question—how averages of the results of the simulations with a coupled three-dimensional model compare with the results of the horizontal averaged model—can only be answered with the help of three-dimensional models. This type of problem is beyond the scope of this chapter, which discusses development and testing of a simple biological model. This simple model, however, could serve as chemical–biological model component in a more complex coupled model system. Coupled models will be discussed in Chapter 6.

## 2.3 SIMPLE PLANKTON MODELS FOR THE OCEAN

In the previous sections, we considered models that are relevant for temperate coastal seas, such as the Baltic Sea. Similar annual cycles of phytoplankton are known in the subarctic North Atlantic; (see, e.g. Miller and Wheeler, 2012). The plankton dynamics in the surface layer of the ocean exhibit a similar pattern, as shown in Fig. 2.8. The main difference is that below the seasonal thermocline, there is a virtual infinite nutrient reservoir. In winter, the surface layer is enriched with nutrients due to vertical mixing, while the phytoplankton stock is low, because of the mixing and of the low illumination. In spring, a thermocline develops and a phytoplankton spring bloom builds up. Because the nutrient supply from below is strongly reduced by the density stratification, the growth of phytoplankton depletes the nutrients in the surface layer.

In summer, the algae are consumed by animals and the phytoplankton stock assumes low levels due to sinking and grazing, while the zooplankton stock increases. In fall, the grazing animals decline and may enter nonfeeding stages for the forthcoming winter. Storms in the fall start to erode the thermocline and enable nutrient pulses from below the thermocline into the upper layer. This triggers brief fall blooms until the stratification is destroyed and vertical mixing resupplies the surface layer with nutrients, which are then available to support the next spring bloom. A more detailed discussion of the annual cycle can be found, for example, in the book of Miller and Wheeler (2012).

### 2.3.1 A Simple NPZ Model for the Ocean Mixed Layer

There are many versions of an NPZD model of the surface layer of the ocean, and we discuss a few, often-quoted examples. We start with the strongly simplified NPZ model of Franks et al. (1986), which takes only the state variables phytoplankton, zooplankton and nutrients, in terms of nitrogen, in account. Ignoring the dynamics of detritus implies that dead plankton, $P$ and $Z$, is directly transformed into nutrients (Fig. 2.21). As an obvious oversimplification, Franks et al. considered the mixed layer as a closed box with no import or exports to the water column below the mixed layer. The equations are,

$$\frac{dN}{dt} = -r_{max}\frac{NP}{k_N + N} + l_{PN}P + l_{ZN}Z + \gamma g(P)Z,$$

$$\frac{dP}{dt} = r_{max}\frac{NP}{k_N + N} - l_{PN}P - g(P)Z,$$

$$\frac{dZ}{dt} = (1 - \gamma)g(P) - l_{ZN}Z$$

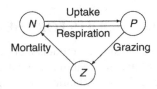

**FIGURE 2.21**   Sketch of the interaction of the state variables in an NPZ model.

we used largely our notations. Phytoplankton grows with a Michaelis–Menton rate, with a maximum $r_{max} = 2/d$ and a half-saturation constant $k_N = 1$ mmol N/m$^3$. The Ivlev grazing function is used, $g(P) = g_{max}(1 - e^{-I_v P})$, with $g_{max} = 1.5/d$ and $I_v = 1$ mmol N/m$^3$. The loss terms, which are called death rates in Franks et al. (1986), are $l_{PN} = 0.1/d$ and $l_{ZN} = 0.2/d$. They introduced a nondimensional rates $\gamma$, to take an unassimilated grazing fraction into account, with the choice $\gamma = 0.3$, but it is mentioned that observational rates vary between 0.3 and 0.8. We note that $1 - \gamma$ is the grazing efficiency. The initial values of the state variables are $N(0) = 1.6$ mmol N/m$^3$, $P(0) = 0.3$ mmol N/m$^3$ and $Z(0) = 0.1$ mmol N/m$^3$, where the choice of the initial nutrient level is obviously underestimated by a factor of 6; see. e.g., Evans and Parslow (1985), Fasham et al. (1990), Miller and Wheeler (2012). A simulation with these sets of parameters gives the rather wildly oscillating state variables as shown in the upper panel of Fig. 2.22(a). These oscillatory behaviour is unrealistic and is obviously caused by the swift recycling of nutrients. To introduce some damping, Franks et al. (1986) modified the grazing term as

$$g^{mod}(P) = g_{max}I_v P(1 - e^{-I_v P}).$$

This approach was motivated by the work of Mayzaud and Poulet (1978). For small phytoplankton concentrations this is equivalent to the Ivlev formula with squared argument, see Equation (2.23),

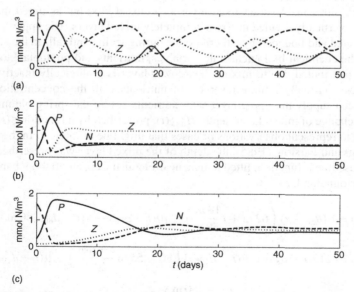

(a)

(b)

(c)

$t$ (days)

**FIGURE 2.22**  Annual cycle of the state variables nutrient, $N$, phytoplankton, $P$, zooplankton, $Z$, with the NPZ model of Franks et al. (1986): (a) The standard run, (b) run with linear growth of zooplankton for high levels of phytoplankton and (c) a run with reduced grazing efficiency.

$$g^{\text{mod}}(P) \approx g_{\text{max}} l_v^2 P^2 \quad \text{for } l_v P \ll 1.$$

But for high phytoplankton concentration, the grazing increases linearly with $P$ without saturation,

$$g^{\text{mod}}(P) \approx g_{\text{max}} l_v P \quad \text{for } l_v P \gg 1.$$

The results of a simulation with the grazing type is shown in Fig. 2.22(b). We find that the oscillations are strongly damped and the state variables quickly reach steady states. However, the zooplankton concentration is obviously much too high. A constant zooplankton stock has been established at a level close to the spring bloom peak.

Another way to damp the oscillations is a change of the value of the unassimilated grazing fraction from $\gamma = 0.3$ to $0.7$. This was pointed out in Miller and Wheeler (2012). With this change, the simulation produces the behaviour shown in Fig. 2.22(c), where the state variables exhibit weak oscillations while approaching a steady state.

### 2.3.2 NPZ and NPZD Models for the Annual Cycle of the Oceanic Mixed Layer

A similar, but more sophisticated, model for the plankton dynamics in the mixed layer of the North Atlantic was proposed by Evans and Parslow (1985). They did not consider the mixed layer as a closed box, but assumed that the ocean is vertically divided into two completely mixed layers. The upper layer is illuminated and bounded by a thermocline during spring and fall. In the surface layer, the chemical biological state variables, nutrients and phytoplankton, are evenly distributed due to mixing. The lower layer is biologically inactive and provides a virtually infinite reservoir of nutrients with the concentration $N_0$, which can supply the upper layer with nutrients before the spring bloom. The rate of change of mixed layer depth, $H_{\text{mix}}(t)$, prescribed by a function $\zeta(t)$, was chosen to represent observations or cases that might be of interest. In the paper of Evans and Parslow (1985), the depth of the mixed layer was prescribed by a piecewise linear function, fitted to data near Flemish Cap, an offshore bank east of Newfoundland, as

$$H_{\text{mix}}(t) = \theta(d_{58} - t)\left(62\,\text{m} + t\frac{18\,\text{m}}{58\,\text{d}}\right) + [\theta(t - d_{58}) - \theta(t - d_{80})]80\,\text{m}$$

$$+[\theta(t - d_{80}) - \theta(t - d_{100})]\left(80 - 55\,\text{m}\frac{t - d_{80}}{20\,\text{d}}\right) + \theta(t - d_{100})25\,\text{m}$$

$$+\theta(t - d_{250})37\,\text{m}\left(\frac{t - 250\,\text{m}}{115\,\text{m}}\right).$$

This amounts to a piecewise constant change of the mixed layer depth,

$$\frac{dH_{mix}}{dt} = \zeta(t) = \left\{ \theta(d_{58} - t)\frac{18}{58} - (\theta(t - d_{80}) - \theta(t - d_{100}))\frac{55}{20} \right.$$
$$\left. + \theta(t - d_{250})\frac{37}{115} \right\} \frac{m}{d}.$$

For some processes, only the deepening of the mixed layer has to be considered. To capture this difference, the quantity $\zeta^+ = \max(\zeta(t), 0)$ was introduced,

$$\zeta^+(t) = \left[ \theta(d_{58} - t)\frac{18}{58} + \theta(t - d_{250})\frac{37}{115} \right] \frac{m}{d}.$$

The annual cycle of the assumed mixed layer is shown in Fig. 2.23(a).

The chemical biological dynamics is represented by the following set of equations:

$$\frac{dH_{mix}}{dt} = \zeta(t),$$

$$\frac{dN}{dt} = r_{max}\frac{NP}{k_N + N} + l_{PN}P + \frac{a_{mix} + \zeta^+(t)}{H_{mix}}(N_0 - N),$$

$$\frac{dP}{dt} = r_{max}\frac{NP}{k_N + N} - l_{PN}P - g_{max}\frac{\theta(P - P_0)(P - P_0)Z}{k_P + (P - P_0)}$$
$$- \frac{a_{mix} + \zeta^+(t)}{H_{mix}}P,$$

$$\frac{dZ}{dt} = fg_{max}\frac{\theta(P - P_0)(P - P_0)Z}{k_P + (P - P_0)} - l_{ZD}Z - \frac{\zeta}{H_{mix}}Z.$$

Here we used again our notations for the state variables and parameters. The respiration rate, loss of phytoplankton to nutrients, is chosen as $l_{PN} = 0.07\,d^{-1}$. For the grazing function, Evans and Parslow (1985) used a Michaelis–Menten function with a grazing threshold at the concentration $P_0$. They introduce a grazing efficiency, $f = 0.5$. The death rate $l_{ZD} = 0.07\,d^{-1}$ corresponds to loss of zooplankton to detritus; although detritus is not represented by a state variable, it is implicitly involved through sinking of dead zooplankton into the lower layer where it is remineralized in the nutrient reservoir. Changes of the mixed-layer thickness affect the state variables in a different way. Deepening of the mixed layer, $\zeta > 0$, implies that water from the inactive zone is mixed with surface water. Then nutrients from the lower layer will mix with nutrient in the surface layer, while phytoplankton and zooplankton will be diluted. If the mixed layer becomes more shallow, no new water is mixed into the surface layer. Consequently, nutrient and phytoplankton concentrations remain unchanged. It is further assumed that the model zooplankton can move vertically to maintain feeding. Thus, in contrast to the phytoplankton, a shallower mixed layer will

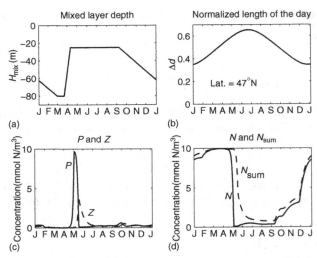

**FIGURE 2.23**     Annual cycles of (a) the mixed-layer depth, (b) the normalized length of the day for 47°N, (c) phytoplankton, $P$, and zooplankton, $Z$, state variables, and (d) of the nutrient (nitrogen) concentration, $N$, and the sum of the nitrogen in all three state variables, with a slightly modified version of the NPZ model of Evans and Parslow (1985).

concentrate the zooplankton. Across the bottom of the surface layer, some diffusion is assumed, which affects phytoplankton and nutrients. Contrary to the model of Franks et al. (1986), in this model the total amount of nutrients is not conserved because of the exchange of material with the infinite reservoir below the upper layer. If we add the model equations, it follows:

$$\frac{d}{dt}(N + P + Z) = \frac{a_{mix} + \zeta^{+}(t)}{H_{mix}}(N_0 - N) - \frac{a_{mix} + \zeta^{+}(t)}{H_{mix}}P - l_{ZD}Z$$

$$- (1 - f)g_{max}\frac{\theta(P - P_0)(P - P_0)Z}{k_P + (P - P_0)} - \frac{\zeta}{H_{mix}}Z.$$

A crucial point is that the regeneration of the initial conditions requires sufficient mixing in winter to resupply the upper layer with nutrients. This can be done with the help of an enhanced mixing, $a_{mix}$, in winter. We use as proxy for the annual cycle of irradiation the normalized length of the day, $\Delta d$, which is described in Chapter 5. An example of $\Delta d$ for the latitude of 47°N is shown in Fig. 2.23(b). We control the onset of the bloom with the assumption that the mixed-layer depth has been reduced to 70 m. The annual cycle of the phytoplankton and zooplankton is shown in Fig. 2.23(c). This cycle recurs every year. The dynamics of the nutrients and the sum of the nitrate in the three state variables is shown in Fig. 2.23(d). During the annual cycle, the mixed layer loses material after the decline of the spring bloom and in summer, but in winter the nutrients are resupplied from the lower part of the ocean. Thus, a stable annual cycle is strongly depends on the winter mixing. The parameters used in this model, which are listed in Table 2.4, are largely the same as in Evans and Parslow (1985), except the maximum uptake rate and the diffusion rates, and we do not explicitly describe the light but use the length of the day as proxy.

These relatively simple approaches were utilized to study the basinwide dynamics of the phytoplankton dynamics in the North Atlantic using steady-state conditions of the climatological mixed-layer-depth data in May (Levitus, 1982) and the latitudinal variation in solar irradiance at the sea surface (Wroblewski et al., 1988). The basinwide impact of the spring bloom on ocean optics for the North Atlantic was investigated by Wroblewski et al. (1988), with this model extended into a time-dependent version to resolve the spring phytoplankton bloom. For such a study, the self-shading of phytoplankton becomes important because the phytoplankton concentration is relevant for the attenuation of light and hence for the transparency of the ocean mixed layer. In this case, the spatial and temporal changes of irradiation cannot be parameterized by the length of the day but must be treated explicitly with irradiation models. Interpolations between the monthly mixed-layer-depth climatologies of Levitus (1982) were used to compute the time-dependent mixing rates. The plankton model combines elements of Evans and Parslow (1985) and Franks et al. (1986). The model estimates the state variables for the North Atlantic in a grid of $1°$ latitude and longitude resolution. There is no lateral connection between the

**TABLE 2.4** Model Parameters

| Parameter | Symbol | Unit | Value |
|---|---|---|---|
| Deep nutrient | $N_0$ | mmol N m$^{-3}$ | 10 |
| Uptake half-saturation | $k_N$ | mmol N m$^{-3}$ | 0.5 |
| Maximum uptake rate | $r_{max}$ | d$^{-1}$ | 0.7 |
| Plant metabolic loss | $l_{PN}$ | d$^{-1}$ | 0.07 |
| Grazing threshold | $P_o$ | mmol N m$^{-3}$ | 0.1 |
| Grazing half-saturation | $k_Z$ | mmol N m$^{-3}$ | 1 |
| Maximum grazing rate | $g_{max}$ | d$^{-1}$ | 1 |
| Grazing efficiency | $f$ | – | 0.5 |
| Loss to carnivores | $l_{ZD}$ | d$^{-1}$ | 0.07 |
| Summer diffusion rate | $a_{mix}$ | m d$^{-1}$ | 0.025 |
| Winter diffusion rate | $a_{mix}$ | m d$^{-1}$ | 0.05 |

The model parameters according to Evans and Parslow (1985), except the maximum uptake rate and the diffusion rates, which are modified.

boxes. Because the distribution of nutrients and the phytoplankton dynamics are affected by horizontal advection and upwelling, these approaches can be considered as early attempts, before fully coupled biological and circulation models were implemented (e.g. Sarmiento et al., 1993).

A further expansion of the model of Evans and Parslow (1985) was provided by Fasham et al. (1990). This model was designed for the coupling into a primitive equation circulation model; (see Sarmiento et al., 1993). There are additional state variables, the detritus, $D$, and the biomass of bacteria, $B$, included. Moreover, nitrate was split into nitrate nitrogen, $N_n$, ammonium nitrogen, $N_r$ and labile dissolved organic nitrogen, $N_d$. An important new aspect of the model compared with the previous ones is that the sinking of detritus, which provides a flux and allows a quantification of material exports from the mixed layer. To estimate a range of realistic sinking rates, Fasham et al. (1990) investigated the sensitivity of the simulation results to different sinking rates.

The additional state variables imply that the number of parameters increases. In particular, because the model zooplankton can feed on phytoplankton, bacteria and detritus, choices or preferences must be defined. This issue will be considered in the next chapter.

# Chapter 3

# More Complex Models

In Chapter 2, we gave a detailed description of a simple model to simulate the yearly cycle of nutrients and plankton in a horizontally averaged two-layer system. Apart from the inclusion of the temperature in several process rates, the physical control was reduced to the formation and the destruction of a thermocline in spring and autumn, respectively. The dynamical ingredients were nutrient-limited growth, food-limited grazing, sinking, loss and recycling. Model runs with constant rates simulate the spring bloom in a reasonable manner but fail to reproduce the observed yearly cycle. An improvement could be achieved by time-dependent sinking rates, which reflect changes of the species composition. During the vernal bloom, diatoms with a relatively high sinking velocity dominate, while the later-developing flagellates and blue-green algae were assumed to have neutral buoyancy or even float towards the sea surface. This assumption has the advantage that we need only one bulk phytoplankton state variable; however, this state variable means different groups at different periods of the year. Important biological processes, such as competition of different groups of phytoplankton for nutrients, remained unresolved. As the next step to improve the model, we will include explicit state variables for the most significant functional groups.

## 3.1 COMPETITION

In natural systems, organisms compete for resources and, according to Darwin's *Theory of Natural Selection*, the competition among genotypes is a driving mechanism of evolution. Within the course of evolution, different species have developed to find an environment that fits best to their demands. Changing environmental conditions—e.g. nutrients, temperature, salinity, light—etc., could result in changing species composition with different dominating species. The natural next step in our model-building exercise is the inclusion of several functional groups, e.g. of phytoplankton, that compete for the same resources, instead of one bulk variable. In this section, we illustrate with a few simple examples of how competition for nutrients can be taken into account. In textbooks, the standard example for the interaction of two competing groups is the Lotka-Volterra equation; see, e.g. Murray (1993). We will, however, use equations close to the model discussed in Chapter 2. We start the discussion with a simple system of two phytoplankton groups, $P_1$ and $P_2$, which compete

Introduction to the Modelling of Marine Ecosystems. http://dx.doi.org/10.1016/B978-0-444-63363-7.00003-9

for the nutrient, $N$. For example, a group becomes dominant when it has an advantage concerning nutrient uptake compared with the other one. For the sake of simplicity, we use a simple model with group specific uptake rates and constant loss rates.

$$\frac{d}{dt}P_1 = \left( r_1 \frac{N^2}{k_1 + N^2} - l_1 \right) P_1, \tag{3.1}$$

$$\frac{d}{dt}P_2 = \left( r_2 \frac{N^2}{k_2 + N^2} - l_2 \right) P_2, \tag{3.2}$$

$$\frac{d}{dt}N = -r_1 \frac{N^2}{k_1 + N^2} P_1 - r_2 \frac{N^2}{k_2 + N^2} P_2 + l_1 P_1 + l_2 P_2, \tag{3.3}$$

where $r_1$ and $r_2$ are the maximum uptake rates, $k_1$ and $k_2$ are the half-saturation constants and $l_1$ and $l_2$ are the loss rates, which recycle the nutrients. Obviously, the group with the larger net growth (uptake–losses) will win the competition. Let us assume that the loss rates are identical for both groups, $l_1 = l_2$, and consider the case where the groups start from the same level, $P_1 = P_2$. Then a critical nutrient level, $N_c$, may exist where both groups have the same chances to develop, i.e.

$$\frac{d}{dt}P_1 = \frac{d}{dt}P_2,$$

implying

$$r_1 \frac{N_c^2}{k_1 + N_c^2} = r_2 \frac{N_c^2}{k_2 + N_c^2}.$$

By simple algebra we find

$$N_c^2 = \frac{r_1 k_2 - r_2 k_1}{r_2 - r_1}.$$

Because $N_c^2$ must be positive, it is clear that both nominator and denominator must either be positive or negative. Thus, only for cases

$$r_2 > r_1 \text{ and } \frac{r_1}{r_2} > \frac{k_2}{k_1}$$

and

$$r_2 < r_1 \text{ and } \frac{r_1}{r_2} < \frac{k_2}{k_1}$$

a critical nutrient level exists. Which of the groups eventually dominates depends on the conditions $N > N_c$ or $N < N_c$. As an example, we chose $r_1 = 2r_2$ and $k_1 = 3k_2$, implying

$$N_c = \sqrt{\frac{k_1}{3}}.$$

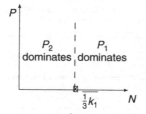

**FIGURE 3.1** Diagram of the areas with respect to the nutrient level, where one the group will win the competition.

If the nutrient concentration $N$ is less than $N_c$, then $P_2$ dominates, while in the opposite case, where the nutrient concentrations exceeds $N_c$, the phytoplankton group $P_1$ will win, as illustrated in the Fig. 3.1.

The reason for this behaviour can be easily understood. For high nutrient concentration, the uptake rates are basically proportional to the maximum rates $r_1$ and $r_2$, while at low nutrient levels, the rates are proportional to $r_1 N^2/k_1$ and $r_2 N^2/k_2$. Hence, a higher maximum rate can be balanced by a larger half-saturation constant. For $r_1 = r_2$ a critical nutrient concentration does not exist; i.e. the competition is completely controlled by the half-saturation constant. The group with the smaller half-saturation constant will win. Similarly, for different maximum rates but equal half-saturation constants, the group with the higher rate will always win.

In this consideration, we assumed that the initial values of the two groups were equal. However, the result of the competition depends also on the ratio of the initial values. In order to illustrate this, we integrate Equations (3.1)–(3.3) numerically for different choices of initial values. The involved parameters are chosen as $k_1 = 3$, $k_2 = 0.15$, $r_1 = 2r_2 = 2/\mathrm{d}$ and $l_1 = l_2 = 0.1/\mathrm{d}$. This choice implies that the group $P_1$ can take up nutrients faster at high nutrient levels but is slightly slower than the group $P_2$ for low-nutrient concentration. A series of experiments for different initial conditions of $P_1$ and $P_2$ is shown in Figs. 3.2 and 3.3. It is found that for the three cases of equal initial values, Fig. 3.2 (top), elevated initial value of $P_1$, Fig. 3.2 (middle), and elevated initial value of $P_2$, the group $P_1$ develops faster and exceeds the group $P_2$. After a further increase of the initial value of $P_2$, Fig. 3.3 (top and middle), both groups reach similar stationary values, while for a significantly higher initial value of $P_2$ this group dominates always, Fig. 3.3 (bottom).

## 3.2 SEVERAL FUNCTIONAL GROUPS

In this section, we expand the box model designed and discussed in Chapter 2 by including a higher process resolution. We consider three components of nutrients: nitrate, ammonium and phosphate. The model phytoplankton comprises three functional groups: diatoms (larger cells, 20–150 $\mu$m), flagellates (smaller

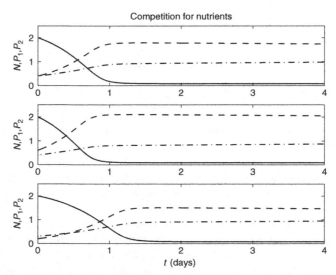

**FIGURE 3.2** Simulation of the competition of two plankton groups, $P_1$ (dashed), $P_2$ (dash-dotted), for a nutrient $N$ (solid), for different initial values. For equal initial values (top), for an elevated initial value of $P_1$ (middle), as well as for a slightly elevated initial value of $P_2$, the group $P_1$ develops faster and dominates.

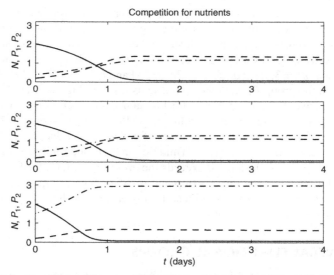

**FIGURE 3.3** Simulation of the competition of two plankton groups, $P_1$ (dashed), $P_2$ (dash-dotted), for a nutrient, $N$ (solid), for different initial values. For subsequently enhanced initial value of $P_2$ (top and middle), both groups tend to approximately the same values, while for a much higher initial value, the group $P_2$ dominates the systems.

cells, 5–20 $\mu$m) and cyanobacteria or blue-green algae (5–15 $\mu$m). The groups have different dynamical signatures. The growth rate of diatoms is considered to be almost independent of temperature variations, while those of flagellates and cyanobacteria are strongly temperature controlled. Only the diatoms are assumed to sink. Both nitrogen and phosphorus can limit the growth of diatoms and flagellates, while cyanobacteria are limited only by phosphate due to their ability to fix atmospheric nitrogen. The process of nitrogen fixation implies the reduction of gaseous nitrogen to ammonium and is an 'expansive', energy-demanding reaction. Therefore, the cyanobacteria can take full advantage of their ability to fix nitrogen at depleted dissolved inorganic nitrogen conditions, i.e. when nitrate and ammonium are no longer available.

For the living cells, we assume the Redfield ratio is valid. We denote the molar ratio of phosphorus to nitrogen as $s_R = 1/16$. Respiration of the model plankton implies the release of ammonium and phosphate in the fixed ratio, $s_R$. Ammonium is either taken up again by phytoplankton or is transformed by a given rate to nitrate via the process of nitrification. As in Chapter 2, the model zooplankton is aggregated in a bulk biomass state variable, $Z$. A list of the state variables is given in Table 3.1.

The use of the Redfield ratio implies a balanced growth of the cells with respect to their internal nutrient pool. Nutrient assimilation includes no deviations during fluctuations in environmental conditions. It is known, e.g. Droop (1974), that the growth rate of phytoplankton also depends on the ratio of the internal nutrient concentrations. There are models that take

**TABLE 3.1** State Variables and Initial Values

| State Variable | Symbol | Initial Values Upper Box (1) | Initial Values Lower Box (2) |
|---|---|---|---|
| Diatoms | $P_1$ | 0.01 | 0.01 |
| Flagellates | $P_2$ | 0.01 | 0.01 |
| Blue-greens | $P_3$ | 0.01 | 0.01 |
| Zooplankton | $Z$ | 0.01 | 0.01 |
| Detritus | $D$ | 0.01 | 0.4 |
| Ammonium, $NH_4$ | $A$ | 0.02 | 0.02 |
| Nitrate, $NO_3$ | $N$ | 1 | 1 |
| Phosphate, $PO_4$ | PO | 0.15 | 0.15 |

Note the state variables and initial values are normalized by $N_{ref} = 4.5 \, \text{mmol m}^{-3}$.

variations in algal content (cell quota), e.g. nitrogen to carbon, or chlorophyll to carbon into account; see, e.g. Geider et al. (1998), Tett (1987), Baretta-Bekker et al. (1997), and references therein. Moreover, it is known that nitrogen fixation by diazotrophic phytoplankton can result in substantial deviations from the Redfield ratio (Karl et al., 1997; Fennel et al., 2002). On the other hand, it is clear that the inclusion of internal nutrient pools requires significantly more state variables and parameters. Therefore, we use the Redfield ratio for simplicity throughout the book.

We consider two vertically arranged boxes, where the upper box represents the euphotic zone, and prescribe the dynamics of the state variables for both boxes. Thus, instead of four state variables considered previously in the box model in Chapter 2 (Section 2.2), we have now to deal with eight variables in each box. This amounts to a total of 16 state variables and the same number of differential equations, which govern the dynamics of the corresponding variables (Fig. 3.4).

We start with the set of equations for the upper box for ammonium $A^{(1)}$, nitrate $N^{(1)}$, phosphate $PO^{(1)}$, diatoms $P_1^{(1)}$, flagellates $P_2^{(1)}$, cyanobacteria $P_3^{(1)}$, detritus $D^{(1)}$ and zooplankton $Z^{(1)}$, where the upper index indicates the box number:

$$\frac{d}{dt}A^{(1)} = -\Pi_A(R_1 P_1^{(1)} + R_2 P_2^{(1)}) + l_{P_1A}(P_1^{(1)} - P_0) + l_{P_2A}(P_2^{(1)} - P_0)$$

$$+l_{P_3A}(P_3^{(1)} - P_0) + l_{ZA}(Z^{(1)} - Z_0)$$

$$-L_{AN}A^{(1)} + L_{DA}D^{(1)}, \tag{3.4}$$

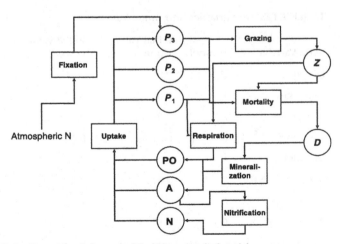

**FIGURE 3.4**   Conceptional diagram of the biogeochemical model.

$$\frac{d}{dt}N^{(1)} = -\Pi_N(R_1 P_1^{(1)} + R_2 P_2^{(1)}) + L_{AN}A^{(1)}$$

$$-A_{\text{mix}}(N^{(1)} - N^{(2)}) + Q_N^{\text{ext}}, \tag{3.5}$$

$$\frac{d}{dt}PO^{(1)} = s_R\Big[-\sum_{n=1}^{3} R_n P_n^{(1)} + l_{P_1A}(P_1^{(1)} - P_0) + l_{P_2A}(P_2^{(1)} - P_0)$$

$$+l_{P_3A}(P_3^{(1)} - P_0) + l_{ZA}(Z^{(1)} - Z_0) + L_{DA}D^{(1)}\Big]$$

$$-A_{\text{mix}}(PO^{(1)} - PO^{(2)}) + Q_{PO}^{\text{ext}}, \tag{3.6}$$

$$\frac{d}{dt}P_1^{(1)} = R_1 P_1^{(1)} - l_{P_1A}(P_1^{(1)} - P_0)$$

$$-G_1 Z^{(1)} - w_1^{\text{sink}}(P_1^{(1)} - P_0), \tag{3.7}$$

$$\frac{d}{dt}P_2^{(1)} = R_2 P_2^{(1)} - (l_{P_2A} + l_{P_2D})(P_2^{(1)} - P_0) - G_2 Z^{(1)}, \tag{3.8}$$

$$\frac{d}{dt}P_3^{(1)} = R_3 P_3^{(1)} - (l_{P_3A} + l_{P_3D})(P_3^{(1)} - P_0) - G_3 Z^{(1)}, \tag{3.9}$$

$$\frac{d}{dt}D^{(1)} = l_{P_2D}(P_2^{(1)} - P_0) - G_{D^{(1)}}Z^{(1)} - L_{DA}D^{(1)}$$

$$+l_{P_3D}(P_3^{(1)} - P_0) + l_{ZD}(Z^{(1)} - Z_0) - w_D^{\text{sink}}D^{(1)}, \tag{3.10}$$

$$\frac{d}{dt}Z^{(1)} = (G_1 + G_2 + G_3 + G_{D^{(1)}})Z^{(1)}$$

$$-(l_{ZA} + l_{ZD})(Z^{(1)} - Z_0). \tag{3.11}$$

We note that the changes in phosphate, as described by Equation (3.6), equal the changes of nitrate and ammonium, scaled by the Redfield ratio, $s_R$,

$$\frac{dPO}{dt} \sim s_R\left(\frac{dN}{dt} + \frac{dA}{dt}\right),$$

apart from external sources and vertical mixing. In the next step, we have to define the processes in terms of rates. We start with the growth rates of the phytoplankton groups. These rates are only needed for the upper box, which represents the euphotic zone. Hereafter, we use the following notation for our limiting function,

$$Y(\alpha, x) = \frac{x^2}{\alpha^2 + x^2}. \tag{3.12}$$

The growth rate for diatoms is chosen as

$$R_1 = r_{\text{max}_1} \min[Y(\alpha_1, A + N), Y(s_R\alpha_1, PO)],$$

where the minimum function ensures that the most limiting nutrient determines the rate. For the flagellates, we use a similar formula but include a temperature-dependent filter

$$R_2 = r_{\text{max}2} \frac{2 + \exp[\beta_{\text{fl}}(T_f - T)]}{1 + \exp[\beta_{\text{fl}}(T_f - T)]} \min[Y(\alpha_2, A + N), Y(s_R \alpha_2, \text{PO})].$$

For the model cyanobacteria (blue-greens), we use a similar expression with a temperature-dependent filter, but we assume that phosphate is the only limiting nutrient. This implies, that the model cyanobacteria biomass can grow at depleted nitrogen and ammonium concentrations as long as phosphate is available. This assumption implicitly involves the process of nitrogen fixation because the blue-green algae take up nitrogen from the atmospheric, along with phosphate according to the Redfield ratio. The rate is described by

$$R_3 = r_{\text{max}3} \frac{1}{1 + \exp[\beta_{\text{bg}}(T_{\text{bg}} - T)]} Y(s_R \alpha_3, \text{PO}).$$

Moreover, we define preference rates for the nutrient uptake as

$$\Pi_A = \frac{A^{(1)}}{A^{(1)} + N^{(1)}}, \quad \Pi_N = \frac{N^{(1)}}{A^{(1)} + N^{(1)}},$$

with $\Pi_A + \Pi_N = 1$. The numerical values of the parameters are listed in Table 3.2. The temperature filter functions for flagellates and cyanobacteria are plotted in Fig. 3.5. Note that the rate for the model diatoms is chosen to be independent of the temperature.

Respiration processes are described as losses to ammonium, $l_{P_1A}$, $l_{P_2A}$, $l_{P_3A}$ and $l_{ZA}$. Similarly, the remineralization of detritus is written as loss of detritus to ammonium. As in Chapter 2, the process rates denoted by capitals, e.g. $L_{DA}$ and $L_{AN}$, are prescribed as functions of state variables and other parameters, while the lowercase letter, '$l$', represents parameters. The mineralization rate depends on the temperature and is chosen as

$$L_{DA} = l_{DA}[1 + \beta_{DA} Y(T_{DA}, T)].$$

Ammonium is either consumed by phytoplankton or is transformed to nitrate at the rate, $L_{AN}$. This process is known as nitrification and is described by

$$L_{AN} = l_{AN} \exp(\beta_{AN} T). \tag{3.13}$$

Next we consider the grazing processes, which are again described by modified Ivlev functions. The zooplankton grazes only in the surface box on flagellates and cyanobacteria, while grazing on diatoms and detritus occurs in both boxes. There are different preferences for the different food resources. The preference rates for grazing are defined by

$$\Pi_1^{(1,2)} = \frac{P_1^{(1,2)}}{P_{\text{sum}}^{(1,2)}}, \quad \Pi_2 = \frac{P_2}{P_{\text{sum}}}, \quad \Pi_3 = \frac{P_3}{P_{\text{sum}}},$$

**TABLE 3.2** Process Rates and Parameters

| Parameter | Diatoms | Flagellates | Blue-greens |
|---|---|---|---|
| Growth | $r_1^{max} = 1\,d^{-1}$ | $r_2^{max} = 0.7\,d^{-1}$ | $r_3^{max} = 0.5\,d^{-1}$ |
| Half-sat. | $\alpha_1 = 0.3$ | $\alpha_2 = 0.15$ | $\alpha_3 = 0.5$ |
| Sinking | $w_1 = 0.1\,d^{-1}$ | $w_2 = 0$ | $w_3 = 0$ |
| Respiration | $l_{P_1A} = 0.01\,d^{-1}$ | $l_{P_2A} = 0.01\,d^{-1}$ | $l_{P_3A} = 0.01\,d^{-1}$ |
| Mortality | $l_{P_1D} = 0.02\,d^{-1}$ | $l_{P_2D} = 0.02\,d^{-1}$ | $l_{P_3D} = 0.02\,d^{-1}$ |
| T-control | – | $T_f = 9°C$ | $T_{bg} = 16°C$ |
|  |  | $\beta_{fl} = 1\,(°C)^{-1}$ | $\beta_{bg} = 1\,(°C)^{-1}$ |
| Grazing | $g_1^{max} = 0.25\,d^{-1}$ | $g_2^{max} = 0.5\,d^{-1}$ | $g_3^{max} = 0.25\,d^{-1}$ |
| Zooplankton |  |  |  |
| Grazing on $D$ | $g_D^{max} = 0.15\,d^{-1}$ |  |  |
| Ivlev parameter | $I_v = 1.5\,e^{a_E T}$ |  |  |
| Respiration | $l_{ZA} = 0.03\,d^{-1}$ |  |  |
| Mortality | $l_{ZD} = 0.01\,d^{-1}$ |  |  |
| Detritus |  |  |  |
| Sinking of $D$ | $w_D = 0.15\,d^{-1}$ |  |  |
| Recycling | $l_{DA} = 0.003\,d^{-1}$ |  |  |
| Nitrification | $l_{AN} = 0.1\,d^{-1}$ |  |  |
| $O_2$ constant | $O_{AN} = 0.01$ |  |  |
| T-control | $\beta_{DA} = 20$ |  |  |
|  | $\beta_{AN} = 0.11\,(°C)^{-1}$ |  |  |
|  | $T_{DA} = 13°C$ |  |  |

Phytoplankton, zooplankton and detritus process rates and parameters.

and

$$\Pi_D^{(1)} = \frac{D^{(1)}}{P_{sum}}, \quad \Pi_D^{(2)} = \frac{D^{(2)}}{P_{sum}}, \tag{3.14}$$

where $P_{sum}^{(1,2)} = \sum_{n=1}^3 P_n^{(1,2)} + D^{(1,2)}$. The grazing rates are prescribed by

$$G_1^{(1,2)} = \Pi_{1,2}\, g_1^{max}[1 - \exp(-I_v P_1^2)], \tag{3.15}$$

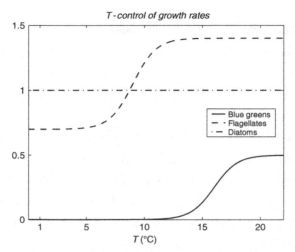

**FIGURE 3.5** Temperature dependence of the growth rates of the model phytoplankton groups: solid – cyanobacteria (blue-greens), dashed – flagellates and dash-dotted – diatoms.

$$G_2 = \Pi_2 \, g_2^{\max} \exp(a_E T)[1 - \exp(-I_v P_2^2)], \qquad (3.16)$$

$$G_3 = \Pi_3 \, g_3^{\max}[1 - \exp(-I_v P_3^2)]. \qquad (3.17)$$

For the grazing on flagellates, we have introduced the Eppley factor, see Equation (2.33), which implies an increasing ingestion with increasing temperature. Because we use the same annual temperature cycle as in Section 2.2, there is no need to limit the activity of zooplankton at very warm conditions. This will be considered in Chapter 6. Moreover, detritus is grazed with different efficiency in the upper box,

$$G_{D^{(1)}} = \Pi_D^{(1)} \, g_{D^{(1)}}^{\max}[1 - \exp(-I_v D^{(1)2})], \qquad (3.18)$$

and in the lower box,

$$G_{D^{(2)}} = \Pi_D^{(2)} \, g_{D^{(2)}}^{\max}[1 - \exp(-I_v D^{(2)2})]. \qquad (3.19)$$

The rates and parameters are listed in Table 3.2. The loss processes of phyto- and zooplankton diminish when small background values, $P_0$ and $Z_0$ for the $P_n$s and $Z$, respectively, are approached. External sources, $Q_N^{\text{ext}}$ and $Q_{\text{PO}}^{\text{ext}}$, are included in the nitrate and phosphate equations (Equations 3.6 and 3.5), respectively.

Finally, we have to formulate the model equations for the lower box. In the lower model layer, there is no growth of phytoplankton. Living flagellates and cyanobacteria stay in the upper layer and their concentrations remain at the background values, $P_0$, below the surface box. However, because of the rapid sinking of the diatoms, we assume a nonzero mortality of diatoms only in the lower box, where zooplankton grazes on diatoms and detritus. For the lower box, the set of equation is

$$\frac{d}{dt}A^{(2)} = l_{ZA}(Z^{(2)} - Z_0) + l_{DA}D^{(2)} - l_{AN}A^{(2)}, \tag{3.20}$$

$$\frac{d}{dt}N^{(2)} = l_{AN}A^{(2)} + A_{mix}(N^{(1)} - N^{(2)}), \tag{3.21}$$

$$\frac{d}{dt}PO^{(2)} = s_R\left[l_{ZA}(Z^{(2)} - Z_0) + l_{DA}D^{(2)}\right]$$

$$+A_{mix}(PO^{(1)} - PO^{(2)}), \tag{3.22}$$

$$\frac{d}{dt}P_1^{(2)} = w_1^{sink}(P_1^{(1)} - P_0) - l_{P_1D}(P_1^{(2)} - P_0) - G_1^{(2)}Z^{(2)}, \tag{3.23}$$

$$\frac{d}{dt}P_2^{(2)} = 0, \tag{3.24}$$

$$\frac{d}{dt}P_3^{(2)} = 0, \tag{3.25}$$

$$\frac{d}{dt}D^{(2)} = l_{P_1D}(P_1^{(2)} - P_0) + l_{ZD}(Z^{(2)} - Z_0) - G_{D^{(2)}}Z^{(2)}$$

$$-l_{DA}D^{(2)} + w_D^{sink}D^{(1)} + S^{ext}, \tag{3.26}$$

$$\frac{d}{dt}Z^{(2)} = (G_{D^{(2)}} + G_1^{(2)})Z^{(2)} - (l_{ZA} + l_{ZD})(Z^{(2)} - Z_0). \tag{3.27}$$

The vertical mixing and, hence the biologically active time interval of the model year are again related to switches, which depend on the length of the day (see Equation 2.31).

After formulating all equations of the model, we can again start with simulations. Parameter values and rates needed are listed in Table 3.2. Here we have used plausible numerical values derived from literature. All of the quoted papers on modelling provide parameter values. For example, extensive tables of parameters can be found in the book of Jørgensen (1994). Throughout this book, we do not discuss methods of parameters estimations.

### 3.2.1 Succession of Phytoplankton

The phytoplankton community in temperate latitudes is characterized by a succession of functional groups. Each functional group describes an important class of phytoplankton in an aggregated manner. As mentioned before, the vernal bloom is dominated by relatively large cells, the diatoms, whereas in summer, smaller cells (flagellates) dominate. Next we conduct a series of experiments with the model sketched in the previous section. In the first exploring runs, we start with the case of no external loads and suppress the development of cyanobacteria; i.e. the cyanobacteria rest at their background value. This implies a closed nitrogen cycle in our model system, similar to the box model discussed in Chapter 2 but with increased process resolution. The initial values of the state variables are listed in Table 3.1. The molar ratio of the initial values of nitrate to phosphate is with $N : PO = 6.6$ clearly below the Redfield ratio, 16:1. Hence, nitrogen is the limiting nutrient in the model. The results of this simulation are shown in Figs. 3.6–3.9. The state variables show a clear annual cycle for nutrients and phytoplankton groups, Figs. 3.6 and 3.7. In the upper box, the nutrients are dissolved during winter and bound in plankton and detritus during summer. In the lower box, the dissolved nutrients accumulate during summer and are mixed over the water column, i.e. both boxes, during winter; i.e. they become equal in the two boxes. The model phytoplankton in the upper box shows the spring bloom peak dominated by diatom and summer values dominated by flagellates. In the lower box, we find substantial concentration of sinking diatoms after the spring bloom. The model phytoplankton decreases to the background level after the onset of mixing in the autumn.

The model zooplankton shows a clear response to the spring bloom but assumes the highest values during summer (see Fig. 3.8). In the lower layer, the cycle of zooplankton is characterized by the response to the sinking model diatoms after the spring bloom. If we add the state variables to bulk nitrate and phytoplankton, we obtain a similar picture to the simple box model with the bulk-phytoplankton state variable (compare, e.g. Fig. 2.11). In the next

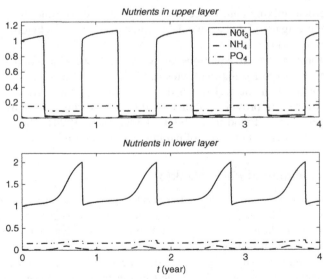

**FIGURE 3.6** Simulation of the nutrients, nitrate $NO_3$ (solid), $NH_4$ (dashed), and phosphate $PO_4$ (dash-dotted), for the upper (top panel) and lower (bottom panel) model boxes.

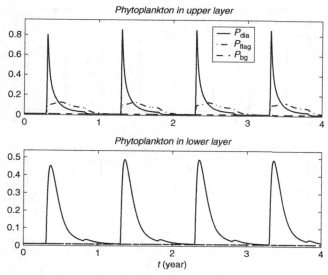

**FIGURE 3.7** Simulation of the model phytoplankton groups, diatoms $P_{dia}$ (solid), $P_{flag}$ (dash-dotted) and cyanobacteria or blue-green algae (not active in this case), $P_{bg}$ (dashed), for the upper model box (top panel). In the lower model box (bottom panel) the model diatoms that sink down accumulate, while the flagellate groups do not sink as living cells.

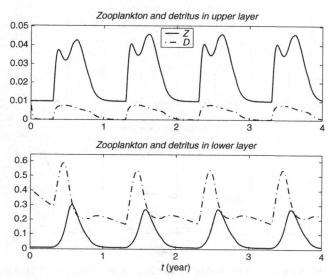

**FIGURE 3.8** Simulation of the model zooplankton, $Z$ (solid), and detritus, $D$ (dashed), for the upper (top panel) and lower (bottom panel) model box.

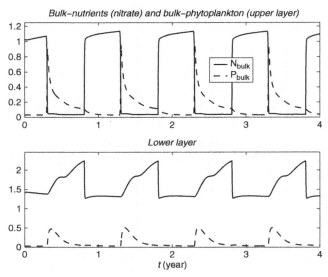

**FIGURE 3.9**    Annual variations of the bulk nutrients (inorganic and detritus nitrogen) (solid lines) and bulk phytoplankton (dashed) in the upper and lower box (top and bottom panels), respectively.

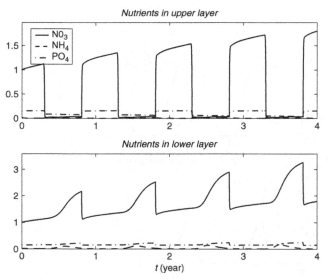

**FIGURE 3.10**    Simulation of the cycling of nutrients, nitrate $NO_3$ (solid), $NH_4$ (dashed), and phosphate $PO_4$ (dash-dotted), for the upper (top panel) and lower (bottom panel) model boxes. Nitrate is added to the surface layer at a constant rate, $Q_N^{ext} = 10^{-3} d^{-1}$.

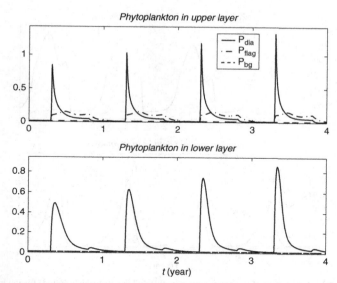

**FIGURE 3.11** Simulation of the phytoplankton, diatoms (solid) and flagellates (dash-dotted), for the upper (top panel) and lower (bottom panel) model boxes. The cyanobacteria (dashed) are not active in this experiment and stay on their background concentration. Nitrate is added to the surface layer at a constant rate, $Q_N^{ext} = 10^{-3} d^{-1}$.

experiment, we add nutrients to the system. It is obvious that an additional input of phosphate to the model system will not change the response because phosphorus is not the limiting nutrient. Hence, we consider only the case of the addition of nitrate at a constant rate, $Q_N^{ext} = 10^{-3} d^{-1}$. Then, in both layers, the winter values of nitrate increase like $\frac{1}{2}Q_N^{ext}t$. The results are shown in Fig. 3.10. The peaks of the spring bloom are increasing in response to the added nutrients, while the phytoplankton concentration remains practically unchanged during the summer season (Fig. 3.11). The excess production fuelled by the nutrient loads is transferred to the model zooplankton as shown in Fig. 3.12. In particular, we find a slight increase of the summer peaks in the top layer and a pronounced enhancement of the zooplankton biomass in the lower box, where the zooplankton responds to the increased diatom biomass, which sinks from the surface layer to the lower box during the spring bloom. We note that after the spring bloom, phosphate in the upper layer decreases. Hence, after several years, the spring bloom peaks will stabilize because phosphate becomes the limiting nutrient. In other words, there is a regime shift with respect to the limiting nutrient from nitrogen to phosphorus.

Next we take the role of cyanobacteria into account. First we repeat the run without external loads. The results are shown in Figs. 3.13 and 3.14. Nitrate increases although there are no external loads, but this increase becomes smaller every year. The nitrogen increases due to nitrogen fixation by cyanobacteria,

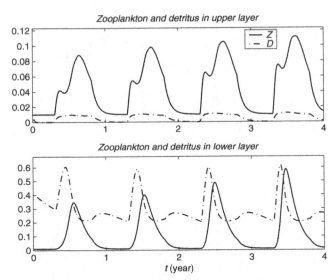

**FIGURE 3.12** Simulation of the zooplankton (solid) and detritus (dash-dotted) for the upper (top panel) and lower (bottom panel) model boxes Nitrate is added to the surface layer at a constant rate, $Q_N^{ext} = 10^{-3} d^{-1}$.

which are only limited by phosphate in the model. As more and more nitrogen gets fixed and diatoms biomass increases in spring, the pool of available phosphate becomes smaller. Therefore, less phosphate is available when conditions become favourable for cyanobacteria.

The model flagellates do not change very much, while the diatoms, in particular the spring bloom peak, increases and the cyanobacteria peak decreases. After several years, the molar ratio of the dissolved inorganic nitrogen to phosphorus approaches the Redfield ratio, 16:1, and the model reaches a stationary cycle.

## 3.3   N₂ FIXATION

The process of nitrogen fixation was included in the model by assuming that the cyanobacteria are limited only through the phosphate. Thus, the cyanobacteria biomass can increase even for depleted nitrate, provided phosphate is still available and other conditions are favourable, i.e. high levels of solar radiation, a warm surface layer and low levels of small-scale turbulence. Then cyanobacteria import nitrate at the Redfield ratio during the uptake of phosphate; hence, nitrogen concentrations can increase even for the case of no external loads. This is shown in Fig. 3.15, where the total nitrogen, sum of nitrogen in all state variables and the total phosphate are plotted for last simulation in the previous section (see Figs. 3.13 and 3.14). While without the activities of cyanobacteria

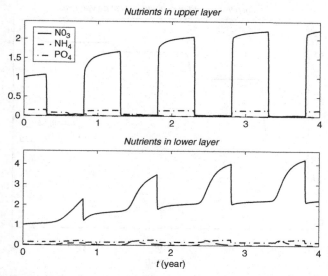

**FIGURE 3.13** Simulation of the nutrients, nitrate $NO_3$ (solid), $NH_4$ (dashed) and phosphate $PO_4$ (dash-dotted), for the upper (top panel) and lower (bottom panel) model boxes.

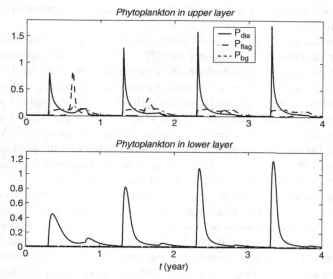

**FIGURE 3.14** Simulation of the model phytoplankton groups, diatoms $P_{dia}$ (solid), flagellates $P_{flag}$ (dash-dotted) and cyanobacteria (blue-green algae) $P_{bg}$ (dashed), for the upper model box (top panel). In the lower model box (bottom panel), the model diatoms accumulate that sink down, while the other to groups do not sink as living cells.

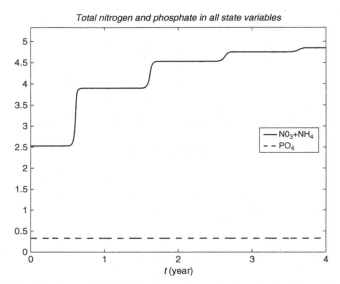

**FIGURE 3.15**  Development of the total nitrogen concentration, dissolved anorganic nutrients, $NH_4$, $NO_3$ and nitrogen in the organic material. After a strong increase of the total nitrogen in the first 3 years, the total nitrogen tends to a saturation level.

the total nitrogen in our model system would be strictly conserved, we find a significant increase of nitrogen in the case of active cyanobacteria due to nitrogen fixation.

In a further simulation, we add nutrients by external loads to the system. We start with the addition of nitrate at a rate $Q_N^{ext} = 10^{-3}\,d^{-1}$. The development of the model nutrients and phytoplankton is shown in Figs. 3.16 and 3.17. Only in the first model year does a substantial cyanobacteria signal, with a strong peak in summer, occur. The added nitrate shifts the molecular ratio of nitrogen to phosphorus towards the Redfield ratio within a 2 year period, and after the second model year, the system is virtually limited by phosphate. After the third year, the addition of nitrate has only a small effect on the phytoplankton development in the model, and the nitrate concentration in the surface layer even shows a slight increase after the spring bloom, Fig. 3.16.

A dramatically different response is found when we add phosphate to the model system with active blue-green algae, say, at a rate of $Q_P^{ext} = 10^{-3}\,d^{-1}$. Then the growth of cyanobacteria is not as strongly restricted by a depletion of phosphate as in the case of strictly constant total phosphorus, and a strong import through fixation of atmospheric nitrogen can occur. The results of the simulation are shown in Figs. 3.18–3.21. The slow increase of phosphate drives an increase of nitrate. This increase is temporarily stopped when phosphate becomes limiting for cyanobacteria development in the summer seasons, Figs. 3.18 and 3.19.

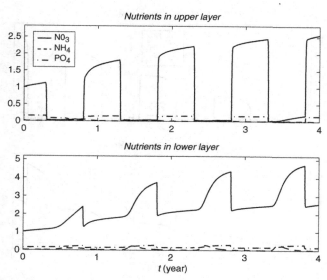

**FIGURE 3.16**   Simulation of the nutrients, nitrate $NO_3$ (solid), $NH_4$ (dashed) and phosphate $PO_4$ (dash-dotted), for the upper (top panel) and lower (bottom panel) model boxes. Nitrate is added at a constant rate, $Q_N^{ext} = 10^{-3} d^{-1}$.

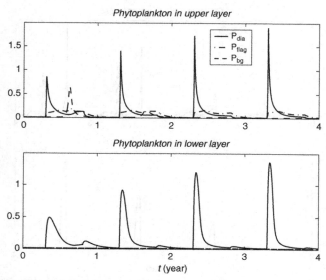

**FIGURE 3.17**   Simulation of the model phytoplankton groups, diatoms $P_{dia}$ (solid), flagellates $P_{flag}$ (dash-dotted) and cyanobacteria (blue-green algae) $P_{bg}$ (dashed), for the upper model box (top panel). In the lower model box (bottom panel), the model sinking diatoms accumulate. Nitrate is added at a constant rate, $Q_N^{ext} = 10^{-3} d^{-1}$.

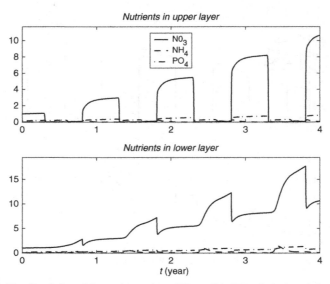

**FIGURE 3.18**   Simulation of the nutrients, nitrate $NO_3$ (solid), $NH_4$ (dashed) and phosphate $PO_4$ (dash-dotted), for the upper (top panel) and lower (bottom panel) model boxes. Phosphate is added at a constant rate, $Q_P^{ext} = 10^{-3} d^{-1}$.

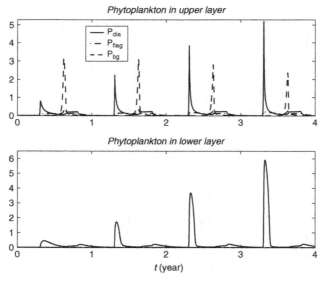

**FIGURE 3.19**   Simulation of the model phytoplankton groups, diatoms $P_{dia}$ (solid), $P_{flag}$ (dash-dotted) and cyanobacteria (blue-green algae) $P_{bg}$ (dashed), for the upper model box (top panel). In the lower model box (bottom panel), the model diatoms that sink down accumulate, while the other two groups do not sink as living cells. Phosphate is added at a constant rate, $Q_P^{ext} = 10^{-3} d^{-1}$.

In the first model year, the cyanobacteria develop rather strongly, but owing to the elevated level of nitrate, imported by the nitrogen fixation of the cyanobacteria, the spring bloom peaks increase and use up a substantial part of the dissolve inorganic nutrients at the Redfield ratio. Thus, a state develops where, due to the increased biomass of diatoms, the model phosphate in the upper box is completely exhausted after the spring bloom and the development of the cyanobacteria in the summer season depends entirely on the external phosphate inputs, added to the system in the time period after the spring bloom.

A substantial increase in the model biomass is found in the next trophic level, the model zooplankton, which ingests the increased food supply, Fig. 3.20. In the middle layer, we also find a substantial increase of the zooplankton biomass in response to the increasing biomass of sinking spring bloom diatoms. Thus, the external inputs of nutrients accumulate eventually in the highest trophic level of our model system. The surprising effects of the cyanobacteria on the budget of our model system can be seen in the total sum of the nutrients in all state variables, which are plotted for nitrogen and phosphorus in Fig. 3.21. Obviously, such an extreme increase of nitrate is unrealistic, implying that important aspects of the biogeochemistry in natural systems are missing in the model.

In particular, the high degree of mineralization implies a high oxygen consumption, which is not yet considered in the model. Low oxygen or even anoxic conditions in turn drive denitrification, which represents a sink of nitrogen. Moreover, biogeochemical processes near the bottom and in the

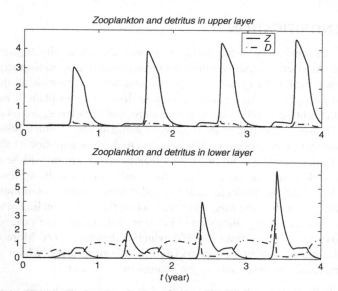

**FIGURE 3.20** Development of the zooplankton (solid) and detritus (dash-dotted) for the upper (top panel) and lower (bottom panel) model boxes with phosphate added to the surface layer at a constant rate, $Q_P^{\text{ext}} = 10^{-3} \text{d}^{-1}$.

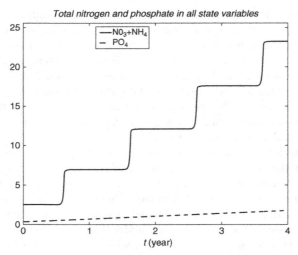

**FIGURE 3.21** Development of the total nitrogen (solid) and phosphate (dashed) concentration, dissolved anorganic nutrients, $NH_4$, $NO_3$ and nitrogen in the organic material. The total nitrogen increases strongly in response to the small addition of phosphate, which is supplied externally at a constant rate, $Q_P^{ext} = 10^{-3} d^{-1}$.

sediments have to be taken into account when dealing with shallow sea areas and high production. These aspects will be considered in the next section.

## 3.4 DENITRIFICATION

The model experiments in the previous sections elucidated the adjustment of the system to external inputs of nutrients through riverine or atmospheric loads and nitrogen fixation by cyanobacteria. Cyanobacteria can provide a substantial import of nitrogen into the system as long as dissolved phosphate is available and other conditions, such as light, temperature, and turbulence, are favourable. There is, however, the important process of denitrification, which results in a sink of nitrogen. This process is closely linked to the concentration of dissolved oxygen. Under anoxic conditions, the oxygen bound in nitrate will be used by bacteria, while molecular nitrogen remains, which can leave the system. Thus, we have to expand the model in such a way that under anoxic conditions nitrate can be used to oxidize the model detritus, while the corresponding amount of molecular nitrogen leaves the system. In order to describe these processes, we include oxygen, $O_2$, as a state variable, which is denoted as $Ox$. Moreover, we add to the model a third box, considered as a bottom layer, which receives sinking material. We take two oxygen sources into account:

- oxygen fluxes through the sea surface, i.e. air sea gas exchange,
- oxygen release during the process of primary production.

The flux through the sea surface is prescribed by

$$Ox^{\text{flux}} = p_{\text{vel}}(O_{\text{sat}} - Ox^{(1)}),$$

where the saturation concentration $O_{\text{sat}}$ depends on the temperature according to

$$O_{\text{sat}} = \frac{a_0}{O_{\text{norm}}}(a_1 - a_2 T),$$

with the coefficients $a_0 = 31.25 \, \text{mmol m}^{-3}$, $a_1 = 14.603$, $a_2 = 0.4025 \, (°\text{C})^{-1}$ and $O_{\text{norm}} = 375 \, \text{mmol m}^{-3}$ (Livingstone, 1993). The photosynthetic oxygen production can be described, in a strongly simplified manner, as

$$h\nu + CO_2 + H_2O \rightarrow CH_2O + O_2,$$

implying a one-to-one molar ratio of assimilated carbon to produced oxygen. The term $h\nu$ indicates the light energy needed to drive the photosynthesis. The sinks of oxygen are represented by the consumption of oxygen due to respiration of phytoplankton, nitrification and microbial decomposition of dead organic material (Fig. 3.22). For the following considerations, we need a consistent description of the chemical transformations, biogeochemical

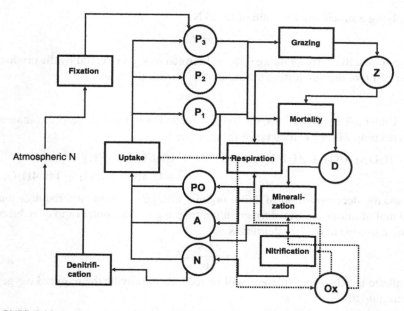

**FIGURE 3.22** A conceptional diagram of the biogeochemical model as in Fig. 3.4, but with inclusion of oxygen as chemical state variable and denitrification.

reactions and the relevant stoichiometric ratios. First we recall that the Redfield ratio follows from the stoichiometric composition of organic matter,

$$(CH_2O)_{106}(NH_3)_{16}(H_3PO_4)_1,$$

i.e. the molar ratio of carbon to nitrogen to phosphorous in living or dead cells is 106:16:1 (Redfield et al., 1963). We use at several places throughout the book the definition $s_R = 1/16$ for the ratio of phosphorous to nitrogen.

In the model, the processes of mineralization of organic material by microbial processes means detritus is oxidized and ammonium and phosphate are reproduced. At oxic conditions, this process is fuelled by dissolved oxygen,

$$(CH_2O)_{106}(NH_3)_{16}H_3PO_4 + 106O_2 \rightarrow 106CO_2 + 16NH_3$$
$$+ H_3PO_4 + 106H_2O.$$

The one-to-one ratio of carbon required to release oxygen (1 mol organic carbon requires 1 mol oxygen) can be utilized to find the relation for ammonia, $NH_3$, with the help of the Redfield ratio of $C/N = 106/16 = 6.625$. This implies the molar relationship

$$s_2 = 6.625.$$

The decomposition of 1 mol organic nitrogen to ammonia requires 6.625 mol oxygen. For the nitrification of ammonia to nitrate, another 2 mol of oxygen are needed,

$$16NH_3 + 32O_2 \rightarrow 16HNO_3 + 16H_2,$$

implying a stoichiometric ratio of $O_2$ to N,

$$s_4 = 2.$$

To estimate the corresponding relation between oxygen released by the production of 1 mol organic nitrogen from nitrate, we use the ratio,

$$s_3 = 8.625.$$

Under anoxic conditions, the oxygen needed to mineralize detritus (anaerobic decomposition) will be taken from nitrate

$$(CH_2O)_{106}(NH_3)_{16}H_3PO_4 + 84.8HNO_3 \rightarrow 106CO_2 + 16NH_3$$
$$+ 42.4N_2 + H_3PO_4 + 148.4H_2O.$$

Thus, the decomposition of 1 mol organic nitrogen to ammonia requires that 5.3 mol of nitrate are transformed into nitrogen gas. The molar ratio of reduced nitrate needed to oxidize detritus is

$$s_1 = \frac{84.8}{16} = 5.3.$$

If nitrate is exhausted, sulphate will be reduced to provide the required oxygen according to

$$(CH_2O)_{106}(NH_3)_{16}H_3PO_4 + 53H_2SO_4 \rightarrow 106CO_2 + 53H_2S$$
$$+ H_3PO_4 + 106H_2O + 16NH_3.$$

By anaerobic decomposition, hydrogen sulphide is synthesized, and nitrogen is recycled with a molar ratio of sulphur to nitrogen,

$$s_5 = \frac{53}{16} = 3.3125,$$

i.e. 3.3 mol of sulphate are reduced to hydrogen sulphide for 1 mol organic nitrogen. To oxidize 1 mol of $H_2S$ to sulphate, 2 mol of $O_2$ are needed.

$$H_2S + 2O_2 + 2H_2O \rightarrow S_2O_4^- + H_2O^+.$$

Because hydrogen sulphide and oxygen do not coexist in natural systems, we count hydrogen sulphide as negative oxygen equivalents. Thus, the negative oxygen equivalent for hydrogen sulphide amounts to the ratio,

$$\frac{H_2S}{O_2} = \frac{1}{2}.$$

Multiplication of hydrogen sulphide values in $mmol/m^3$ by 2 gives the value of 'negative oxygen' in $mmol/m^3$, which can be used to describe the hydrogen sulphide concentration (Fonselius, 1981).

With the help of the stoichiometric ratios, listed in Table 3.3, we can now formulate the equations for the state variable oxygen in the three model boxes. The equation for oxygen in the upper model box is

$$\frac{d}{dt} Ox^{(1)} = \frac{N_{norm}}{O_{norm}} \Bigg[ (s_2 \Pi_A + s_3 \Pi_N) \sum_{n=1}^{3} R_n (P_n^{(1)} - P_0) \tag{3.28}$$

$$-s_2 (l_{P_1A}(P_1^{(1)} - P_0) + l_{P_2A}(P_2^{(1)} - P_0)$$

$$+l_{P_3A}(P_3^{(1)} - P_0) + l_{ZA}(Z^{(1)} - Z_0))$$

$$-s_4 L_{AN} A^{(1)} \Bigg] - A_{mix}(Ox^{(1)} - Ox^{(2)}) + Ox^{flux}.$$

**TABLE 3.3** Stoichiometric Ratios

| Parameter | Symbol | Value |
|---|---|---|
| Redfield-ratio (P/N) | $s_R$ | 1/16 |
| Reduced nitrate/oxydized detritus | $s_1$ | 5.3 |
| Oxygen demand/recycled nitrogen | $s_2$ | 6.625 |
| Oxygen production related to $N$ | $s_3$ | 8.825 |
| Nitrification | $s_4$ | 2/3 |
| Hydrogen sulphide, nitrogen | $s_5$ | 3.3125 |

The first group of terms refers to the production of oxygen during primary production, while the remaining terms in the bracket refer to the loss of oxygen due to respiration and nitrification. The last term prescribes the oxygen flux through the air–sea interface. For the middle box we assume that, similar to the upper box, the oxygen concentration is always positive, while the temperature is practically constant on the winter level. Then we have

$$\frac{d}{dt} Ox^{(2)} = \frac{N_{norm}}{O_{norm}} \left[ -s_2 l_{ZA}(Z^{(2)} - Z_0) - s_4 L_{AN} A^{(2)} \right. \tag{3.29}$$

$$\left. -s_2 l_{DA} D^{(2)} \right] + A_{mix}(Ox^{(1)} - Ox^{(2)}).$$

In the bottom box, the oxygen can be positive or negative,

$$\frac{d}{dt} Ox^{(3)} = \frac{N_{norm}}{O_{norm}} \left\{ -s_4 L_{AN}^{(3)} A^{(3)} \right. \tag{3.30}$$

$$\left. -s_2(\theta(Ox^{(3)}) + [1 - \theta(Ox^{(3)})][1 - \theta(N^{(3)})]) l_{DA} D^{(3)} \right\}.$$

Note that according to our definition of the step function (2.24) it follows that $[1 - \theta(N^{(3)})] = 0$ for $N^{(3)} > 0$, but $[1 - \theta(N^{(3)})] = 1$ for $N^{(3)} = 0$. The same applies for the switch $[1 - \theta(Ox^{(3)})]$, which is zero for oxic conditions, but equals one for zero or negative oxygen concentration. The expression $\theta(Ox^{(3)}) + [1 - \theta(Ox^{(3)})][1 - \theta(N^{(3)})]$ in Equation (3.30) is one for both positive oxygen concentration, due to the first term, and zero or 'negative' oxygen concentrations, if nitrate is depleted. However, for anoxic conditions, the expression can be zero for nonzero nitrate concentration. Then denitrification is possible and oxygen is taken from the nitrate. If all nitrate is reduced, microbial decomposition of detritus is connected to the production of hydrogen sulphide, i.e. 'negative oxygen'.

Moreover, the nitrification process depends on the availability of oxygen, and we have taken this into account by the rate $L_{AN}^{(3)}$, which is defined as

$$L_{AN}^{(3)} = l_{AN} \frac{Ox^{(3)}}{O_{AN} + Ox^{(3)}}; \tag{3.31}$$

see Stigebrandt and Wulff (1987). The numerical value of the parameter $O_{AN}$ is given in Table 3.2.

For the other state variables, the equations in the upper box are the same as in the set (3.4)–(3.11). The model equations for the middle box are

$$\frac{d}{dt} A^{(2)} = l_{ZA}(Z^{(2)} - Z_0) + l_{DA} D^{(2)} - l_{AN} A^{(2)}, \tag{3.32}$$

$$\frac{d}{dt} N^{(2)} = l_{AN} A^{(2)} + A_{mix}(N^{(1)} - N^{(2)}), \tag{3.33}$$

$$\frac{d}{dt} PO^{(2)} = s_R \left[ l_{ZA}(Z^{(2)} - Z_0) + l_{DA} D^{(2)} \right]$$

$$+ A_{mix}(PO^{(1)} - PO^{(2)}), \tag{3.34}$$

$$\frac{d}{dt}P_1^{(2)} = w_1^{sink}(P_1^{(1)} - P_0) - w_1^{sink}(P_1^{(2)} - P_0) \tag{3.35}$$
$$-l_{P_1D}(P_1^{(2)} - P_0),$$

$$\frac{d}{dt}P_2^{(2)} = 0, \tag{3.36}$$

$$\frac{d}{dt}P_3^{(2)} = 0, \tag{3.37}$$

$$\frac{d}{dt}D^{(2)} = l_{P_1D}(P_1^{(2)} - P_0) + l_{ZD}(Z^{(2)} - Z_0) - G_{D^{(2)}}Z^{(2)}$$
$$-l_{DA}D^{(2)} + w_D^{sink}D^{(1)} - w_D^{sink}D^{(2)}, \tag{3.38}$$

$$\frac{d}{dt}Z^{(2)} = (G_{D^{(2)}} + G_1^{(2)})Z^{(2)} - (l_{ZA} + l_{ZD})(Z^{(2)} - Z_0), \tag{3.39}$$

and, for the bottom box,

$$\frac{d}{dt}A^{(3)} = l_{DA}D^{(3)} - L_{AN}^{(3)}A^{(3)}, \tag{3.40}$$

$$\frac{d}{dt}N^{(3)} = L_{AN}^{(3)}A^{(3)} - s_1[1 - \theta(Ox^{(3)})]\theta(N^{(3)})l_{DA}D^{(3)}, \tag{3.41}$$

$$\frac{d}{dt}PO^{(3)} = s_R l_{DA}D^{(3)}, \tag{3.42}$$

$$\frac{d}{dt}P_1^{(3)} = w_1^{sink}(P_1^{(2)} - P_0) - l_{P_1D}(P_1^{(3)} - P_0), \tag{3.43}$$

$$\frac{d}{dt}P_2^{(3)} = 0, \tag{3.44}$$

$$\frac{d}{dt}P_3^{(3)} = 0, \tag{3.45}$$

$$\frac{d}{dt}D^{(3)} = l_{P_1D}(P_1^{(3)} - P_0)$$
$$-l_{DA}D^{(3)} + w_D^{sink}D^{(2)} + S^{ext}, \tag{3.46}$$

$$\frac{d}{dt}Z^{(3)} = 0. \tag{3.47}$$

Note that in the second equation for the bottom box (3.41) the process of denitrification is taken into account, which amounts to a sink of nitrate under anoxic conditions (negative oxygen). As already mentioned in the discussion of the differential equation for the model oxygen in the bottom box (3.30), it is implicitly assumed that at the rate of the microbial use of oxygen, which is bound in nitrate to oxidize detritus, the resulting molecular nitrogen leaves the system. This process can only work as long as nitrate exists. Thus, as nitrate is depleted and $N^{(3)}$ becomes zero then the step function switches off the process. At the same time, microbial use of oxygen bound in sulphate starts and produces hydrogen sulphide, counted as negative oxygen. This process is activated at zero level of nitrate and zero or negative oxygen.

In the bottom box it is assumed, for simplicity, that no living phyto- or zooplankton exists. All organic material enters the bottom box as detritus, which is subject to microbial transformations, as described in Equation (3.46) by the rate $l_{P_1 D}$.

### 3.4.1 Numerical Experiments

As in the previous two sections, we conduct a series of experiments with the expanded model: first, with no external imports, combined with the two cases of no active and active cyanobacteria; second, with external nitrogen import, combined with the two cases of no active and active cyanobacteria; and, third, with external phosphorous import and active cyanobacteria, see Table 3.4. The initial conditions are listed in Table 3.1, where the detritus value is chosen for the bottom box, while in the upper and middle boxes the initial value of detritus is zero. The initial values of oxygen, in model units, are chosen as $Ox^{1,2} = 1.3$ and $Ox^{1,2} = 0.1$. We note, that the oxygen values in ml l$^{-1}$ follows from the model value by multiplication with the factor 8.4. The MATLAB programs for these experiments are list in the appendix and are available on the website of the book.

#### 3.4.1.1 Experiment 1

In the first experiment, we set external inputs of nutrients to zero and deactivate the dynamics of cyanobacteria; i.e. the growth rate of cyanobacteria vanishes and there is no grazing on cyanobacteria. The results of the simulation for

**TABLE 3.4** List of the Model Experiments

| |
| --- |
| **Experiment 1:** |
| No imports of nutrients |
| (a) Deactivated cyanobacteria |
| (b) Activated cyanobacteria |
| **Experiment 2:** |
| Import of nitrogen |
| (a) Deactivated cyanobacteria |
| (b) Activated cyanobacteria |
| **Experiment 3:** |
| Import of phosphorous |
| (a) activated cyanobacteria |

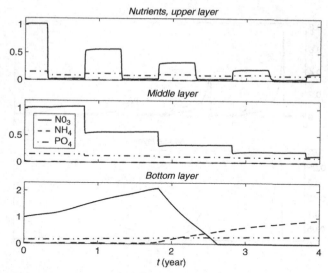

**FIGURE 3.23** Simulation of the cycling of nutrients, nitrate $NO_3$ (solid), $NH_4$ (dashed) and phosphate $PO_4$ (dash-dotted), for the upper (top panel), middle (middle panel) and lower (bottom panel) model boxes.

4 years are shown in Figs. 3.23–3.27. First, we look at the dissolved nutrients in the three layers (Fig. 3.23).

Because phosphate is not a limiting nutrient, it shows no significant changes. The winter values of nitrate in the upper box are decreasing. Due to the sinking of the model diatoms and detritus, there is a downward flux of material which accumulates in the bottom box. In the lower box, nitrate increases during the first two model years, but then the dissolved nitrate concentration decreases until it is completely depleted. This is due to the process of denitrification, which is controlled by the oxygen concentration in the bottom layer (see Fig. 3.26).

While the oxygen in the upper and middle box shows an annual cycle, mainly in response to the temperature and mixing during the winter season, the oxygen in the bottom box approaches zero after two model years and becomes negative later. At the same time, ammonium starts to increase because nitrification stops when the oxygen is exhausted. The decrease of dissolved nitrogen in the upper layer implies decreasing levels of phytoplankton (Fig. 3.24). In particular, the model diatoms, which dominate the spring bloom, are reduced to smaller levels while the flagellates are only weakly affected. The decreasing food resources results in a diminishing zooplankton biomass (Fig. 3.25). The detritus in the bottom layer diminishes also because the supply through sinking plankton decreases while the microbial driven mineralization of detritus to ammonium continues by consuming oxygen bound in sulphate. Thus, hydrogen sulphide is produced, which is counted as negative oxygen (see Fig. 3.26).

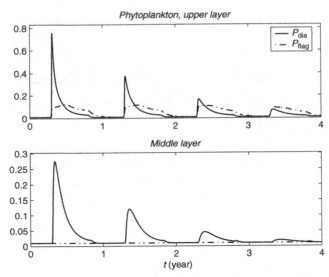

**FIGURE 3.24**    Simulation of the model phytoplankton groups, diatoms $P_{dia}$ (solid) and flagellates $P_{flag}$ (dash-dotted) for the upper model box (top panel). The model diatoms sink down into the middle model box (bottom panel), while the living flagellates do not sink.

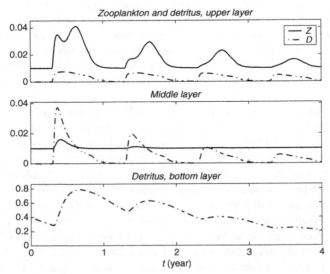

**FIGURE 3.25**    Simulation of the model zooplankton (solid) and detritus (dash-dotted) in experiment 1 for the upper (top panel), middle (middle panel) and (bottom panel) model box.

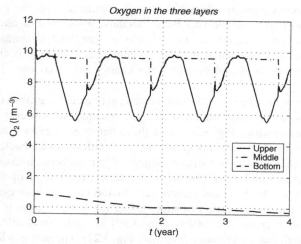

**FIGURE 3.26** Simulated development of the model oxygen in the upper (solid), middle (dash-dotted) and bottom (dashed box). Hydrogen sulphide is expressed by 'negative oxygen'.

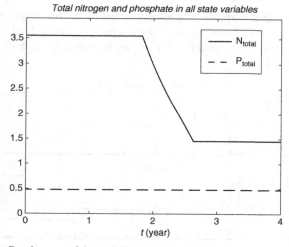

**FIGURE 3.27** Development of the total nitrogen, dissolved anorganic nutrients, $NH_4$ and $NO_3$, and nitrogen in the organic material (solid) and the total phosphate concentration (dashed). The total nitrogen is strongly reduced due to denitrification.

The total amount of the nutrients, nitrogen and phosphorous in all state variables is plotted in Fig. 3.27. These figures demonstrates that a substantial amount, more than 50%, of the initially available nitrogen is lost through denitrification. Thus we find that without external supply of nitrate, the model system becomes oligotrophic after a couple of years.

These results show a substantial change compared with the previous experiments with the one or two boxes models, where stable annual cycles have established. Because the molar ratio of dissolved inorganic nutrients ($N : P$) is significant smaller than the Redfield ratio (16:1), it is clear that an external addition of phosphate would not change the behaviour of the model system.

However, if the dynamics of cyanobacteria activated, i.e. growth of and grazing on cyanobacteria are 'switched on', the dynamics of nutrients (Fig. 3.28), model phytoplankton (Fig. 3.29), and the other state variables are strongly modified. In the first two model summers, significant cyanobacteria blooms are simulated with the associated import of atmospheric nitrogen through fixation. We also find a decrease of the winter concentrations of the nutrients in the upper model box, but slower than in the case without cyanobacteria (Fig. 3.23). The accumulation of nutrients in the bottom layer is faster than without cyanobacteria. The budget (Fig. 3.30) is significantly affected by the import through nitrogen fixation (compare Fig. 3.27). The nitrogen loss through denitrification is about balanced by the gain through fixation of nitrogen.

### 3.4.1.2 Experiment 2

In the second experiment, we retain the settings of the first experiment except that an external supply of nitrate added into the upper layer at constant rate, $Q_N^{ext} = 10^{-3}\,d^{-1}$, is considered. First we consider the case of deactivated

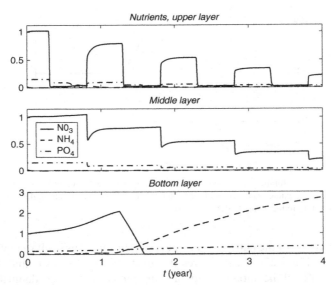

**FIGURE 3.28** Simulation of the cycling of nutrients as in Fig. 3.23, but with active-model cyanobacteria.

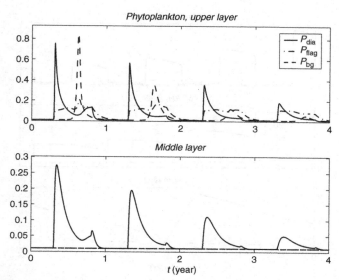

**FIGURE 3.29** Simulation of the cycling of the phytoplankton groups as in Fig. 3.24, but with active-model cyanobacteria.

cyanobacteria (i.e. growth of and grazing on cyanobacteria are 'switched off' in the model). The results of a 4 year simulation are shown in Figs. 3.31–3.33. The development of nutrients in the three boxes (Fig. 3.31) shows a decrease of the winter values of nitrate in the upper layer and an eventual accumulation of the dissolved inorganic nutrients in the bottom layer, as is the case without import (Fig. 3.23). The main difference in the case of import is that the levels for

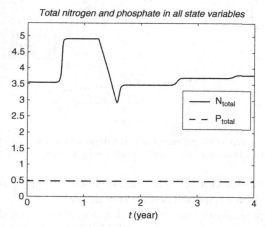

**FIGURE 3.30** Development of the total nitrogen and phosphate as in Fig. 3.27, but with active-model cyanobacteria.

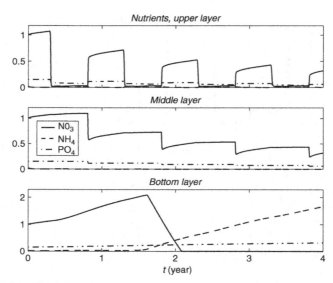

**FIGURE 3.31**   Simulation of the cycling of nutrients, nitrate $NO_3$ (solid), $NH_4$ (dashed) and phosphate $PO_4$ (dash-dotted), for the upper (top panel), middle (middle panel), and lower (bottom panel) model boxes. Nitrate is added to the surface layer at a constant rate, $Q_N^{ext} = 10^{-3}\,d^{-1}$.

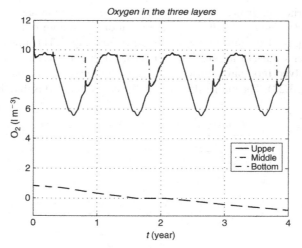

**FIGURE 3.32**   Simulated development of the model oxygen Experiment 2 in the upper (solid), middle (dash-dotted) and bottom (dashed box). Negative oxygen stands for hydrogen sulphide.

nitrogen are somewhat elevated. The increased nutrient supply is passed along to the phytoplankton groups and detritus (not shown).

The hydrogen sulphide concentration, i.e. negative oxygen, develops faster because more organic material is mineralized (Fig. 3.32). The total amount of the nutrients, nitrogen and phosphorus in all state variables (see Fig. 3.33), is

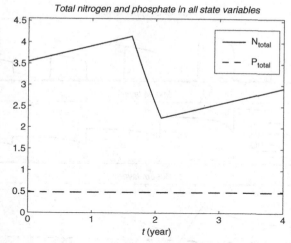

**FIGURE 3.33** Development of the total nitrogen (solid) and phosphate (dashed) concentration, dissolved anorganic nutrients, NH, NO$_3$ and nitrogen in the organic material. The total nitrogen increases in response to the small addition of nitrate, which is externally supplied at a constant rate, $Q_N^{ext} = 10^{-3}\,d^{-1}$.

different compared to the previous example (Fig. 3.27) because the nitrogen in the system increases linearly due to the external supply, except during the phase of a strong loss driven by denitrification. The denitrification phase starts earlier, but is of shorter duration. These differences from the first experiment are due to the increased concentration of detritus, which accelerates oxygen consumption. Note that the expressions that describe oxygen consumption, second term in Equation (3.30), and the denitrification, second term in Equation (3.41), are related to the detritus concentration in the bottom layer.

We repeat the simulation with active-model cyanobacteria. Then in the first two model years, we find substantial blooms of the cyanobacteria and a corresponding input of nitrogen due to the fixation (Figs. 3.34 and 3.35). After the second model year, the summer phosphate concentration in the surface layer is significantly reduced; thereafter, the general development is similar to the case without cyanobacteria. The total nitrogen budget looks similar to the case without external supply except that the inputs yield a linear increase of the total nitrogen, Fig. 3.36.

In both simulations, the upper layer becomes oligotrophic after 2–3 years because the sinking material is eventually trapped in the lower layer. The import of nitrate alone cannot compensate for the losses due to the downward flux because the molar ratio of the nutrients is also affected and phosphate is drawn from the surface layer as well. However, an interesting effect emerges in the fourth summer of the simulation. The system in the surface box becomes phosphate limited, and the external nitrate supply accumulates with high increasing concentrations of nitrate.

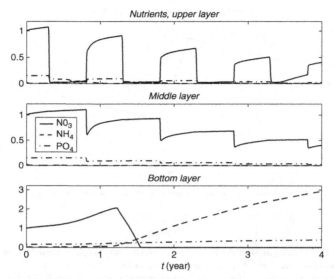

**FIGURE 3.34** Simulation of the cycling of nutrients as in Fig. 3.31, but with active-model cyanobacteria.

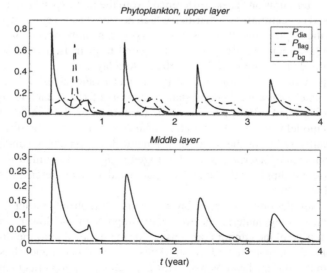

**FIGURE 3.35** Simulation of the model phytoplankton groups, diatoms $P_{dia}$ (solid), flagellates $P_{flag}$ (dash-dotted) and cyanobacteria (dash) in the upper and middle model box. Nitrate is added to the surface layer at a constant rate, $Q_N^{ext} = 10^{-3}\,d^{-1}$.

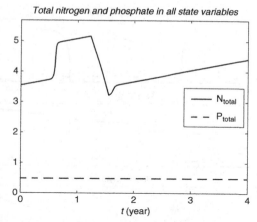

**FIGURE 3.36** Development of the total nitrogen (solid) and phosphate (dashed) concentration as in Fig. 3.33, but with active-model cyanobacteria.

### 3.4.1.3 Experiment 3

In the third experiment, we add phosphate to the upper layer at a constant rate, $Q_P^{ext} = 10^{-3} d^{-1}$, and include cyanobacteria. The results of the 4 year simulation are shown in Figs. 3.37–3.41. The nutrients in the upper layer show a clear increase in the winter values for both phosphate, due to the external supply, and, nitrate, owing to the nitrogen fixation of the cyanobacteria (see Fig. 3.37). Within the annual cycles, we find that nitrate in the upper box is depleted after the spring bloom, while phosphate is depleted after the summer bloom of the cyanobacteria (Fig. 3.38). The nutrient concentrations in the middle box clearly shows an increasing trend. In the bottom layer, the concentration of nitrate indicates that the denitrification starts earlier than in the previous experiments. This is due to the elevated level of detritus in the bottom layer, which uses up the oxygen earlier, as shown in Figs. 3.39 and 3.40.

The summer peaks of the cyanobacteria blooms (Fig. 3.38) stabilize after a few years, while the spring bloom peak of the model diatoms increases. Thus, the functional groups of diatoms and flagellates are limited by nitrogen after the spring bloom, while phosphate is available until the onset of the bloom of cyanobacteria, which use up the phosphate. The nitrogen input provided by the cyanobacteria cycles through the food web and is used in the next spring bloom, implying increasing spring bloom peaks. Phosphate is sufficiently available due to the external supply; however, the increasing phosphate concentration is balanced by the higher concentration of model diatoms, which dominate the spring bloom.

The model zooplankton shows a slight increase in the surface box and also an increasing concentration in response to the sinking diatoms after the spring

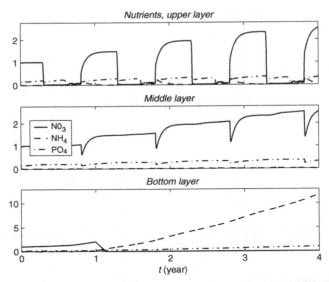

**FIGURE 3.37** Simulation of the cycling of nutrients, nitrate $NO_3$ (solid), $NH_4$ (dashed) and phosphate $PO_4$ (dash-dotted), for the upper (top panel), middle (middle panel), and lower (bottom panel) model boxes. Phosphate is added to the surface layer at a constant rate, $Q_P^{ext} = 10^{-3}$ $d^{-1}$.

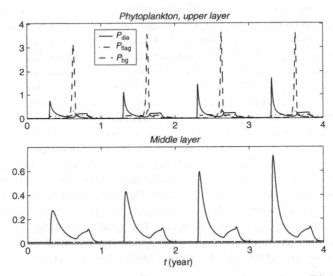

**FIGURE 3.38** Simulation of the model phytoplankton groups, diatoms $P_{dia}$ (solid), $P_{flag}$ (dash-dotted) and cyanobacteria (blue-green algae) $P_{bg}$ (dashed), for the upper model box (top panel). The model diatoms can sink into the middle box (lower panel), while the other two groups do not sink as living cells. Phosphate is added at a constant rate, $Q_P^{ext} = 10^{-3}$ $d^{-1}$.

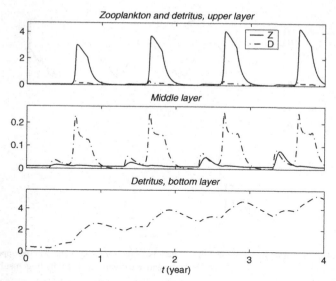

**FIGURE 3.39** Simulation of the model zooplankton (solid) and detritus (dash-dotted) in Experiment 3 for the upper (top panel), middle (middle panel) and bottom (lower panel) model box.

bloom. The strongest response to the external supply of phosphate is found in the detritus concentration in the bottom layer. This is shown in Fig. 3.39.

The total amount of nitrogen and phosphorus in all state variables is shown in Fig. 3.41. The denitrification has a substantial effect but only for a limited period. After the nitrate is depleted in the bottom layer, the system is dominated by the nitrogen import through nitrogen fixation. Comparing Figs. 3.41 and 3.21, we find basically a similar behaviour of the total nitrogen in the system except that the inclusion of the denitrification process leads to a substantial reduction in the first model year until nitrate in the bottom layer is completely reduced.

The most striking effects seen in the experiments with the two-box models in Section 3.3 and the three-box model is that the inclusion of the bottom box leads to significantly lower levels of the spring bloom peaks (compare Figs. 3.38 and 3.19) and, consequently, a much smaller response of the model zooplankton biomass in response to the addition of phosphate. The substantial flux of matter, expressed as detritus, into the bottom box limits the increase of the state variables in the surface box.

## 3.4.2 Processes in Sediments

So far we have considered processes in the water column only and described the impact of processes outside the water column via boundary conditions. In the previous sections, we included the flux of oxygen through the surface boundary.

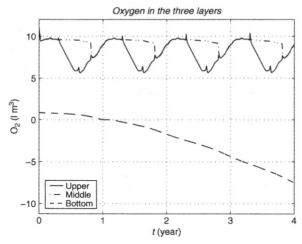

**FIGURE 3.40**   Simulated development of the model oxygen for Experiment 3 in the upper (solid), middle (dash-dotted), and bottom (dashed box). Negative oxygen stands for hydrogen sulphide.

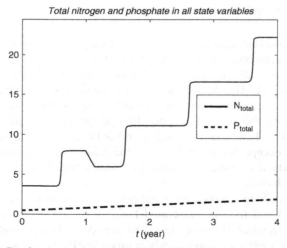

**FIGURE 3.41**   Development of the total nitrogen (solid) and phosphate (dashed) concentration, dissolved anorganic nutrients, $NH_4$, $NO_3$ and nitrogen, in the organic material. The total nitrogen increases strongly in response to the small addition of phosphate, which is supplied externally at a constant rate, $Q_P^{ext} = 10^{-3}\ d^{-1}$.

We assumed also that atmospheric nitrogen is exchanged through the air–sea interface, providing the molecular nitrogen fluxes needed for nitrogen fixation. We also considered a near-bottom model box that can become anaerobic and in which biogeochemical transformations may occur. So far we have not explicitly included benthic–pelagic interactions and biogeochemical processes in the sediments. The importance of the impact of benthic–pelagic coupling and of sedimentary processes on a system obviously depends on depth of the system at hand. The deeper the ocean, the smaller the importance of the effects of process at the seabed and the longer the time scales of the feedback of biogeochemical transformation in sediments to the pelagic system. Thus, sinking material that accumulates at the seafloor in deep oceans is virtually lost for the pelagic ecosystem for long time periods. In shallow systems, where the light reaches the bottom, the benthic flora and fauna can be as complex as in the pelagic domain. In such systems, the benthic–pelagic interaction must be an integral part of the model system.

In this consideration, we assume that the term *sediment* refers to solid material that has been, or is potentially subject to erosion, transport and deposition. In a biogeochemical model, we can assign a certain amount of the detritus, which accumulates at the bottom, to a state variable 'sediment-detritus'. We confine the following discussion on the biogeochemical active upper part of sediments, which consists of a layer of recently deposited sedimentary material. This layer of sediments has the capacity to store nutrients, which can either be released into near-bed waters or can be removed from the system through burial in deeper sediment layers. The flux of dissolved nutrients into the water column and resuspension of sediment-detritus is controlled by the turbulence level in a bottom boundary layer. Buried sedimentary material is diagenetically consolidated and has a very low biogeochemical activity. This material is practically removed from the model system and represents a sink for nutrients.

For phosphorus, the burial in the sediment represents the most important sink of the system, whereas for oxidized nitrogen there is also the microbial driven denitrification, which represents another important path for leaving the system. In relatively shallow marginal seas, a significant loss of nitrate occurs due to denitrification in the sediments. Hence, the removal of nitrogen from the system by processes in the sediments is, especially in oxygenated waters, an important factor in the nutrient budget. In summary, the most important processes that should be included into a sediment model are:

- accumulation and storage of organic matter (sediment-detritus),
- release of dissolved nutrients and resuspension of matter,
- description of biogeochemical and diagenetic processes, such as burial, denitrification and oxidation.

The nitrogen cycle in the sediments, relevant for biogeochemical processes in the water column, is not very different from that in the water column. For a simple model approach, we can assume that mineralization of nitrogen amounts to the production of dissolved inorganic nitrogen in the form of ammonium. The ammonium is released from the sediment into the near-bottom water and is then again involved in the marine biogeochemical cycle. Mineralization needs an equivalent of oxygen that might be drawn from the near-bottom water. If oxygen is depleted, the available nitrate will be reduced, and molecular nitrogen is released (denitrification) and leaves the system. If nitrate is exhausted, the microbial driven sulphate respiration is favoured, which leads to the formation of hydrogen sulphide.

Under aerobic conditions in the near-bottom water, a redoxcline will be formed in the sediments. Near this redoxcline, processes like mineralization, nitrification and denitrification occur. In a simple approach, we can parameterize these processes. A common approach is to prescribe a rate at which a certain part of mineralized and nitrified nitrogen will be denitrified.

Compared to nitrogen compounds, a considerable amount of phosphate can be bound more or less permanently within the sediment. In estuarine sediments, a pool of oxidized iron exists that binds part of the phosphate. In well-oxidized sediment, the pool of oxidized iron increases, while under low oxygen conditions the capability to bind phosphate, diminishes and mineralized phosphate is released and can leave the sediment. In the presence of hydrogen sulphide the iron-phosphate complexes are transferred into ferrous sulphide, and consequently, large amounts of phosphate are liberated.

$$Fe_{ox}PO_4{}^{3-} + H_2S \rightarrow FeS + PO_4{}^{3-}$$

The redox conditions at the sediment–water interface control the biogeochemical functioning of the sediment in the nitrogen cycle. In aerobic sediments, denitrification takes place and burial of phosphate is favoured. Anaerobic sediments, on the other hand, are likely to increase the amount of mineralized nutrients. However, low oxygen or anaerobic conditions may slow down the mineralization processes in sediments.

In principle, we can construct box models to describe the outlined processes in the sediment in a manner similar to that used to describe the pelagic system. However, the fluxes through the water–sediment interface are greatly controlled by the dynamics of the bottom boundary layer, which can be described adequately only with a three-dimensional model approach. In Chapter 6, we will give the mathematical formulations to simulate some of the previously discussed properties of the role of sedimentary processes.

Diagenetic models, which are used to explicitly resolve diagenetic processes in sediments, require a large number of state variables and equations. Their vertical resolution is very fine, say, in the order of a few millimetres. Often, these components are used as standalone models, which try to find steady-state solutions. This is not appropriate if the components are part of an ecosystem model system. A reduction of complexity and computational demands require aggregated state variables, which are essential for the description of early diagenetic processes. A successful approach was developed in the frame of the ERSEM system as described in Ruardij and van Raaphorst (1995) and Ebenhöh et al. (1995). This model shows a good compromise between a reasonable vertical resolution and a sufficient representation of key processes.

Another sediment-related model type concerns sedimentation, resuspension and transportation of material in the bottom boundary layer. The relevance for ecosystem models exists in the ability to describe near-bottom transport of biogeochemical active matter that support an important pathway for organic carbon and nutrients. The general mechanism can briefly be described as follows: Near-bottom currents and, in shallow areas, surface waves generate bottom shear stress acting on the seabed. If the bottom shear stress exceeds a critical value for bed load transport or resuspension, then sedimentary material can be mobilized from the seabed. Sedimentary particles can be transported in suspension or as bed load. When the bottom shear stress falls below a critical value, particles can be deposited at the sediment surface. The critical values of bottom shear stress are material properties depending on grain size, compactness and stickiness.

Several model approaches of different complexity were developed to simulate these processes (see Puls et al., 1994; Holt and James, 1999; Nielsen, 1994; Black and vincent, 2001; Gerritsen et al., 2001). Originally developed for sediment transport simulations, these models can also be used for the simulation of the transports of particulate organic matter.

There are also biological processes, known as bioresuspension and biodeposition, that affect fluxes through the water–sediment interface. These biologically mediated fluxes can be divided into direct and indirect processes: Direct effects are due to the activities of benthic animals, which catch particles from the water column or remove particles from the sediment. Indirect effect are related to modifications of the bottom roughness, and hence, changes of the bottom shear stress by animals living at the sea floor. A review of biologically mediated fluxes is given in Graf and Rosenberg (1997).

# Chapter 4

# Modelling Life Cycles of Copepods and Fish

In the preceding chapters, the relationships between the different trophic levels of the lower part of the food web were expressed through state variables without explicit consideration of life cycles of animals. This may be an appropriate simplification for the modelling of unicellular organisms, such as phytoplankton, as long as an explicit description of the formation of resting stages, such as spores or cysts, is not needed. Otherwise, sexual reproduction and formation of dormant stages must be included by additional state variables and process formulations. For animals such as copepods or fish, an explicit description of life cycles with different stages becomes important.

This chapter is devoted to a theoretical description of life cycles in models. We start the discussion with the trophic level of zooplankton and use copepods as example. The theoretical approach will also be applied to fish, however, with some modifications. In the models discussed so far, the role of zooplankton was implicitly included by phytoplankton mortality or by a bulk zooplankton variable, which provided a dynamically developing grazing pressure on the model phytoplankton. Many problems require a more detailed description of the variations of zooplankton, including growth, development and reproduction, as well as the feedback to the lower and higher levels of the food web. For example, studies of the recruitment success of fish involve stage-resolving descriptions of zooplankton in order to address size-selective feeding of larvae, juvenile and adult fish.

Models of zooplankton biomass, including several species and stages, were proposed by Vinogradov et al. (1972). Although several stages were included as state variables, the development—i.e. the transfer from one stage into another one—driven by growth, was not considered. Stage-resolving population models were used by Wroblewski (1982) and Lynch et al. (1998). The development of stages was included by using experimentally estimated stage duration times. A link of population models with individual growth was proposed by Carlotti and Sciandra (1989), where the rates describing the development were calculated by a growth equation for mean individuals. Individual-based modelling of population dynamics has been considered for example by Batchelder and Miller (1989) and Miller and Tande (1993). An overview of existing zooplankton models can also be found in the review article by Carlotti et al. (2000).

Introduction to the Modelling of Marine Ecosystems. http://dx.doi.org/10.1016/B978-0-444-63363-7.00004-0

## 4.1   GROWTH AND STAGE DURATION

The development of copepods is characterized by the propagation of the animals through several stages: eggs, nauplii, copepodites and adults. Each stage can be defined by a range of body weight, or more precisely, body mass. The bounds of mass intervals are defined by the mass levels where molting from the previous stage into the current one, and molting of the current stage to the next one, occur. For example, let $m$ be the mass of a certain stage, '$i$', then this stage is characterized by the interval $X_{i-1} \leq m \leq X_i$, where $X_{i-1}$ is the molting mass of the previous stage $i - 1$ and $X_i$ is the molting mass at which stage $i$ molts into the next stage $i + 1$.

The dynamics of growth of an individual animal, i.e. the increase of mass with time, is usually governed by an equation of the type

$$\frac{d}{dt}m = (g - l)m^p, \tag{4.1}$$

where $g$ is a grazing or ingestion rate that depends on food supply and temperature, $l$ is a loss or egestion rate that can be expressed as a certain amount of the ingested food and $p$ is an allometric exponent which has typically a value of about $p = 0.75$. An allometric exponent smaller than 1 reflects the fact that the higher stages grow slower relative to their body mass than the smaller stages. This equation was derived from experiments in rearing tanks (Carlotti and Sciandra, 1989).

The integration of Equation (4.1) with the initial condition $m = m_e$ for $t = 0$, where $m_e$ is the mass of the eggs, can be done by separating the variables as

$$\frac{dm}{m^p} = (g - l)dt.$$

Integration gives

$$\frac{m^{1-p}}{1-p} = \int_0^t dt'[g(t') - l(t')] + C.$$

The constant $C$ is determined by the initial condition, i.e. $C = m_e^{1-p}/(1 - p)$. For $p = 3/4$, it follows

$$m(t) = \left\{ m_e^{1/4} + \frac{1}{4}\int_0^t dt'[g(t') - l(t')] \right\}^4. \tag{4.2}$$

A further important observable quantity is the duration time of stages, $D_i$, which prescribes how long individuals stay in a certain stage. Experimental estimates of stage duration time are usually represented by so-called Belehradek formulas of the type

$$D = \frac{a}{(b + T)^c},$$

where $T$ is the temperature in degrees Celsius and $a$, $b$ and $c$ are empirical coefficients. Examples of duration times be found in Corkett and McLaren (1978).

The stage duration time, $D_i$, can also be calculated by integration of Equation (4.1) with the initial condition $m = X_{i-1}$ for $t = 0$. Similar to Equation (4.2) we find

$$m(t) = \left[ X_{i-1}^{1/4} + \frac{1}{4} \int_0^t \mathrm{d}t' \widehat{\alpha}(t') \right]^4$$

where $\widehat{\alpha} = g - l$. In particular, for $t = D_i$ it follows

$$m(D_i) = X_i = \left[ X_{i-1}^{1/4} + \frac{1}{4} \int_0^{D_i} \mathrm{d}t' \widehat{\alpha}(t') \right]^4 .$$

For strictly constant conditions, as in a rearing experiment with constant food supply and fixed temperature, the integration can be performed and we find

$$m(D_i) = X_i = \left( X_{i-1}^{1/4} + \frac{1}{4} \widehat{\alpha} D_i \right)^4 .$$

Then the duration is given by

$$D_i = \frac{4}{\widehat{\alpha}} (X_i^{1/4} - X_{i-1}^{1/4}). \tag{4.3}$$

As shown in the following subsections, there are good reasons to avoid a allometric exponent less than one. We can choose an equation for growth as

$$\frac{\mathrm{d}}{\mathrm{d}t} m_i(t) = [g_i(t) - l_i(t)] m_i, \tag{4.4}$$

where $m_i$ refers to the mass within a certain stage, i.e. $X_{i-1} \le m \le X_i$. However, the effect of the allometric scaling must now be described by stage-dependent grazing rates, $g_i(t)$, and loss rates, $l_i(t)$. We calculate the stage duration time, $D_i$, by integration of Equation (4.4) with the initial condition $m = X_{i-1}$ for $t = 0$:

$$m_i(t) = X_{i-1} \exp \left\{ \int_0^t \mathrm{d}t' [g_i(t') - l_i(t')] \right\} .$$

In particular, for $t = D_i$, we find the maximum mass of stage $i$ as

$$m(D_i) = X_i = X_{i-1} \exp \left[ \int_0^{D_i} \mathrm{d}t' \widehat{\beta}_i(t') \right] ,$$

where, $\widehat{\beta}_i = g_i - l_i$. Thus, the stage duration is implicitly given by

$$\int_0^{D_i} \mathrm{d}t' \widehat{\beta}_i(t') = \ln \left( \frac{X_i}{X_{i-1}} \right) .$$

For strictly constant conditions, it follows

$$D_i = \frac{1}{\widehat{\beta}_i} \ln \left( \frac{X_i}{X_{i-1}} \right) . \tag{4.5}$$

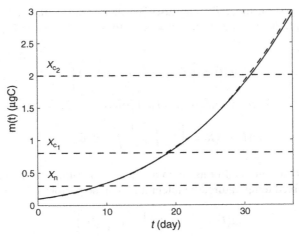

**FIGURE 4.1** Development of a model copepod through different stages as described with an allometric exponent $p = 0.75$ (solid) and $p = 1$ (dashed). The plots demonstrate the equivalence of the two growth equations (4.1) and (4.4).

Because the durations must be independent of the used description, we can express the $\widehat{\beta_i}$'s through $\widehat{\alpha}$ by combining Equations (4.3) and (4.5). Then, we find

$$\widehat{\beta_i} = \frac{\widehat{\alpha}}{4} \frac{\ln\left(\dfrac{X_i}{X_{i-1}}\right)}{X_i^{1/4} - X_{i-1}^{1/4}}. \tag{4.6}$$

An example for the development of the body mass of a model copepod with four aggregated stages, nauplii, $m_e < m < X_n$, two groups of copepodites, $X_n < m < X_{c_1}$ and $X_{c_1} < m < X_{c_2}$, and adults, $X_{c_2} < m < X_a$, is shown in Fig. 4.1. The numerical values for the $X_i$'s are listed in Table 4.3. The solid line in Fig. 4.1 refers to Equation (4.1) and the dashed line to Equation (4.4). The effective growth rate, $\widehat{\alpha}$, is chosen as $\widehat{\alpha} = 0.081 \, \mathrm{d}^{-1}(\mu gC)^{-1/4}$ and is the same value for all stages, while the stage-dependent rate, $\widehat{\beta_i}$, is determined by the relationship (4.6). This consideration shows that the growth equations (4.1) and (4.4) provide equivalent descriptions of the development of body mass.

## 4.2 STAGE-RESOLVING MODELS OF COPEPODS

The considerations in the following sections aim at a consistent stage-resolving description of zooplankton in a chemical-biological model that can be embedded in a ocean circulation model. The theory follows partly the paper of Fennel (2001). First, we specify the state variables for the species model. In order to keep the model simple, we use aggregated state variables. We include five stages of the copepod model eggs, nauplii, copepodites 1 and 2 and adults. In other words, we merge all nauplii stages into one state variable model nauplii. Similarly, we merge the copepodites stages into the state variables model copepods 1 and 2, and partly into model adults. The corresponding variables of biomass per unit of volume are $Z_e$, $Z_n$, $Z_{c_1}$, $Z_{c_2}$ and $Z_a$. The corresponding numbers of individuals per unit of volume are $N_e$, $N_n$, $N_{c_1}$, $N_{c_2}$ and $N_a$. A more detailed resolution of stages is possible by expanding the number of state variables and equations. In order to specify the dynamic signatures of our copepods model, we choose the example of *Pseudocalanus*, which is described in detail in the review article of Corkett and McLaren (1978). *Pseudocalanus* is one of the important prey for planktivorous fish, also in the Baltic Sea, and has a mean generation time of 30 days.

## 4.2.1 Population Density

To develop the theory, we start with the population density of copepods, $\sigma^{cop}(m, t)$, see Equation (1.4), which describes the number of individuals having the mass $m$. Because we exclusively consider copepods, we drop the superscript, 'cop' in the following sections. By definition, $\sigma(m, t)\mathrm{d}m$ is the number of individuals in the interval $(m, m + \mathrm{d}m)$ at time $t$. The state variables $Z_i$ and $N_i$, $(i = \mathrm{e, n, c_1, c_2, a})$ are related to the population density. The total zooplankton biomass and the total number of individuals can be expressed by integrals of $\sigma(m, t)$ as

$$Z_{tot} = \int_{m_e}^{X_a} \sigma(m)m\,\mathrm{d}m \quad \text{and} \quad N_{tot} = \int_{m_e}^{X_a} \sigma(m)\mathrm{d}m,$$

where $m_e$ is the egg mass and $X_a$ the maximum mass of matured adults. For a certain stage, $i$, the individual mass, $m$, is confined to the interval $X_{i-1} \leq m \leq X_i$, where $X_{i-1}$ and $X_i$ are the molting mass of the previous and the current stage. The biomass, $Z_i$, and number of individuals, $N_i$, of the stage $i$ are obviously

$$Z_i = \int_{X_{i-1}}^{X_i} \sigma(m)m\,\mathrm{d}m, \tag{4.7}$$

and

$$N_i = \int_{X_{i-1}}^{X_i} \sigma(m)\mathrm{d}m. \tag{4.8}$$

The dynamics, i.e. the total change in time, of $\sigma(m)$ is controlled by mortality as the loss term and egg production as the source term,

$$\frac{\mathrm{d}\sigma}{\mathrm{d}t} = -\mu\sigma + T_{ae}\delta(m - m_e), \tag{4.9}$$

where $T_{ae}$ is an egg production rate, which prescribes the transfer rate of adults to eggs, and $\delta(m - m_e)$ is Dirac's $\delta$ function. Owing to the growth of the individuals, the distribution density $\sigma(m, t)$ will propagate along the $m$-axis with the 'propagation speed', $\frac{dm}{dt}$, which is controlled by the effective growth (grazing minus losses) of the individuals according to Equation (4.4). Because the mass of a copepod, $m$, changes with time according to Equation (4.4), we may write Equation (4.9) as

$$\frac{\partial \sigma}{\partial t} + \frac{\partial \sigma}{\partial m}\frac{dm}{dt} = -\mu\sigma + T_{ae}\delta(m - m_e).$$

With Equation (4.4), it follows

$$\frac{\partial \sigma}{\partial t} + (g - l)m\frac{\partial \sigma}{\partial m} = -\mu\sigma + T_{ae}\delta(m - m_e), \tag{4.10}$$

where

$$g - l = \sum_i (g_i - l_i)H(m, X_i),$$

with $H = 1$ for $X_{i-1} \leq m \leq X_i$ and $H = 0$ otherwise. Equation (4.10) for the population density, $\sigma(m)$, is equivalent to the von Foerster equation; see Murray (1993). A visualization of the behaviour of the population density for an idealized example is sketched in Fig. 4.2, which shows how a coherent cohort propagates along the $m$-axis, while the peak value decreases due to mortality. This corresponds, for example, to a rearing tank experiment, where at $t_0$ a number of eggs are put into the tank and then propagate through the stages.

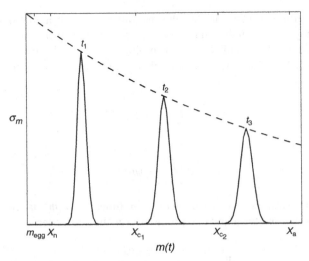

**FIGURE 4.2**    Propagation of a cohort along the $m$-axis.

Let $N_0$ be the initial concentration of eggs; then the equation for the population density is

$$\frac{d}{dt}\sigma(m, t) = \frac{\partial \sigma}{\partial t} + \frac{\partial \sigma}{\partial m}\frac{dm}{dt} = -\mu\sigma(m, t),$$

with the initial condition

$$\sigma(m, 0) = N_0\delta(m(0) - m),$$

where $m(0) = m_e$. The solution is obviously

$$\sigma = N_0\delta(m(t) - m)e^{-\mu t},$$

with $m(t) = m_e \exp[\int_0^t dt'(g - l)]$. However, in general the individuals within a cohort may develop differently, and if there are already adults, the adults may lay eggs more or less continuously. Hence, the distribution will be smeared out, and $\sigma(m, t)$ becomes a more or less continuous function of $m$ comprising many overlapping cohorts.

For the population density, $\sigma_i$, of a certain stage, $i$, there must be a similar equation as Equation (4.9),

$$\frac{d\sigma_i}{dt} = -\mu_i\sigma_i + T_{(i-1),i}\delta(m - X_{i-1}) - T_{i,(i+1)}\delta(m - X_i), \tag{4.11}$$

where $T_{(i-1),i}\delta(m - X_{i-1})$ and $T_{i,(i+1)}\delta(m - X_i)$ prescribe the rates at which individuals of the stage $i - 1$ are transferred to the stage $i$, and individuals of the stage $i$ molt into the stage $i + 1$, respectively. In order to specify the transfer rates, we look at a certain stage $i$, and note that the temporal change of the number of individuals in the stage is described by

$$\frac{\partial}{\partial t}\int_{X_{i-1}}^{X_i} dm\sigma_i(m).$$

This change is driven by the fluxes in and out of the mass interval of the given stage and by the decay due to mortality,

$$\frac{\partial}{\partial t}\int_{X_{i-1}}^{X_i} dm\sigma_i(m) = -\frac{dm}{dt}\sigma_i|_{m=X_i} + \frac{dm}{dt}\sigma_i|_{m=X_{i-1}} - \mu_i\int_{X_{i-1}}^{X_i}\sigma_i(m)dm.$$

This can be rewritten as

$$\int_{X_{i-1}}^{X_i} dm\left[\frac{\partial}{\partial t}\sigma_i(m) + \frac{dm}{dt}\sigma_i(m)\delta(m - X_i) + \mu_i\sigma_i(m)\right.$$
$$\left. - \frac{dm}{dt}\sigma_i(m)\delta(m - X_{i-1})\right] = 0,$$

or, equivalently,

$$\frac{\partial}{\partial t}\sigma_i(m) = -\mu_i\sigma_i(m) - (g_i - l_i)X_i\sigma_i(X_i)\delta(m - X_i)$$
$$+ (g_{i-1} - l_{i-1})X_{i-1}\sigma_i(X_{i-1})\delta(m - X_{i-1}), \tag{4.12}$$

where Equation (4.4) was used.

Comparing Equation (4.12) with Equation (4.11), we find that the transfer rates are determined by the 'propagation speed' at the molting masses and the number of individuals, having the molting masses $X_i$ and $X_{i-1}$,

$$T_{(i-1),i} = (g_{i-1} - l_{i-1})X_{i-1}\sigma_{i-1}, \tag{4.13}$$

$$T_{i,(i+1)} = (g_i - l_i)X_i\sigma_i. \tag{4.14}$$

We can interpret this as follows

$$T_{(i-1),i} = \frac{dm}{dt}\sigma(m)|_{m\approx X_{i-1}} \approx \frac{\Delta m\sigma(m)}{\Delta t}|_{m\approx X_{i-1}}, \tag{4.15}$$

where $\Delta m\sigma(m)|_{m\approx X_{i-1}}$ is, by definition, the number of individuals in a small interval, $\Delta m$, around $X_{i-1}$ and $\Delta t = \Delta t_{i-1,i}$ is the time needed for the molting process.

## 4.2.2 Stage-Resolving Population Models

Population models describe the dynamics of the number of individuals, $N_i$, of the different stages. The dynamic equations for $N_i$ follow from Equations (4.8) and (4.11) as

$$\frac{dN_i}{dt} = \int_{X_{i-1}}^{X_i} \frac{d\sigma(m)}{dt}dm \tag{4.16}$$

$$= -\mu_i N_i + T_{(i-1),i} - T_{i,(i+1)}.$$

The difficulty is to express the transfer rates by known state variables $N_{i-1}$ or $N_i$ and other parameters. A simple approach is $T_{i,(i+1)} \sim N_i/D_i$, where the $D_i$'s are stage duration times. Then, the corresponding equations for the number of individuals can be written as

$$\frac{d}{dt}N_e = q_e - \mu_e N_e - \frac{1}{D_e}N_e, \tag{4.17}$$

$$\frac{d}{dt}N_n = \frac{1}{D_e}N_e - \mu_n N_n - \frac{1}{D_n}N_n, \tag{4.18}$$

$$\frac{d}{dt}N_{c_1} = \frac{1}{D_n}N_n - \mu_{c_1}N_{c_1} - \frac{1}{D_{c_1}}N_{c_1}, \tag{4.19}$$

$$\frac{d}{dt}N_{c_2} = \frac{1}{D_{c_1}}N_{c_1} - \mu_{c_2}N_{c_2} - \frac{1}{D_{c_2}}N_{c_2}, \tag{4.20}$$

$$\frac{d}{dt}N_a = \frac{1}{D_{c_2}}N_{c_2} - \mu_a N_a. \tag{4.21}$$

Such a set of equations was used in several papers, e.g. Gupta et al. (1994), and Lynch et al. (1998). In order to illustrate the properties of such a model, we discuss as an example the development of an initial pulse of eggs, $N_e(0) = 3000$ ind m$^{-3}$, and $N_i(0) = 0$ for $i = (n, c_1, c_2, a)$. The duration times, which are take from experimental data (Klein-Breteler et al., 1994) are $D_e = 5$ d, $D_n = 8.8$ d, $D_{c_1} = 10.16$ d, $D_{c_2} = 12$ d. The mortalities are chosen as $\mu_e = 0.1$ d$^{-1}$, $\mu_n = 0.05$ d$^{-1}$, $\mu_{c_1} = 0.05$ d$^{-1}$, $\mu_{c_2} = 0.03$ d$^{-1}$ and $\mu_a = 0.01$ d$^{-1}$. A list of symbols used in this chapter is given in Table 4.1. The solution, which describes the propagation through the different stages from eggs to adults is shown in Fig. 4.3. After about 30 days, all surviving individuals have reached the adult stage. This reproduce, the experimental data. However, because the data were used to determine the numerical values of the rates, this result was more or less forced.

An obviously unrealistic feature is the early start of development in the higher stages. This becomes more obvious when egg laying is included in the model run. Let us assume that half of the adults are female and lay one egg per day. Setting the source term $q_e = 0.5 N_a$ in Equation (4.17), we obtain the result shown in Fig. 4.4. After about 10 days, the initially, starting population of eggs has reached the adult stage and then the female adults begin to lay eggs. Obviously, in this model approach the propagation of the copepods through the different stages is too fast.

### 4.2.3 Population Model and Individual Growth

As alternative approach we can use the evolution of the mean individual mass to control the onset of molting, i.e. the transfer to the higher stages. This can be done in a consistent manner for a coherent cohort, where all individuals develop in the same way. This approach was proposed by Carlotti and Sciandra (1989) and Carlotti and Nival (1992). From Equation (4.17), it follows that the molting of the individuals is controlled by the amount of ingested food. Part of the energy provided by the ingested food is used for the molting process rather than for the increase in body mass. Moreover, it is known that the molting process is already initiated when the mass of the individuals approaches the molting mass. In order to take this into account, we can introduce filter functions that slow down the growth and smoothly activate the transfer rate when approaching the molting mass. If we look, for example, at the transfer term $T_{(i-1),i}\sigma_{i-1}\delta(m - X_{i-1})$ in Equation (4.11), we can write the function, $\delta(m - X_{i-1})$, as derivative of the step function

$$\delta(m - X_{i-1}) = \frac{d}{dm}\theta(m - X_{i-1}).$$

To avoid the abrupt change at the step, we can replace the step function by a function, which provides a representation of the step function with a smooth transition. We choose in particular the Fermi function,

$$f(m - X_{i-1}) = \frac{1}{1 + \exp[\frac{\kappa}{X_{i-1}}(m - X_{i-1})]}. \tag{4.22}$$

**TABLE 4.1** Symbols Used in the Stage-Resolving Theory of Copepods

| Variable | Meaning | Dimension |
|---|---|---|
| *Expressions independent of stages* | | |
| $p$ | Allometric exponent | – |
| $g$ | Grazing rate | time$^{-1}$ |
| $l$ | Loss rate | time$^{-1}$ |
| $\hat{\alpha} = g - l$ | Effective growth rate | time$^{-1}$ |
| $\sigma$ | Population density | ind (mass)$^{-1}$ |
| *Stage-dependent expressions* | | |
| $i = e$ | Subscript for eggs | – |
| $i = n$ | Subscript for model nauplii | – |
| $i = c_1$ | Subscript for model copepodites 1 | – |
| $i = c_2$ | Subscript for model copepodites 2 | – |
| $i = a$ | Subscript for model adults | – |
| $q_e$ | Egg source term | ind (vol. time)$^{-1}$ |
| $h$ | Hatching rate (eggs) | time$^{-1}$ |
| $g_i$ | Stage dependent grazing rate | time$^{-1}$ |
| $l_i$ | Stage dependent loss rate | time$^{-1}$ |
| $\beta_i = g_i - l_i$ | Stage dependent effective growth rate | time$^{-1}$ |
| $\mu_i$ | Stage dependent mortality rate | time$^{-1}$ |
| $X_i$ | Upper mass limit of stage $i$ | mass |
| $\langle m \rangle_i$ | Mass of initiation of molting | mass |
| $D_i$ | Stage duration time | time |
| $\Delta t_{i-1,i}$ | Duration of the molting phase | time |
| $N_i$ | Abundance of stage $i$ | ind (volume)$^{-1}$ |
| $Z_i$ | Concentration of stage $i$ | mass(volume)$^{-1}$ |
| $m_i = \frac{Z_i}{N_i}$ | Mean individual mass | mass |
| $T_{i-1,i}$ | General transfer rate | ind (time)$^{-1}$ |
| $\tau_{i-1,i}$ | Specified transfer rate | time$^{-1}$ |
| $f(m - X_i)$ | Fermi function | – |
| $\kappa$ | Control parameter (Fermi function) | – |

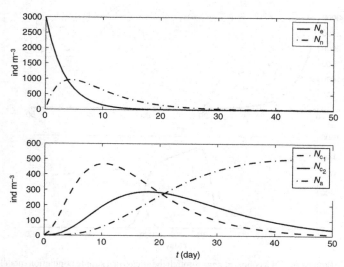

**FIGURE 4.3**  Propagation of a cohort of eggs through the stages in a simple population model with prescribed transition rates.

The steepness of the transition from one to zero is controlled by the auxiliary parameter, $\kappa$. For large values of the parameter $\kappa$, the function $f(m - X_{i-1})$ becomes a step function, $\theta(m - X_{i-1})$. In the following, we set the parameter $\kappa$ to $\kappa = 20$. Let $\langle m \rangle_{i-1}$ be a mass somewhat smaller than the molting mass $X_{i-1}$, at which the initiation of the molting process commences. Then we can use the

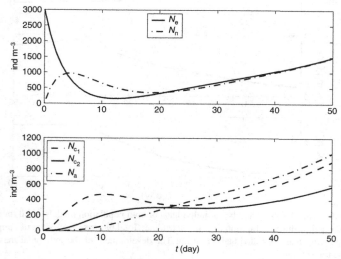

**FIGURE 4.4**  As Fig. 4.3, but with reproduction. It is assumed that half of the adults are female and lay one egg per day.

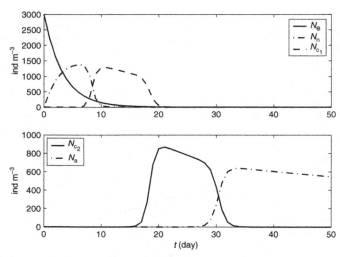

**FIGURE 4.5**    Propagation of a cohort of eggs through the stages in a simple population model with prescribed transition rates, which are activated when the individuals approach the molting mass.

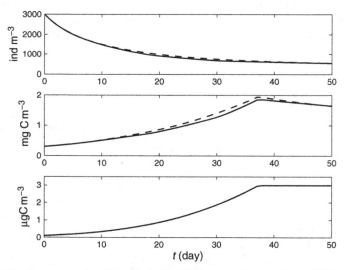

**FIGURE 4.6**    Top: The total number of individuals of both population models, solid for activated transfers, dashed for inverse duration time rates. Middle: Total biomass as product of numbers and the corresponding mass of individuals. Bottom: The development of the individual mass, which controls the transfer rates.

Fermi function with the arguments $f(\langle m \rangle_{i-1} - m)$ to provide a smooth transition of the transfer to the following stage due to molting. Noting the property of the Fermi functions, $f(x - y) = 1 - f(y - x)$, the transfer rates, $T_{i-1,i}$, can be defined as

$$T_{i-1,i} = (g_{i-1} - l_{i-1})X_{i-1}f[\langle m \rangle_{i-1} - m(t)]\sigma(X_{i-1}). \qquad (4.23)$$

Because the Fermi function is only activated when the cohort approaches the molting mass, we can replace $\sigma(X_{i-1})$ by $N_{i-1}$ in Equation (4.23). The typical structures of the Fermi functions are shown in Fig. 4.7 (lower panel). A conceptional diagram of the stage-resolving zooplankton model is shown in Fig. 4.8, where we also indicate how a bulk state variable for zooplankton is replaced by a model component with enhanced process resolution. In order to illustrate this model approach, we use a set of equations similar as Equations (4.17)–(4.21) with transfer rates

$$T_{i-1,i} = \frac{1}{\Delta t} f(0.95X_{i-1} - m)N_{i-1};$$

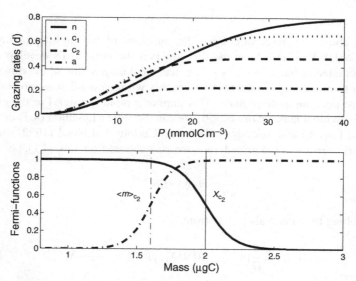

**FIGURE 4.7**    Top: Grazing rates as functions of the available food for $T = 10\,°C$. For the lower stages, the ratio of ingested food to body mass is larger than for the higher stages. In food shortage, the higher stages are better at catching food due to a higher mobility. Bottom: Fermi functions with properties of low-pass and high-pass filter $f(Z_i/N_i - X_i)$ (solid) and $f(\langle m \rangle_i - Z_i/N_i)$ (dash-dotted), respectively, to control the development for $i = c_2$.

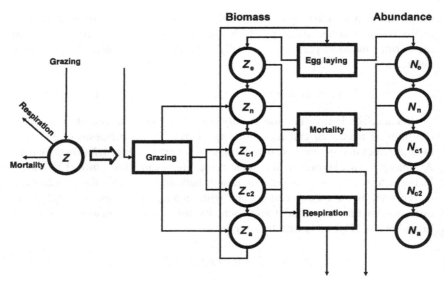

**FIGURE 4.8**    A conceptual diagram of the stage-resolving zooplankton model, which replaces the bulk zooplankton biomass concentration in NPZD models, as, e.g. in Fig. 2.9.

i.e. we chose the mass values for the initiation of the molting process as $\langle m \rangle_{i-1} = 0.95 X_{i-1}$ and $\Delta t$ is the time interval the individuals develop along the mass interval $\langle m \rangle_{i-1}$ to $X_{i-1}$ rather than the duration time of the stage. The transfer from one stage to the next occurs only within a small mass interval near the corresponding molting mass. This implies a rapid transfer. For the growth of the individual mass in the cohort, we can use either Equation (4.1) or (4.4). Because Carlotti and Sciandra (1989) and Carlotti and Nival (1992) used an allometric version of the growth rate, we may choose Equation (4.1), i.e.

$$\frac{\mathrm{d}}{\mathrm{d}t} m = (g - l) m^{0.75}.$$

The molting time intervals follow from

$$\Delta t_{i-1,i} = \frac{4}{\widehat{\alpha}} [X_{i-1}^{1/4} - (0.95 X_{i-1})^{1/4}] \approx \frac{0.012}{\widehat{\alpha}} X_{i-1}^{1/4}, \qquad (4.24)$$

as $\Delta t_{nc_1} = 0.46\,\mathrm{d}$, $\Delta t_{c_1 c_2} = 0.60\,\mathrm{d}$ and $\Delta t_{c_2 a} = 0.75\,\mathrm{d}$. For the hatching time scale, we use the same value as in the previous section, $\Delta t_{en} = 5\,\mathrm{d}$. If we apply the growth equation (4.4), it follows analogously

$$\Delta t_{i-1,i} = \frac{1}{\widehat{\beta}_{i-1}} \ln \frac{X_{i-1}}{0.95 X_{i-1}} \approx \frac{0.05}{\widehat{\beta}_{i-1}}.$$

Introducing

$$\tau_{i-1,i} = \frac{f(\langle m \rangle_{i-1} - m)}{\Delta t_{i-1,i}}, \tag{4.25}$$

we find in our notations the model equations as

$$\frac{d}{dt} N_e = q_e - \mu_e N_e - \tau_{en} N_e, \tag{4.26}$$

$$\frac{d}{dt} N_n = \tau_{en} N_e - \mu_n N_n - \tau_{nc_1} N_n, \tag{4.27}$$

$$\frac{d}{dt} N_{c_1} = \tau_{nc_1} N_n - \mu_{c_1} N_{c_1} - \tau_{c_1 c_2} N_{c_1}, \tag{4.28}$$

$$\frac{d}{dt} N_{c_2} = \tau_{c_1 c_2} N_{c_1} - \mu_{c_2} N_{c_2} - \tau_{c_2 a} N_{c_2}, \tag{4.29}$$

$$\frac{d}{dt} N_a = \tau_{c_2 a} N_{c_2} - \mu_a N_a. \tag{4.30}$$

Thus, the transfer rates are no longer prescribed by empirical rates but are computed within the model. However, the egg production is still set by a source term, $q_e$, and the model is restricted to the development of coherent cohorts where all individuals grow more or less synchronously. The structure of the model equations (4.26)–(4.30) is qualitatively equivalent to the theory of Carlotti and Sciandra (1989), although we use aggregated state variables and somewhat different mathematical formulations for the process rates in our model.

The simulation of the development of a cohort, shown in Fig. 4.5, is more realistic than the example corresponding case with transfer rates set to duration times, shown in Fig. 4.3. If we look at the development of the total biomass and abundance of both models, i.e. with transfer rates controlled by duration time and by individual growth, we find that the results for the integrated values are practically identical, see Fig. 4.6. Thus, the major improvement achieved by growth-controlled transfer rates basically results in a better description of the development of the different stages.

### 4.2.4 Stage-Resolving Biomass Model

An equivalent way to achieve a consistent model description for copepods can be formulated by a stage-resolving description of biomass (Fennel, 2001) as sketched in Fig. 4.8. The dynamic equations for the biomass concentration, $Z_i$, of the stage, $i$, can be derived by combining Equation (4.12), multiplied by $m$, and Equation (4.4) multiplied by $\sigma_i$,

$$m\frac{\partial}{\partial t}\sigma_i(m) = -\mu_i m\sigma_i(m) - (g_i - l_i)X_i m\sigma_i(X_i)\delta(m - X_i)$$
$$+(g_{i-1} - l_{i-1})X_{i-1}m\sigma_i(X_{i-1})\delta(m - X_{i-1}),$$

$$\sigma_i(m)\frac{\partial}{\partial t}m = (g_{i-1} - l_{i-1})m\sigma_i(m).$$

Adding both equations yields,

$$\frac{\partial}{\partial t}m\sigma_i(m) = (g_{i-1} - l_{i-1})m\sigma_i(m) - \mu_i m\sigma_i(m)$$
$$+(g_{i-1} - l_{i-1})X_{i-1}m\sigma_i(X_{i-1})\delta(m - X_{i-1}),$$
$$-(g_i - l_i)X_i m\sigma_i(X_i)\delta(m - X_i).$$

Integration from $X_{i-1}$ to $X_i$ then, gives according to Equation (4.7),

$$\frac{dZ_i}{dt} = \int_{X_{i-1}}^{X_i} dm\frac{d}{dt}m\sigma(m)$$
$$= (g_i - l_i - \mu_i)Z_i + T_{(i-1),i}X_{i-1} - T_{i,(i+1)}X_i, \qquad (4.31)$$

where the transfer rates are introduced as in Equation (4.23). Note that this kind of equation can only be derived for a growth equation of the type (4.4). For an allometric exponent different from 1 we would find, instead of expressions of the form $m\sigma$, terms like $m^{0.75}\sigma$, which cannot be expressed as zooplankton biomass by integration over $m$. Because the filter function is activated only if the individual mass approaches the corresponding molting mass, we can replace terms such as $dm\sigma(m)m|_{m\approx X_{i-1}}$ by $Z_{i-1}$. Then, the equations for the stage dependent biomass can be written as

$$\frac{d}{dt}Z_e = \tau_{ae}Z_a - \tau_{en}Z_e - \mu_e Z_e, \qquad (4.32)$$

$$\frac{d}{dt}Z_n = \tau_{en}Z_e + (g_n - \mu_n - l_n)Z_n - \tau_{nc_1}Z_n, \qquad (4.33)$$

$$\frac{d}{dt}Z_{c_1} = \tau_{nc_1}Z_n + (g_{c_1} - \mu_{c_1} - l_{c_1})Z_{c_1} - \tau_{c_1c_2}Z_{c_1}, \qquad (4.34)$$

$$\frac{d}{dt}Z_{c_2} = \tau_{c_1c_2}Z_{c_1} + (g_{c_2} - \mu_{c_2} - l_{c_2})Z_{c_2} - \tau_{c_2a}Z_{c_2}, \qquad (4.35)$$

$$\frac{d}{dt}Z_a = \tau_{c_2a}Z_{c_2} + (g_a - \mu_a - l_a)Z_a - \tau_{ae}Z_a, \qquad (4.36)$$

where the $\tau_{i-1,i}$'s are given by Equation (4.25). The dynamics of the state variables are controlled by transfer rates, $\tau_{i,i+1}$, grazing rates, $g_i$, loss rates, $l_i$, and mortality rates, $\mu_i$, where $i = (e, n, c_1, c_2, a)$. The process rates will be specified later. In Equations (4.32) and (4.36), we have introduced the term $\tau_{ae}$, which means 'transfer of adults to eggs'. This term can be prescribed with the Fermi function $f(\langle m \rangle_a - m)$, where $\langle m \rangle_a$ is a maturing mass, which must be reached before the female adults can lay eggs. The number of eggs produced per day can then be expressed as a certain fraction of the ingested food. Thus, instead of a source term for eggs, $q_e$, which was needed in population models, we have a term that describes the production of the biomass of eggs. The abundance of eggs follows by division of the egg biomass by the mass of a egg, $m_e$. The corresponding equations for the number of individuals are,

$$\frac{d}{dt}N_e = \frac{1}{m_e}\tau_{ae}Z_a - \mu_e N_e - \tau_{en}N_e, \tag{4.37}$$

$$\frac{d}{dt}N_n = \tau_{en}N_e - \mu_n N_n - \tau_{nc_1}N_n, \tag{4.38}$$

$$\frac{d}{dt}N_{c_1} = \tau_{nc_1}N_n - \mu_{c_1}N_{c_1} - \tau_{c_1 c_2}N_{c_1}, \tag{4.39}$$

$$\frac{d}{dt}N_{c_2} = \tau_{c_1 c_2}N_{c_1} - \mu_{c_2}N_{c_2} - \tau_{c_2 a}N_{c_2}, \tag{4.40}$$

$$\frac{d}{dt}N_a = \tau_{c_2 a}N_{c_2} - \mu_a N_a. \tag{4.41}$$

The death rates, $\mu_i$, are the same as in the set (4.32)–(4.36). These two sets of equations can be solved without a parallel integration of the growth equation for individuals. For the example of the development of an initial number of eggs, which was considered in previous sections, we can estimate the mean individual mass $\overline{m}_i$, of the stage $i$, by $\overline{m}_i = Z_i/N_i$. Note that this variable is needed for the Fermi function to control the transfer terms. The mass $\overline{m}$ in the Fermi function (4.22) must be replaced by the mean individual mass for the stages, $\overline{m}_i$,

$$f(\overline{m}_i - X_i) = \frac{1}{1 + \exp(\frac{\kappa}{X_i}[\overline{m}_i - X_i])}, \tag{4.42}$$

Multiplying the growth term by the filter function ensures that the growth decreases when the individuals are approaching the molting mass. In a statistical sense, the function (4.42) provides the connection of the individual level to the bulk biomass of the stage. The function drops from one to zero if the mean individual mass, $\overline{m}_i = Z_i/N_i$, of a stage approaches the molting mass $X_i$. The grazing rate, $g_i$, will be multiplied by this function in order to channel the development from growth into molting (see Fig. 4.7). In the same way, we change the Fermi function in the transfer rate (4.25),

$$\tau_{i-1,i} = \frac{f(\langle m \rangle_{i-1} - \overline{m}_i)}{\Delta t_{i-1,i}}. \tag{4.43}$$

The transfer rate to the next stage is activated when the mean individual mass, $\bar{m}_i$, exceeds $\langle m \rangle_i$. Again we choose the mass values, $\langle m \rangle_i$, somewhat smaller than the molting mass of the corresponding stages.

Because the number of equations have now increased, the question may arise as to what we have achieved. The most important progress lies in the ability to now include reproduction. That is, we can now consider the development of various overlapping cohorts in response to a more or less continuously laying of eggs. An attractive feature of including the biomass state variables is the explicit conservation of mass in the model. This provides a convenient way to check the consistency of models by monitoring the total biomass, which must be constant or balanced by external inputs minus outputs. Population models are generally not constrained in such an easy way because there is no law for the conservation of the number of individuals.

## 4.3 EXPERIMENTAL SIMULATIONS

In the preceding section, we constructed the model by defining state variables, constructing the dynamic equations and formulating of the processes. In order to perform model runs, we have to specify the numerical values of all parameters. In the following subsections we provide a detailed list of all process rates and choices of parameter. The copepod model is characterized by mass intervals and process parameters, which are summarized in Tables 4.2 and 4.3. We adopt parameter values corresponding to *Pseudocalanus*, however, it is clear that the model applies also to several other species, provided that the state variables and process parameters are accordingly adjusted.

### 4.3.1 Choice of Parameters

#### 4.3.1.1 Grazing Rates

The grazing rates describe the amount of food ingested per day by an animal in relation to its biomass. Thus, in general, the lower stages have higher grazing

**TABLE 4.2** Grazing Parameters

| Stage | $\beta_i|_{T=0\,°C}$ | $\beta_i|_{T=10\,°C}$ | $\beta_i|_{T=15\,°C}$ | $I_i$ |
|---|---|---|---|---|
| nauplii | 0.5 d$^{-1}$ | 0.93 d$^{-1}$ | 1.3 d$^{-1}$ | $2.5 \times 10^3$ (m$^3$ mmolC$^{-1}$)$^2$ |
| copepodites 1 | 0.35 d$^{-1}$ | 0.66 d$^{-1}$ | 0.9 d$^{-1}$ | $4.7 \times 10^3$ (m$^3$ mmolC$^{-1}$)$^2$ |
| copepodites 2 | 0.25 d$^{-1}$ | 0.47 d$^{-1}$ | 0.64 d$^{-1}$ | $7 \times 10^3$ (m$^3$ mmolC$^{-1}$)$^2$ |
| adults | 0.12 d$^{-1}$ | 0.22 d$^{-1}$ | 0.31 d$^{-1}$ | $10.1 \times 10^3$ (m$^3$ mmolC$^{-1}$)$^2$ |

Stage-dependent grazing parameter, $\beta_i$, for different temperatures according to Equation (4.45) and stage-dependent Ivlev parameters, $I_i$.

**TABLE 4.3** Mass Parameters of the Copepod Model

| Egg mass | $m_e = 0.1\,\mu gC$ | | |
|---|---|---|---|
| Stage | Molting | Maturation | Mean |
| nauplii | $X_n = 0.3\,\mu gC$ | – | $\langle m \rangle_n = 0.22\,\mu gC$ |
| copepodites 1 | $X_{c_1} = 0.8\,\mu gC$ | – | $\langle m \rangle_{c_1} = 0.6\,\mu gC$ |
| copepodites 2 | $X_{c_2} = 2\,\mu gC$ | – | $\langle m \rangle_{c_2} = 1.6\,\mu gC$ |
| Adults | – | $X_a = 3\,\mu gC$ | $\langle m \rangle_a = 2.6\,\mu gC$ |

rates than the higher stages. However, for declining resources, it is assumed that the higher stages have an advantage due to higher mobility to capture food. These features are formulated in terms of a modified Ivlev formula,

$$g_i(P) = \beta_i[1 - \exp(-I_i P^2)]f(\overline{m}_i - X_i), \qquad (4.44)$$

where $P$ is the food concentration, i.e. phytoplankton, and the $I_i$s are stage-dependent Ivlev parameters. The function $f(\overline{m}_i - X_i)$ in Equation (4.44) was defined by Equation (4.42) and represents a stage-dependent Fermi function, which works as a 'low-pass filter' and ensures that the growth decreases if the mean individual mass, $\overline{m}_i$, reaches the maximum mass, $X_i$, of the corresponding stage and may molt to the next stage. The filter property of this function is activated, i.e. its value drops from one to zero, if the mean individual mass, $\overline{m}_i$, of a stage approaches the molting mass, $X_i$. The maximum grazing rate, $\beta_i$, depends on the temperature. We choose an Eppley factor, $\exp(a_E T)$ with $a_E = 0.063(°C)^{-1}$,

$$\beta_i = b_i \exp(a_E T). \qquad (4.45)$$

The numerical values are listed in Table 4.2. The decrease of $b_i$ for the higher stages reflects that higher stages with more biomass ingest a smaller amount of food, expressed as percentage of the body mass, than the lower stages. The mass parameters listed in Table 4.3 were derived from data of the Baltic Sea (Hernroth, 1985).

### 4.3.1.2 Loss Rates

The ingested food is partly used for growth and partly for the metabolism of the animals. We follow Corkett and McLaren (1978) and assume that 35% of the ingestion is lost by egestion, 10% as excretion and 10% by respiration. Moreover, about 15% of the ingestion is assumed to be needed for the molting processes. These losses are expressed by the rates $l_i = 0.7\,g_i(P, T)$, for $(i = n, c_1, c_2)$. For the adults, we assume $l_a = 0.8\,g_a(P, T)$.

### 4.3.1.3 Reproduction

The rate of reproduction describes the egg production of the female adults after reaching the maturation mass. This amounts to a transfer rate from adults to eggs. We assume that half of the adults are female and 30% of the ingested food is transferred into egg biomass:

$$\tau_{ae} = \left(\frac{1}{2}\right) 0.3 \, g_i. \tag{4.46}$$

Then the number of new eggs per day is given by

$$\tau_{ea} Z_a / m_e. \tag{4.47}$$

This approach allows the egg-laying rate to depend on food availability and temperature—here, through the grazing rates—but ignores the discontinuous release of clutches of eggs with time intervals of a couple of days, see, e.g. Hirche et al. (1997). The transfer from eggs to nauplii (hatching) is set by the embryonic duration, about 10 days at low temperatures, (Corkett and McLaren, 1978). The effect of temperature is accounted for by an Eppley factor,

$$\tau_{en} = h \, \theta(T - T_0) \, \exp[a_E(T - T_0)], \tag{4.48}$$

with $h = 0.124$, $T_0 = 2.5\,°C$ and $\theta(x)$ being the step function, $\theta(x) = 1$ for $x > 0$ and $\theta(x) = 0$ for $x < 0$. Thus, only at temperatures exceeding $2.5\,°C$ do the model eggs hatch to model nauplii.

### 4.3.1.4 Mortality and Overwintering

Because the food chain model is truncated at the level of the model zooplankton, the mortality rates involve natural death and death by predation by planktivorous fish and other predators. The choice of the mortality rates is difficult because the basic factors are not well known. In the literature, stage-dependent mortalities with some seasonal variation are often used. The typical orders of magnitude vary from $\mu \sim 0.01 \ldots 0.1 \, d^{-1}$. In our model, we choose constant mortality rates for eggs and nauplii model, $\mu_e = 0.2 \, d^{-1}$ and $\mu_n = 0.033 \, d^{-1}$. Because constant rates for all stages are certainly an oversimplification, we relate the mortality of the copepodites and adults to a time-dependent function, which is proportional to a constant minus the normalized length of the day, $\Delta d(t)$. The normalized length of the day are plotted in Fig. 5.1. As reported from the Baltic Monitoring Programme (HELCOM, 1996, p. 96), a low abundance in autumn and winter can be expected. Thus, we choose $\mu_{c_1} = \mu_{c_2} = 0.1[1 - \Delta d(t)] \, d^{-1}$ and $\mu_a = 0.05[1 - \Delta d(t)] \, d^{-1}$. Because $\Delta d(t)$ varies between 0.2 and 0.8, we have $\mu_{c_1} = \mu_{c_2} = 0.02 \ldots 0.08 \, d^{-1}$ and $\mu_a = 0.01 \ldots 0.04 \, d^{-1}$.

We further assume that only the adults overwinter. In order to mimic the overwintering, we put the mortality of adults to zero if their biomass concentration reaches the threshold of $0.3\,\text{mmolC}\,\text{m}^{-3}$, while the other state variables tend towards zero in wintertime due to corresponding mortality rates.

### 4.3.2 Rearing Tanks

We start with a simulation of the development of a cohort of eggs in a rearing tank with constant food and temperature condition. To this end, we integrate the model equations (4.32)–(4.36) and (4.37)–(4.41) with the initial conditions at $t_0 = 0$: $Z_i = 0$ and $N_i = 0$ for $i = (\text{n}, c_1, c_2, a)$ and $N_e(0) = 3000\,\text{ind}\,\text{m}^{-3}$ and $Z_e = N_e\,m_{\text{egg}}$. First we consider the case of no reproduction. The temperature is chosen as $T = 8\,°\text{C}$ and the food, expressed in nitrogen units, as $P = 4.5\,\text{mmol}\,\text{Nm}^{-3}$. The results of the numerical integration, which is done with MATLAB, are shown in Figs. 4.9 and 4.10. The development of eggs to adults is shown for the stage-resolving biomass, Fig. 4.9, as well as for the stage-resolving number of individuals, Fig. 4.9. Without reproduction (no new eggs), the biomass accumulates eventually in the adult stage. The control of the high-pass filters in the transfer rates (4.43) can clearly be recognized in Fig. 4.9(a) and (b). Owing to the overlapping of the high- and low-pass filters, the transition between stages is relatively smooth. The duration times of the stages can be estimated from Fig. 4.9(a) and (b). A full generation time, i.e. the development of eggs to mature adults, takes about 25 days.

The total number of individuals, $N_{\text{tot}}$, decreases while the total biomass and the mean value of the individual mass of the whole population, $m_{mean} = Z_{\text{tot}}/N_{\text{tot}}$, increase, see Fig. 4.10. The development of the $i$th stage to the next stage occurs when the mean individual mass, $\overline{m}_i$, approaches the molting mass, $X_i$. Comparing Figs. 4.3 and 4.5 with Figs. 4.9(b) and 4.10 shows that the stage-resolving biomass model gives virtually the same results as a population model with growth-controlled transfer rates. Thus, in the case of the development of a coherent cohort without reproduction, the evolution of the mean individual mass and the number of individuals can be used to describe the dynamics of the model zooplankton, as in Carlotti and Sciandra (1989) and Carlotti and Nival (1992).

In order to show the influence of the temperature on the development of the copepod model, we repeat the simulation for a higher temperature, e.g. $T = 15\,°\text{C}$. The result, which is shown for the abundance in Fig. 4.11, gives a significantly shorter generation time of about 18 days. This is in agreement with the observational findings of Klein-Breteler et al. (1994).

In the next series of experiments, we repeat the simulations, but with reproduction. The results, which are plotted in Figs. 4.12–4.14, show a similar development as in the previous case until the adults start to lay eggs. After the onset of reproduction, there is a clear increase in biomass and number of individuals of the model zooplankton. It is obvious that the time evolution of the mean individual mass, $\overline{m} = Z_{\text{tot}}/N_{\text{tot}}$, cannot be used to describe the system in an unambiguous way. For example, the mean individual mass of the population

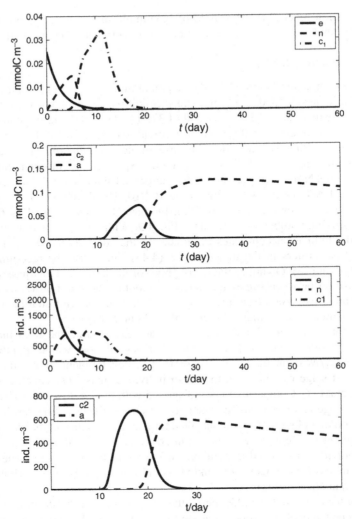

**FIGURE 4.9**    Stage resolving simulation of the development of zooplankton biomass (a) and abundance (b) in a rearing tank without reproduction at a constant food level, $4.5\,\mathrm{mmolN\,m^{-3}}$, and a constant temperature, $T = 8\,^{\circ}\mathrm{C}$.

decreases after the onset of reproduction, because it no longer represents a more or less coherent cohort, but is the mean value of all existing stages. In other words, the evolution of $\overline{m}$ is no longer equivalent to a solution of a growth equation of the mass of a mean individual. Reproduction cannot be simulated by a population model controlled by a growth equation for $\overline{m}$, but requires a description by a stage-resolving biomass model

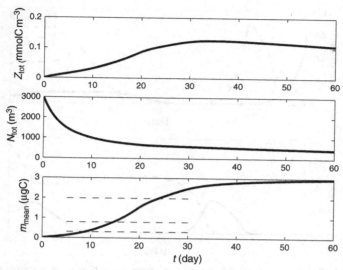

**FIGURE 4.10** Total zooplankton biomass (top), numbers of individuals (middle) and mean individual mass (bottom) in a rearing tank without reproduction. In the bottom panel, the molting mass of nauplii and copepodites model $c_1$ and $c_2$ are indicated by dashed lines.

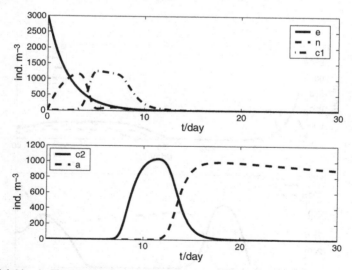

**FIGURE 4.11** As Fig. 4.9, but for 15 °C.

## 4.3.3 Inclusion of Lower Trophic Levels

After the simulations in rearing tanks for constant environmental conditions, we wish now to approach marine *in situ* conditions and take the dynamic links to the lower trophic levels into account. As a first step, we link an NPD model to

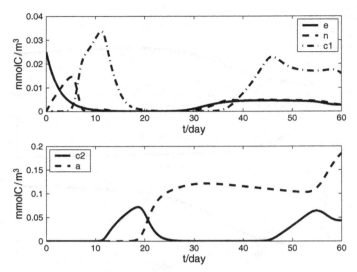

**FIGURE 4.12**    Stage-resolving simulation of the zooplankton biomass in a rearing tank with reproduction, at a constant food level, $4.5\,\mathrm{mmolN\,m^{-3}}$, and, constant temperature $T = 8\,°\mathrm{C}$.

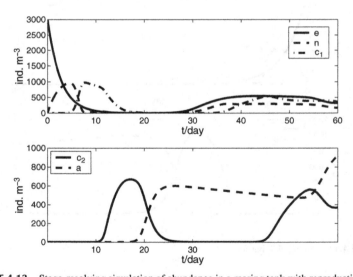

**FIGURE 4.13**    Stage-resolving simulation of abundance in a rearing tank with reproduction.

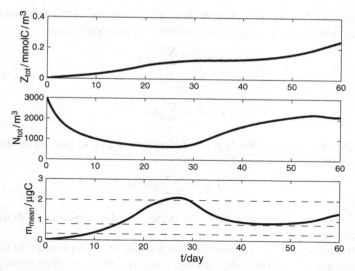

**FIGURE 4.14** Total zooplankton biomass (top), numbers of individuals (middle) and mean individual mass (bottom), in a rearing tank without reproduction. In the bottom panel, the molting mass of nauplii and copepodites model $c_1$ and $c_2$ are indicated by dashed lines.

the zooplankton model, given by Equations (4.32)–(4.36) and (4.37)–(4.41), to marine systems.

The simple model of the lower trophic levels has the state variables limiting nutrients, $N$, phytoplankton, $P$, and detritus, $D$.

$$\frac{d}{dt}N = -M(N)P + l_{PN}P + L_{ZN}Z$$

$$+ l_{DN}D + Q^{\text{ext}}, \tag{4.49}$$

$$\frac{d}{dt}P = M(N)P - G(P)Z - l_{PD}P - l_{PN}P, \tag{4.50}$$

$$\frac{d}{dt}D = L_{ZD}Z - l_{DN}D + l_{PD}P - Q^{\text{ext}}. \tag{4.51}$$

The uptake is controlled by a modified Michaelis-Menten formula,

$$M(N) = \frac{rN^2}{\alpha^2 + N^2}. \tag{4.52}$$

The role of daylight is included qualitatively in the rate, $r$, by means of an astronomical standard formula of the normalized length of the day for temperate latitudes, $\phi = 54°N$, see Fig. 5.1. The effect of temperature is included by an Eppley factor. Moreover, as a crude reflection of mixed-layer formation, we set the uptake rate to zero for temperatures below 2.5 °C, i.e.

$$r = r_{\max}\theta(T - T_o)\Delta d(t)\exp(aT).$$

The loss rates $L_{ZN}$ and $L_{ZD}$ describe how much biomass of the different stages of the zooplankton is transferred to nutrients, $N$, and detritus, $D$.

$$L_{ZN}Z = \sum_i l_i Z_i,$$

$$L_{ZD}Z = \sum_i \mu_i Z_i,$$

while $G(P)Z$ represents the loss of phytoplankton due to grazing of different stages of the copepods model,

$$G(P)Z = \sum_i g_i(P)Z_i,$$

where, $i = (n, c_1, c_2, a)$. The rate $l_{DN}$ prescribes the recycling of detritus to nutrients, see Table 4.4.

The mortality rate, $\mu_P$, prescribes the transfer of phytoplankton to detritus, including a crude approach for sinking, while $l_{PN}$ describes how much phytoplankton biomass is transferred directly to nutrients through respiration losses. Finally, the rate $Q^{\text{ext}}$ prescribes an external input of nutrients. In Table 4.4, the numerical values of the various parameters are listed in carbon units, where the Redfield ratio, C:N = 106:16, is assumed.

### 4.3.4 Simulation of Biennial Cycles

The model box can be envisaged as a part of the upper mixed layer with no horizontal advection and restricted vertical exchange. The box could also be considered as a grid box of a circulation model with no interaction of the

**TABLE 4.4** Parameters of a Simple Phytoplankton Model

| Symbol | Parameter | Value |
|---|---|---|
| $L_{DN}$ | Recycling $D$ to $N$ | $0.01 \cdot \exp(a_E T)\,\text{d}^{-1}$ |
| $l_{PD}$ | Mortality | $0.02\,\text{d}^{-1}$ |
| $l_{PN}$ | Transfer $P$ to $N$ | $0.1\,M(N)$ |
| $r_{\max}$ | Maximum uptake | $0.5\,\text{d}^{-1}$ |
| $\alpha^2$ | Half-saturation | $40\,\text{mmolC}^2\,\text{m}^{-6}$ |
| $Q^{\text{ext}}$ | Nutrient input | $0.2 \times \Delta d\,\text{mmolC}\,\text{m}^{-3}\text{d}^{-1}$ |

Note that the function $M(N)$ is given by Equation (4.52) and $\Delta d$ refers to the normalized length of the day defined by Equation (5.2).

neighbouring grid boxes. We consider a minimum of physical control, i.e. we use the annual cycle of the surface temperature in the Arkona Basin of the Baltic Sea and simulate the seasonal variation of irradiation by a standard formula of the normalized length of the day for 54°N, as shown in Figs. 2.12 and 5.1, respectively. The lower part of the food web is described by Equations (4.49)–(4.51).

As the standard experiment, we look at overwintering copepods that start to lay eggs after the onset of the vernal phytoplankton bloom. The dynamics of the state variables are shown in Figs. 4.15–4.17. The development of nutrients, phytoplankton, bulk zooplankton (i.e. the sum of all stages) and detritus is shown in Fig. 4.15. In this model, the onset of the spring bloom is controlled by temperature. The development of thermal stratification, which is not resolved in a box model, starts if the water temperature exceeds 2.5 °C. The peak of the phytoplankton bloom is controlled by the mortality rate of phytoplankton, which also includes sinking (transfer to detritus), and by zooplankton grazing. The total or bulk biomass of the zooplankton model, where all stages are lumped together, shows a pronounced maximum in the summer. The reduced grazing pressure in autumn allows a secondary bloom of the phytoplankton. During winter the detritus is recycled to nutrients.

The stage-resolving dynamics of the copepods model are shown in Figs. 4.16 and 4.17 for both biomass and the number of individuals of the model stages. After the onset of the spring bloom, the overwintering adults start to lay eggs. A significant increase of the zooplankton model follows after the 'egg signal' has

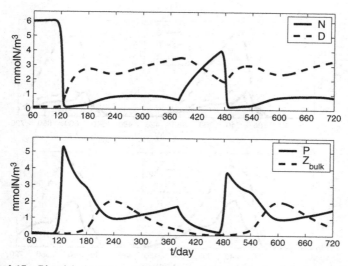

**FIGURE 4.15** Biennial simulation of nutrients and detritus (top), phytoplankton and total zooplankton biomass (bottom).

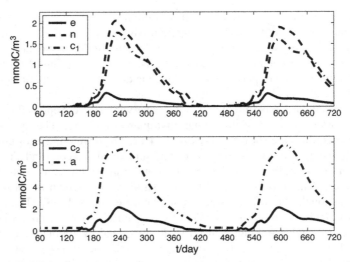

**FIGURE 4.16**    Biennial simulation of the biomass of the different stages.

'propagated' through the stages (time scale about 50 days) and newly developed adults start reproduction. The biomass of the zooplankton model accumulates in the adult stage. The highest number of individuals is found for the nauplii model.

We also show the dynamics of the total number of individuals and the egg-laying rate in Fig. 4.18. The egg rate increases rapidly after the onset of

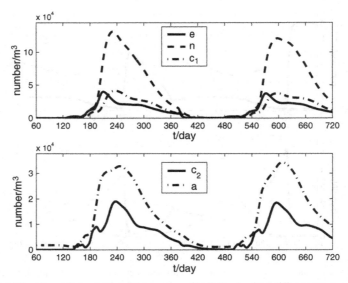

**FIGURE 4.17**    Biennial simulation of the number of individuals of the different stages.

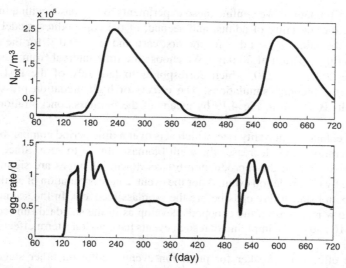

**FIGURE 4.18**  Biennial simulation of the total number of individuals (top) and the variation of the egg rates (bottom), with indications of three distinct generations of copepods.

the spring bloom and later shows some oscillations that can be related to the decrease of adults due to mortality until the next generation arrives in the adult stage and starts to lay eggs. The next generation of eggs propagate through the different stages and generate a further peak in the egg rate. The time variations of the egg rate indicate the development of three distinct generations of our copepods, until smaller values are reached in late summer, when the zooplankton abundance decreases. The egg rate drops to zero when the transfer to the dormant phase starts.

We may also look at the effect of 'bottom-up' control by adding nutrients to the box model. We can introduce an external supply of nutrients, prescribed by the source term, $Q^{ext}$, in the nutrient equation (4.49), and remove the equivalent amount of detritus to keep the mass in the system constant. The results of the simulations (not shown) reveal that the increase of nutrients has only a small effect on the model phytoplankton because the excess phytoplankton is grazed rapidly by zooplankton. Only the peak of the spring bloom increases in the second year of the simulation, in response to the excess nutrients, which accumulate during the biologically inactive wintertime. The signal of the added nutrient propagates through the food web and finally accumulates in the adults' biomass. The total abundance and the egg rate show a similar behaviour, as shown in Fig. 4.18.

In our next simulations, we study the effects of a 'top-down' control by means of a temporary, increase in mortality, which may be caused by, episodes of higher predation. We confine these experiments to two cases with a increase of, first, the mortality of adults, and second, of the copepodites model 2. We assume the values $\mu_a = 1\,\mathrm{d}^{-1}$ for the first case and $\mu_{c_2} = 1\,\mathrm{d}^{-1}$ for the second case, during a period of 30 days. We choose the time interval from the model days $t = 200$ to $t = 230$, which corresponds to late July of the first model year of the biennial simulations. The effects of high predation pressure on the adults is shown in Fig. 4.19 by means of the biomass concentration of all stages.

The enhanced mortality rate, which acts over a time period comparable with the generation time, decreases the adult biomass almost to zero. Since during this period of time the reproduction breaks down, all stages are significantly affected by the predation event. After the event, a new population of adults must develop and start to lay eggs before the biomass recovers. In the second year, after the winter season, the copepods develop as in the standard run (compare Figs. 4.16 and 4.19), implying that such events have no long-term effects on the system.

The effects are weaker for predation events acting on other stages than the adults, e.g. copepodites 2. The results of the biennial simulation of the second experiment are shown in Fig. 4.20. Although the biomass of the $c_2$ stage decreases almost to zero during the predation event, the biomass of the adults remains at a relatively high level and maintains reproduction. Obviously, a predation event on adults has the severest effects on the zooplankton model because reproduction is directly affected.

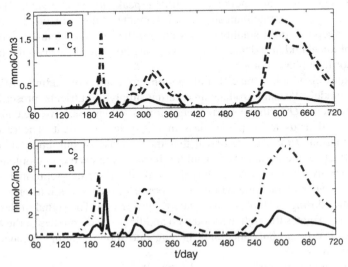

**FIGURE 4.19**  Stage-resolving simulation of the biomass for a temporary increase in mortality of the adults.

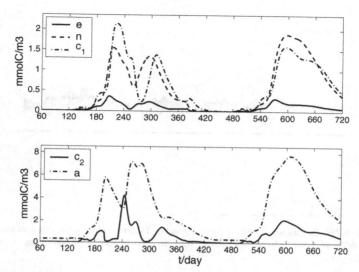

**FIGURE 4.20** Stage-resolving simulation of the biomass for a temporary increase mortality (predation event) of the copepodites model 2, $c_2$.

## 4.4 A FISH MODEL

In Chapters 2 and 3, we discussed lower food web models (biogeochemical models) with a bulk zooplankton variable, which was used to truncate the higher trophic levels. In this chapter, we included life-cycle-resolving zooplankton models, in particular for copepods. As a next natural step, we wish to add fish into the model system. To this end, we need a consistent theoretical description of fish. We use again the example of the Baltic Sea, which has the advantage that we can deal with a relatively simple fish stock structure, dominated by two prey species (sprat and herring) and one predator species (cod). The central idea is to construct a model with size (mass-class)-dependent predator–prey dynamics and a two-way interaction among the NPZD and fish model components through feeding of prey fish on zooplankton and recycling of fish biomass to nutrients and detritus.

### 4.4.1 Formulation of the Theory

A size-structured fish model can be formulated similar to the stage-resolving zooplankton model, but the interaction between the different mass classes of two prey species and one predator is more complex. The model includes two sets of evolution equations: one for biomass and one for number of individuals within each mass class. Similar to for the zooplankton model, the fish biomass can increase by consumption, while the abundance increases only during the reproduction phase through a pulselike addition of newborns; otherwise, the

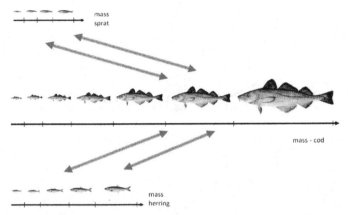

**FIGURE 4.21**   Illustration of the mass-class structured model, showing the different mass axes for the different fishes. The arrows indicate the size-class structured predator–prey interaction. (See color plate.)

number of individuals is decreasing through natural and fishing mortality. A control parameter is the average individual mass, $m$, defined by the ratio of biomass to number of individuals. A schematic of the structure of the fish model is shown in Fig. 4.21. During their life cycle, the fishes 'propagate' along their mass axis, passing successively through the different mass intervals. The 'propagation speeds' are defined by the corresponding growth rates, which will be specified later. As indicated by the arrows, the predator consumes only prey of smaller mass classes.

### 4.4.2   Structure of the Fish Model Equations

The mass axis of each species 'x' [$x = (\textbf{cod}, \textbf{herring}, \textbf{sprat})$], starts with $m_0$ and ends with $m_{\text{end}}$. We choose seven mass classes for cod, six for herring and five for sprat. The state variables can be written as vectors. For cod, the state vectors of the biomass, $\textbf{B}^{\text{cod}}$, and number of individuals, $\textbf{N}^{\text{cod}}$, are

$$
\textbf{B}^{\text{cod}} = \begin{pmatrix} B_1^{\text{cod}} \\ B_2^{\text{cod}} \\ B_3^{\text{cod}} \\ B_4^{\text{cod}} \\ B_5^{\text{cod}} \\ B_6^{\text{cod}} \\ B_7^{\text{cod}} \end{pmatrix} \quad \text{and} \quad \textbf{N}^{\text{cod}} = \begin{pmatrix} N_1^{\text{cod}} \\ N_2^{\text{cod}} \\ N_3^{\text{cod}} \\ N_4^{\text{cod}} \\ N_5^{\text{cod}} \\ N_6^{\text{cod}} \\ N_7^{\text{cod}} \end{pmatrix}.
$$

The elements are the biomass and number of the seven mass classes per unit volume. The dynamic changes of the state vectors in time is governed by the evolution equations:

$$\frac{d}{dt}\mathbf{B}^{cod} = (\mathbf{G}^{cod} + \mathbf{M}^{cod} + \mathbf{O}^{cod} + \tau^{cod})\mathbf{B}^{cod}, \tag{4.53}$$

and

$$\frac{d}{dt}\mathbf{N}^{cod} = \mathbf{M}^{cod}\mathbf{N}^{cod} + (\tau_N^{cod} + \mathbf{O}_N^{cod})\mathbf{B}^{cod}, \tag{4.54}$$

The matrices $\mathbf{G}^{cod}$, $\mathbf{M}^{cod}$ and $\mathbf{O}^{cod}$ describe the dynamic changes due to effective growth, mortality and reproduction (production of offspring), respectively. The matrices $\tau^{cod}$ and $\tau_N^{cod}$ define the transfer rates at which the fish of one mass class are promoted to the next higher one when the mean individual mass approaches the end of the mass class at hand. The structure of the matrices is as follows:

$$\mathbf{G}^{cod} = \begin{pmatrix} g_{C1}^{eff} & 0 & 0 & 0 & 0 & 0 & 0 \\ 0 & g_{C2}^{eff} & 0 & 0 & 0 & 0 & 0 \\ 0 & 0 & g_{C3}^{eff} & 0 & 0 & 0 & 0 \\ 0 & 0 & 0 & g_{C4}^{eff} & 0 & 0 & 0 \\ 0 & 0 & 0 & 0 & g_{C5}^{eff} & 0 & 0 \\ 0 & 0 & 0 & 0 & 0 & g_{C6}^{eff} & 0 \\ 0 & 0 & 0 & 0 & 0 & 0 & g_{C7}^{eff} \end{pmatrix},$$

$$\mathbf{O}^{cod} = \begin{pmatrix} 0 & 0 & 0 & 0 & 0 & O_{C6} & O_{C7} \\ 0 & 0 & 0 & 0 & 0 & 0 & 0 \\ 0 & 0 & 0 & 0 & 0 & 0 & 0 \\ 0 & 0 & 0 & 0 & 0 & 0 & 0 \\ 0 & 0 & 0 & 0 & 0 & 0 & 0 \\ 0 & 0 & 0 & 0 & 0 & -O_{C6} & 0 \\ 0 & 0 & 0 & 0 & 0 & 0 & -O_{C7} \end{pmatrix},$$

$$\mathbf{M}^{cod} = \begin{pmatrix} -\mu_{C1} & 0 & 0 & 0 & 0 & 0 & 0 \\ 0 & -\mu_{C2} & 0 & 0 & 0 & 0 & 0 \\ 0 & 0 & -\mu_{C3} & 0 & 0 & 0 & 0 \\ 0 & 0 & 0 & -\mu_{C4} & 0 & 0 & 0 \\ 0 & 0 & 0 & 0 & -\mu_{C5} & 0 & 0 \\ 0 & 0 & 0 & 0 & 0 & -\mu_{C6} & 0 \\ 0 & 0 & 0 & 0 & 0 & 0 & -\mu_{C7} \end{pmatrix},$$

and

$$
\tau^{\text{cod}} = \begin{pmatrix}
-\tau_1 & 0 & 0 & 0 & 0 & 0 & 0 \\
\tau_1 & -\tau_2 & 0 & 0 & 0 & 0 & 0 \\
0 & \tau_2 & -\tau_3 & 0 & 0 & 0 & 0 \\
0 & 0 & \tau_3 & -\tau_4 & 0 & 0 & 0 \\
0 & 0 & 0 & \tau_4 & -\tau_5 & 0 & 0 \\
0 & 0 & 0 & 0 & \tau_5 & -\tau_6 & 0 \\
0 & 0 & 0 & 0 & 0 & \tau_6 & 0
\end{pmatrix}.
$$

For the number of individuals, $\mathbf{N}^{\text{cod}}$, the transfer matrix $\tau_N^{\text{cod}}$ and $\mathbf{O}_N^{\text{cod}}$ are slightly modified in comparison with $\tau^{\text{cod}}$ and $\mathbf{O}^{\text{cod}}$,

$$
\tau_N^{\text{cod}} = \begin{pmatrix}
-\frac{\tau_1}{m_1^c} & 0 & 0 & 0 & 0 & 0 & 0 \\
\frac{\tau_1}{m_1^c} & -\frac{\tau_2}{m_2^c} & 0 & 0 & 0 & 0 & 0 \\
0 & \frac{\tau_2}{m_2^c} & -\frac{\tau_3}{m_3^c} & 0 & 0 & 0 & 0 \\
0 & 0 & \frac{\tau_3}{m_3^c} & -\frac{\tau_4}{m_4^c} & 0 & 0 & 0 \\
0 & 0 & 0 & \frac{\tau_4}{m_4^c} & -\frac{\tau_5}{m_5^c} & 0 & 0 \\
0 & 0 & 0 & 0 & \frac{\tau_5}{m_5^c} & -\frac{\tau_6}{m_6^c} & 0 \\
0 & 0 & 0 & 0 & 0 & \frac{\tau_6}{m_6^c} & 0
\end{pmatrix},
$$

and

$$
\mathbf{O}_N^{\text{cod}} = \begin{pmatrix}
0 & 0 & 0 & 0 & 0 & \frac{O_{C6}}{m_0^c} & \frac{O_{C7}}{m_0^c} \\
0 & 0 & 0 & 0 & 0 & 0 & 0 \\
0 & 0 & 0 & 0 & 0 & 0 & 0 \\
0 & 0 & 0 & 0 & 0 & 0 & 0 \\
0 & 0 & 0 & 0 & 0 & 0 & 0 \\
0 & 0 & 0 & 0 & 0 & 0 & 0 \\
0 & 0 & 0 & 0 & 0 & 0 & 0
\end{pmatrix}.
$$

A comparison of Equations (4.53) and (4.54) shows that the equation for the abundances contains biomass variables. This is required to describe the reproduction in a consistent mass conserving way. Reproduction is controlled by the 'offspring' rates, which transfer a certain amount of biomass from the matured fish to the smallest mass interval that includes the newborns.

For the prey species, herring and sprat, we have similar equations and process matrices. The differences in the mathematical structure are that the state vectors of herring and sprat contain six and five elements, respectively. The corresponding processes are governed by (6,6) and (5,5) matrices.

For herring, the state vectors are,

$$\mathbf{B}^{her} = \begin{pmatrix} B_1^{her} \\ B_2^{her} \\ B_3^{her} \\ B_4^{her} \\ B_5^{her} \\ B_6^{her} \end{pmatrix}, \text{ and } \mathbf{N}^{her} = \begin{pmatrix} N_1^{her} \\ N_2^{her} \\ N_3^{her} \\ N_4^{her} \\ N_5^{her} \\ N_6^{her} \end{pmatrix}.$$

The evolution equation for the biomass, $\mathbf{B}^{her}$, is,

$$\frac{d}{dt}\mathbf{B}^{her} = (\mathbf{G}^{her} + \mathbf{M}^{her} + \mathbf{O}^{her} + \tau^{her})\mathbf{B}^{her},$$

and for the number of individuals, $\mathbf{N}^{her}$,

$$\frac{d}{dt}\mathbf{N}^{her} = \mathbf{M}^{her}\mathbf{N}^{her} + (\tau_N^{her} + \mathbf{O}_N^{her})\mathbf{B}^{her}.$$

The process matrices have the structures,

$$\mathbf{G}^{her} = \begin{pmatrix} g_{H1}^{eff} & 0 & 0 & 0 & 0 & 0 \\ 0 & g_{H2}^{eff} & 0 & 0 & 0 & \\ 0 & 0 & g_{H3}^{eff} & 0 & 0 & 0 \\ 0 & 0 & 0 & g_{H4}^{eff} & 0 & 0 \\ 0 & 0 & 0 & 0 & g_{H5}^{eff} & 0 \\ 0 & 0 & 0 & 0 & 0 & g_{H6}^{eff} \end{pmatrix},$$

$$\mathbf{O}^{her} = \begin{pmatrix} 0 & 0 & 0 & 0 & O_{H5} & O_{H6} \\ 0 & 0 & 0 & 0 & 0 & 0 \\ 0 & 0 & 0 & 0 & 0 & 0 \\ 0 & 0 & 0 & 0 & 0 & 0 \\ 0 & 0 & 0 & 0 & -O_{H5} & 0 \\ 0 & 0 & 0 & 0 & 0 & -O_{H6} \end{pmatrix},$$

$$\mathbf{M}^{her} = \begin{pmatrix} -\mu_{H1} & 0 & 0 & 0 & 0 & 0 \\ 0 & -\mu_{H2} & 0 & 0 & 0 & 0 \\ 0 & 0 & -\mu_{H3} & 0 & 0 & 0 \\ 0 & 0 & 0 & -\mu_{H4} & 0 & 0 \\ 0 & 0 & 0 & 0 & -\mu_{H5} & 0 \\ 0 & 0 & 0 & 0 & 0 & -\mu_{H6} \end{pmatrix},$$

$$\tau^{\text{her}} = \begin{pmatrix} -\tau_1 & 0 & 0 & 0 & 0 & 0 \\ \tau_1 & -\tau_2 & 0 & 0 & 0 & 0 \\ 0 & \tau_2 & -\tau_3 & 0 & 0 & 0 \\ 0 & 0 & \tau_3 & -\tau_4 & 0 & 0 \\ 0 & 0 & 0 & \tau_4 & -\tau_5 & 0 \\ 0 & 0 & 0 & 0 & \tau_5 & 0 \end{pmatrix}$$

$$\tau_N^{\text{her}} = \begin{pmatrix} -\frac{\tau_1}{m_1^h} & 0 & 0 & 0 & 0 & 0 \\ \frac{\tau_1}{m_1^h} & -\frac{\tau_2}{m_2^h} & 0 & 0 & 0 & 0 \\ 0 & \frac{\tau_2}{m_2^h} & -\frac{\tau_3}{m_3^h} & 0 & 0 & 0 \\ 0 & 0 & \frac{\tau_3}{m_3^h} & -\frac{\tau_4}{m_4^h} & 0 & 0 \\ 0 & 0 & 0 & \frac{\tau_4}{m_4^h} & -\frac{\tau_5}{m_5^h} & 0 \\ 0 & 0 & 0 & 0 & \frac{\tau_5}{m_5^h} & 0 \end{pmatrix},$$

and

$$\mathbf{O}^{\text{her}} = \begin{pmatrix} 0 & 0 & 0 & 0 & \frac{O_{H5}}{m_0^h} & \frac{O_{H6}}{m_0^h} \\ 0 & 0 & 0 & 0 & 0 & 0 \\ 0 & 0 & 0 & 0 & 0 & 0 \\ 0 & 0 & 0 & 0 & 0 & 0 \\ 0 & 0 & 0 & 0 & 0 & 0 \\ 0 & 0 & 0 & 0 & 0 & 0 \end{pmatrix}.$$

For the vectors of sprat biomass, $\mathbf{B}^{\text{spra}}$, and number of individuals, $\mathbf{N}^{\text{spra}}$, we have the evolution equations,:

$$\frac{d}{dt}\mathbf{B}^{\text{spra}} = (\mathbf{G}^{\text{spra}} + \mathbf{M}^{\text{spra}} + \mathbf{O}^{\text{spra}} + \tau^{\text{spra}})\mathbf{B}^{\text{spra}},$$

and

$$\frac{d}{dt}\mathbf{N}^{\text{spra}} = \mathbf{M}^{\text{spra}}\mathbf{N}^{\text{spra}} + (\tau_N^{\text{spra}} + \mathbf{O}_N^{\text{spra}})\mathbf{B}^{\text{spra}},$$

where

$$\mathbf{B}^{\text{spra}} = \begin{pmatrix} B_1^{\text{spra}} \\ B_2^{\text{spra}} \\ B_3^{\text{spra}} \\ B_4^{\text{spra}} \\ B_5^{\text{spra}} \end{pmatrix} \quad \text{and} \quad \mathbf{N}^{\text{spra}} = \begin{pmatrix} N_1^{\text{spra}} \\ N_2^{\text{spra}} \\ N_3^{\text{spra}} \\ N_4^{\text{spra}} \\ N_5^{\text{spra}} \end{pmatrix},$$

The process matrices are

$$\mathbf{G}^{\mathrm{spra}} = \begin{pmatrix} g_{S1}^{\mathrm{eff}} & 0 & 0 & 0 & 0 \\ 0 & g_{S2}^{\mathrm{eff}} & 0 & 0 & 0 \\ 0 & 0 & g_{S3}^{\mathrm{eff}} & 0 & 0 \\ 0 & 0 & 0 & g_{S4}^{\mathrm{eff}} & 0 \\ 0 & 0 & 0 & 0 & g_{S5}^{\mathrm{eff}} \end{pmatrix},$$

$$\mathbf{O}^{\mathrm{spra}} = \begin{pmatrix} 0 & 0 & 0 & O_{S4} & O_{S5} \\ 0 & 0 & 0 & 0 & 0 \\ 0 & 0 & 0 & 0 & 0 \\ 0 & 0 & 0 & -O_{S4} & 0 \\ 0 & 0 & 0 & 0 & -O_{S5} \end{pmatrix},$$

$$\mathbf{M}^{\mathrm{spra}} = \begin{pmatrix} -\mu_{S1} & 0 & 0 & 0 & 0 \\ 0 & -\mu_{S2} & 0 & 0 & 0 \\ 0 & 0 & -\mu_{S3} & 0 & 0 \\ 0 & 0 & 0 & -\mu_{S4} & 0 \\ 0 & 0 & 0 & 0 & -\mu_{S5} \end{pmatrix},$$

$$\tau^{\mathrm{spra}} = \begin{pmatrix} -\tau_1 & 0 & 0 & 0 & 0 \\ \tau_1 & -\tau_2 & 0 & 0 & 0 \\ 0 & \tau_2 & -\tau_3 & 0 & 0 \\ 0 & 0 & \tau_3 & -\tau_4 & 0 \\ 0 & 0 & 0 & \tau_4 & 0 \end{pmatrix}$$

$$\tau_N^{\mathrm{spra}} = \begin{pmatrix} -\frac{\tau_1}{m_1^s} & 0 & 0 & 0 & 0 \\ \frac{\tau_1}{m_1^s} & -\frac{\tau_2}{m_2^s} & 0 & 0 & 0 \\ 0 & \frac{\tau_2}{m_2^s} & -\frac{\tau_3}{m_3^s} & 0 & 0 \\ 0 & 0 & \frac{\tau_3}{m_3^s} & -\frac{\tau_4}{m_4^s} & 0 \\ 0 & 0 & 0 & \frac{\tau_4}{m_4^s} & 0 \end{pmatrix},$$

and

$$\mathbf{O}_N^{\mathrm{spra}} = \begin{pmatrix} 0 & 0 & 0 & \frac{O_{S4}}{m_0^s} & \frac{O_{S5}}{m_0^s} \\ 0 & 0 & 0 & 0 & 0 \\ 0 & 0 & 0 & 0 & 0 \\ 0 & 0 & 0 & 0 & 0 \\ 0 & 0 & 0 & 0 & 0 \end{pmatrix}.$$

The definition of the various matrix elements will be given in the following subsections.

### 4.4.3 Predator–Prey Interaction and Effective Growth

The effective growth rates are defined by the differences between food consumption and metabolic losses. The maximum growth rates are taken from Fennel (2008), where data of the growth of average individuals of the three species were derived from Bertalanffy formulas by mapping the continuous growth formulas with allometric terms onto piecewise linear growth equations. The parameters of the Bertalanffy growth formulas, which were estimated by Horbowy (1989), represent observations and accumulated knowledge in a condensed manner.

In the model, the fish of the smallest mass class of cod eat zooplankton, while fish of the second mass class consume both zooplankton and small prey fish. The other classes of cod eat fish only. The prey fishes, sprat and herring are zooplanktivores, i.e. they eat zooplankton. For simplicity, we do not include cannibalism of cod or, as a more complex interaction, that prey fish may eat cod eggs. The feeding of sprat and herring on zooplankton is food limited through an Ivlev-type limiting function, $G(Z) = [1 - \exp(-I_Z Z)]$, with the Ivlev constant $I_Z = 1 \text{ m}^3/\text{mmolC}$. Thus, the feeding terms are

$$g_i^{S,H} = g_{H_i,S_i}^{\max} G(Z)A(T)Q(t), \tag{4.55}$$

where the effect of temperature is described by the factor $A(T)$,

$$A(T) = \Theta(T_{\text{ref}} - T)\exp(aT) + \Theta(T - T_{\text{ref}})\exp(bT - cT^2), \tag{4.56}$$

and where $T_{\text{ref}} = 18\,°\text{C}$, $a = 0.063/\,°\text{C}$, $b = 0.106/\,°\text{C}$, $c = 0.002/\,°\text{C}^2$ and $\Theta(x)$ is the step function. The term $Q(t)$ reflects the annual cycle of food quality and was chosen to mimic the time variation of food quality in the course of the year. A similar piecewise linear term was proposed by Megrey et al. (2007). An experimental motivation for this approach was provided by the findings of Arrhenius (1998b), Arrhenius (1998a) and Arrhenius and Hansen (1996),

$$Q(t) = [0.1 + \Theta(d_{100} - t)]\frac{d_{100} - t}{d_{145}} + \Theta(t - d_{320})\left[0.9 - \frac{(t - d_{320})}{d_{145}}\right]$$
$$+ [\Theta(t - d_{100}) - \Theta(t - d_{320})](t - d_{100})\frac{0.9}{d_{220}},$$

where $d_x$ refers to the day of the year.

The metabolic losses consist of a certain part of food uptake as defined by Equation (4.55) and a rate due to basic metabolism, which is independent of food consumption (Table 4.5). The choice of loss terms describing the fluxes from herring and sprat to nutrients (egestion and respiration) are explicitly,

$$L_{H_i N} = 0.0625\left[g_i^H + \frac{1}{2}A(T)Q(t)g_{H_i}^{\max}\right],$$

**TABLE 4.5** Parameters of the Fish Model

| Species | Mass Class | Mass (g) | $g_{x_i}^{max}$ (d⁻¹) | $g_{C_i}^{max}$ (d⁻¹) | $\mu_{x_i}^b$ (d⁻¹) | $\mu_{x_i}^s$ (d⁻¹) | $f_{x_i}$ (d⁻¹) | $o_{x_i}$ (d⁻¹) |
|---|---|---|---|---|---|---|---|---|
| Cod | $m_0^C$ | 2 | 0.0082 | – | 0.01 | 0 | – | – |
| | $m_1^c$ | 5 | 0.0066 | 0.0079 | $10^{-4}$ | 0 | – | – |
| | $m_2^c$ | 30 | – | 0.0092 | $10^{-4}$ | 0.002 | – | – |
| | $m_3^c$ | 100 | – | 0.0059 | $5 \times 10^{-5}$ | 0.002 | – | – |
| | $m_4^c$ | 200 | – | 0.0049 | $5 \times 10^{-5}$ | 0.002 | – | – |
| | $m_5^c$ | 800 | – | 0.0036 | $10^{-5}$ | 0.002 | $6.85 \times 10^{-4}$ | $10^{-5}$ |
| | $m_6^c$ | 1500 | – | 0.0023 | $10^{-11}$ [a] | 0.001 | $6 \times 10^{-3}$ | $5 \times 10^{-5}$ |
| | $m_{end}^C$ | $10^4$ | – | – | – | – | – | – |
| Herring | $m_0^H$ | 2 | 0.0160 | – | 0.01 | 0.1 | – | – |
| | $m_1^h$ | 5 | 0.0134 | – | $10^{-4}$ | 0 | – | – |
| | $m_2^h$ | 10 | 0.0118 | – | $10^{-4}$ | 0 | – | – |
| | $m_3^h$ | 30 | 0.0051 | – | $10^{-4}$ | 0 | – | – |
| | $m_4^h$ | 60 | 0.0024 | – | $5 \times 10^{-5}$ | 0 | – | $3 \times 10^{-5}$ |
| | $m_5^h$ | 150 | 0.0008 | – | $10^{-12}$ [a] | 0.01 | $2.74 \times 10^{-4}$ | $6 \times 10^{-5}$ |
| | $m_{end}^H$ | 240 | – | – | – | – | – | – |
| Sprat | $m_0^S$ | 1 | 0.0139 | – | 0.01 | 0.1 | – | – |
| | $m_1^s$ | 5 | 0.0075 | – | $10^{-4}$ | 0 | – | – |
| | $m_2^s$ | 10 | 0.0037 | – | $10^{-4}$ | 0 | – | – |
| | $m_3^s$ | 15 | 0.0032 | – | $5 \times 10^{-5}$ | 0 | – | $3 \times 10^{-4}$ |
| | $m_4^s$ | 20 | 0.0021 | – | $10^{-12}$ [a] | 0.005 | $2.74 \times 10^{-4}$ | $5 \times 10^{-4}$ |
| | $m_{end}^S$ | 35 | – | – | – | – | – | – |

A list of the basic characteristic parameters of the fish model for each mass class of cod (C), herring (H), and sprat (S). The consumption rates $g_{x_i}^{max}$ and $\mathbf{g}_{C_i}^{max}$ refer to eating of zooplankton and fish, respectively. The index $x$ refers to C, H and S.

[a] Some values of the mortality, $\mu_{x_i}^b$ refer to density-dependent terms; i.e. they are multiplied by the corresponding biomass concentration and have thus the unit $km^3\,g^{-1}\,d^{-1}$.

and

$$L_{S_i N} = 0.0625 \left[ g_i^S + \frac{1}{2} A(T) Q(t) g_{S_i}^{max} \right].$$

The losses governing the flux to detritus (excretion) are given by

$$L_{H_i D} = 0.0625 g_i^H \quad \text{and} \quad L_{S_i D} = 0.0625 g_i^S. \tag{4.57}$$

Thus, the effective growth rates in the matrices $\mathbf{G}^{her}$ and $\mathbf{G}^{spra}$ are determined by

$$g_{H_i}^{eff} = g_i^H - L_{H_i N} - L_{H_i D}, \tag{4.58}$$

and

$$g_{S_i}^{eff} = g_i^S - L_{S_i N} - L_{S_i D}. \tag{4.59}$$

The growth of cod is driven by a size-structured consumption of sprat and herring. The predator–prey terms are defined by

$$\mathbf{P}_{HS_k} = \mathbf{g}_{C_k}^{max} \frac{\sum_{i=1}^{6} p_{k,i} B_i^{her} G(B_i^{her}) + \sum_{i=1}^{5} p_{k,i} B_i^{spra} G(B_i^{spra})}{\sigma_k} A(T) Q(t), \tag{4.60}$$

for $2 \leq k \leq 7$. Here, $\mathbf{g}_{C_k}^{max}$ is the maximum consumption rate of the mass class $k$ of the prey that can be consumed by cod, and $G(B_i^{her}) \sim [1 - exp(I_C B_i^{her})]$ is an Ivlev function for cod that controls the food limitation for small prey concentrations (Fennel, 2010). Preferences of food are expressed through the preference coefficients $p_{k,i}$, listed in Table 4.6, and the $\sigma_k$ sums are defined by,

$$\sigma_k = \sum_{i=1}^{5} p_{k,i} B_i^{spra} + \sum_{i=1}^{6} p_{k,i} B_i^{her}. \tag{4.61}$$

The values of the elements in the matrix $p_{i,k}$ are chosen to ensure that cod consumes only prey smaller than the predator. The resulting effective growth rates of cod are then,

$$g_{C_1}^{eff} = g_1^C - L_{C_1 N} - L_{C_1 D} \tag{4.62}$$

$$g_{C_2}^{eff} = \mathbf{P}_{HS_2} + g_2^C - L_{C_2 N} - L_{C_2 D}, \tag{4.63}$$

$$g_{C_k}^{eff} = \mathbf{P}_{HS_k} - L_{C_k N} - L_{C_k D}, \quad \text{for } 3 \leq k \leq 7. \tag{4.64}$$

Similarly to $L_{H_i N}$, etc., we define loss rates of cod to the nutrient and detritus pool,

$$L_{C_k N} = L_{C_k D} = L_{C_k}^{zoo} g_k^C + L_{C_k}^{fish} \mathbf{P}_{HS_k} + L_{C_k}^{basic} \mathbf{g}_{C_k}^{max} Q(t) A(T). \tag{4.65}$$

The factors are given in Table 4.7. With these expressions, all matrix elements of $\mathbf{G}^{cod}$, $\mathbf{G}^{her}$ and $\mathbf{G}^{spra}$ are defined.

**TABLE 4.6** The Elements of the Preference Matrix, $p_{k,i}$

| $k/i$ | $H_1/S_1$ | $H_2/S_2$ | $H_3/S_3$ | $H_4/S_4$ | $H_5/S_5$ | $H_6$ |
|---|---|---|---|---|---|---|
| $C_1$ | 0 | 0 | 0 | 0 | 0 | 0 |
| $C_2$ | 1 | 0 | 0 | 0 | 0 | 0 |
| $C_3$ | 1 | 1 | 0 | 0 | 0 | 0 |
| $C_4$ | 1 | 1 | 1 | 0 | 0 | 0 |
| $C_5$ | 0.2 | 1 | 1 | 1 | 0/1 | 0 |
| $C_6$ | 0.1 | 0.3 | 0.6 | 1 | 1 | 1 |
| $C_7$ | 0.05 | 0.3 | 0.6 | 1 | 1 | 1 |

**TABLE 4.7** Loss Rates of Cod

| Mass Class $k$ | $L_{C_k}^{zoo}$ [1/d] | $L_{C_k}^{fish}$ [1/d] | $L_{C_k}^{basic}$ [1/d] |
|---|---|---|---|
| 1 | 0.125 | 0 | 0 |
| 2 | 0.125 | 0.15 | 0.015 |
| 3 | 0 | 0.15 | 0.015 |
| 4 | 0 | 0.15 | 0.015 |
| 5 | 0 | 0.2 | 0.015 |
| 6 | 0 | 0.2 | 0.015 |
| 7 | 0 | 0.45 | 0.005 |

## 4.4.4 Modelling Reproduction and Mortality

In our model, the reproduction rates, i.e. the matrix elements of the matrices $\mathbf{O}^x$, listed in Table 4.5, describe the mass loss of the matured mass classes during the spawning periods. The corresponding mass flow is channelled to the recruits; i.e. it controls the increase of the number of individuals in the smallest mass class with the start value $m_0$. The reproduction (offspring rates) is restricted to certain time windows, spawning periods, which can be expressed in terms of $\Theta$ functions,

$$T_{window}^C = \Theta(t - d_{60}) - \Theta(t - d_{150}), \text{ and}$$
$$T_{window}^H = T_{window}^S = \Theta(t - d_{60}) - \Theta(t - d_{120}),$$

The offspring rates are

$$O_{S_i} = o_{S_i} T^S_{window},$$

$$O_{H_i} = o_{H_i} T^H_{window},$$

$$O_{C_i} = o_{C_i} T^C_{window}.$$

The mortality of cod consists of three parts: a background mortality $\mu^b_{C_i}$, starvation mortality $\mu^{starv}_{C_i}$ which is acting at scarce food resources and fishing mortality, $F_{C_i}$. The background level of mortality is listed in Table 4.5. For cod, the starvation mortality is related to the ratio of prey biomass to the biomass of the predators; i.e. if the sum of the suitable food biomass is less than the tenfold of the predator biomass, then the starvation mortality rate for cod is activated. The fishing mortality, $F_{C_i}$, is zero during the spawning period and equals $f_{C_i}$ otherwise. Thus, the elements in the mortality matrix are given by,

$$\mu_{C_i} = \mu^b_{C_i} + \mu^{starv}_{C_i} + F_{C_i}, \ 1 \le i \le 6, \ \text{and} \ \mu_{C_7} = \mu^b_{C_7} B^{cod}_7 + \mu^{starv}_{C_7} + F_{C_7}.$$

For the largest mass class, the background mortality is proportional to the biomass, which implies a quadratic loss term in the corresponding equations. This choice helps to increase the stability of the model. For the fishing rates, we assume that the fishery is closed during the reproduction period, implying,

$$F_{C_i} = f_{C_i} (1 - T^C_{window}),$$

$$F_{H_i} = f_{H_i} (1 - T^H_{window}),$$

and

$$F_{S_i} = f_{S_i} (1 - T^S_{window}).$$

For sprat and herring, the predation mortality plays an important role in addition to the background, starvation, and fishing mortality rates, which are specified in Table 4.5. The consumption of prey implies a reduction of both the prey biomass and abundance. This is taken into account in the prey–predator interaction terms,

$$\Pi_{S_i} = \sum_{k=1}^{7} \frac{g^{max}_{C_k} p_{k,i} B^{cod}_k}{\sigma_k} A(T) Q(t), \ \text{and} \tag{4.66}$$

$$\Pi_{H_i} = \sum_{k=1}^{7} \frac{g^{max}_{C_k} p_{k,i} B^{cod}_k}{\sigma_k} A(T) Q(t). \tag{4.67}$$

These terms correspond to the predator–prey interaction term (4.60) and prescribe the loss of biomass of herring and sprat due to cod predation.

The starvation mortality rates of sprat and herring are activated if the effective growth rates become negative.

To give an impression of the order of magnitude of the fishing mortality with respect to the usual scale of the fishing rates, we note that $f_{H_5} = 2.7397 \times 10^{-4}/d = 0.1/year$ and $f_{C_6} = 6.8493 \times 10^{-4}/d = 0.25/year$. The total mortalities are thus given by

$$\mu_{S_i} = \mu_{S_i}^b + \mu_{S_i}^{starv} + \Pi_{S_i}, \quad 1 \le i \le 4,$$

$$\mu_{S_5} = \mu_{S_5}^b B_5^{spra} + \mu_{S_5}^{starv} + F_{S_5} + \Pi_{S_5},$$

and

$$\mu_{H_i} = \mu_{H_i}^b + \mu_{H_i}^{starv} + \Pi_{H_i}, \quad 1 \le i \le 5,$$

$$\mu_{H_6} = \mu_{H_6}^b B_6^{her} + \mu_{H_6}^{starv} + F_{H_6} + \Pi_{H_6}.$$

With these terms, all matrix elements of $\mathbf{M}^x$ are defined for the three groups of fish. Higher-order interactions, like cannibalism of cod or prey fish eating newborns of the predator are not yet taken into account, but this can be done in a straightforward manner.

Finally we have to define the matrix elements of the transfer matrices $\tau^x$. The transfer function is enabled if the mass of the mean individual, $m$, reaches the upper limit of the corresponding mass interval. Instead of Fermi's function used for the stage-resolving zooplankton, we use a simpler switch-on with a linear ramp.

$$\tau_i(m, m_i) = \theta(m - m_i) + \frac{m - s\, m_i}{m_i(1 - s)}(\theta(m - s\, m_i) - \theta(m - m_i), \quad (4.68)$$

where the parameter $s = 0.99$ controls the slope of the ramp.

## 4.4.5 Coupling Fish and Lower Trophic Levels

In order to connect the lower and upper food web in the model system, we use a simple NPZD model, as described in Section 4.3.3, but for simplicity, with a bulk zooplankton state variable. The coupling of the NPZD model and the fish model can be implemented through three channels:

1. consumption of zooplankton by the zooplanktivores, expressed by the mass flux $\mathbf{G_F}$,
2. respiration, i.e. transfer of matter from fish to dissolved nutrients, by the mass flux $\mathbf{L_{FN}}$,
3. recycling of fish to detritus, where the corresponding mass flux, $\mathbf{L_{FD}}$, is maintained by excretion rates $L_{x_iD}$ and mortalities, $\mu_{x_i}$.

The resulting equations for the NPZD, then, are

$$\frac{dN}{dt} = -M(N)P + l_{PN}P + l_{DN}D + l_{ZN}Z + Q_N^{import} + \mathbf{L_{FN}}, \quad (4.69)$$

$$\frac{dP}{dt} = M(N)P - l_{PN}P - g(P)Z - l_{PD}P, \qquad (4.70)$$

$$\frac{dZ}{dt} = g(P)Z - l_{ZN}Z - l_{ZD}Z - \mathbf{G_F}, \qquad (4.71)$$

$$\frac{dD}{dt} = l_{ZD}Z + l_{PD}P - (l_{DN} + l_{DD_{\text{sed}}})D + \mathbf{L_{FD}} \qquad (4.72)$$

$$\frac{dD_{\text{sed}}}{dt} = l_{DD_{\text{sed}}}D. \qquad (4.73)$$

The mass flux, $\mathbf{G_F}$, which represents the removal of zooplankton biomass through sprat, herring and the two smallest mass classes of cod, depends on the biomass concentration of the mass classes $B_i^x$, and food-limited consumption rates $g_i^X(Z)$, where $x = (\text{sprat, herring, cod})$ and the index $i$ refers to the corresponding mass class. Moreover, we use a conversion factor, $f_{\text{con}}$, to transform biomass concentration in $g/km^3$ to carbon units in $mmolC\,m^{-3}$, where $1\,g\,km^{-3} \sim f_{\text{con}}mmolC\,m^{-3}$ with $f_{\text{con}} = 10^{-8}$. Then, we have,

$$\mathbf{G_F} = f_{\text{con}}\left[\sum_{i=1}^{6} g_i^{her}(Z)B_i^{her} + \sum_{i=1}^{5} g_i^{spra}(Z)B_i^{spra} + \sum_{i=1}^{2} g_i^{cod}(Z)B_i^{cod}\right].$$

Part of the consumption flows into respirational losses, described by rates $L_{S_iN}$, $L_{H_iN}$ and $L_{C_iN}$ for sprat, herring and cod, respectively. The coupling of the fish model and the NPZD model with respect to nutrient state variable amounts to the addition of the flux term, $\mathbf{L_{FN}}$ (loss of fish to nutrients), in Equation (4.69),

$$\mathbf{L_{FN}} = f_{\text{con}}\left(\sum_{i=1}^{5} L_{S_iN}B_i^{spra} + \sum_{i=1}^{6} L_{H_iN}B_i^{her} + \sum_{i=1}^{7} L_{C_iN}B_i^{cod}\right).$$

Losses of fish biomass through excretion at the rates $L_{S_iD}$, $L_{H_iD}$ and $L_{C_iD}$ and mortality of fish at rates $\mu_{x_i}$ constitute a further pathway from the fish model to the NPZD model, implying an input term, $\mathbf{L_{FD}}$ (loss of fish to detritus), in the detritus equation (4.72),

$$\mathbf{L_{FD}} = f_{\text{con}}\left[\sum_{i=1}^{5}(L_{S_iD} + \mu_{S_i})B_i^{spra} + \sum_{i=1}^{6}(L_{H_iD} + \mu_{H_i})B_i^{her}\right.$$
$$\left. + \sum_{i=1}^{7}(L_{C_iD} + \mu_{C_i})B_i^{cod}\right]. \qquad (4.74)$$

The rates were described in the previous section and the numerical values specified Table 4.5. To describe the export of matter by accumulation in the sediment, we have introduced an equation for the state variable, $D_{\text{sed}}$, which represents the detritus buried in sediments. Together with the set (4.69)–(4.73), our fish model forms a nutrient to fish model system with a true two-way coupling of the upper and lower part of the food web.

### 4.4.6 Example Scenarios

After the outline of the nutrient to fish model, we will now show some example simulations with this box model that are motivated by observations. The fish catches in the Baltic exhibit significant interannual variability, as shown in Fig. 4.22.

The data are available from the International Council for the Exploration of the Sea (ICES)[1]; an overview on the fish dynamics in the Baltic can be found, e.g. in Hammer et al. (2008). While the variation of herring catches are relatively weak, the catches of cod and sprat vary rather strongly in opposite phase—low sprat catches correspond to high cod catches and visa verse. The possible reasons for the variations are:

- fishery pressure, in particular, on cod,
- eutrophication due to the nutrients loads of the rivers discharging into the Baltic,
- variations in the reproduction of cod due to anoxic conditions in the spawning areas.

In order to qualitatively study the role of these factors, we simulate different scenarios with the aid of multipliers that mimic variations in fishery, loads, and

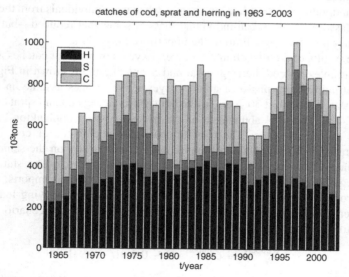

**FIGURE 4.22** Annual variations of the catches of sprat, herring and cod (data source: ICES). (See color plate.)

1. www.ices.dk

**TABLE 4.8** Multipliers for the Simulated Scenarios 1–6

| | Loads | Fishery Mortality | Reproduction |
|---|---|---|---|
| 1 | 1 | 0.3 | 1 |
| 2 | $1 - 0.3 \times \Theta(t - y_{18})$ | 0.3 | 1 |
| 3 | 1 | $0.15 + 0.5 \times \Theta(t - y_{12})$ $+0.35\Theta(t - y_{30})$ | 1 |
| 4 | 1 | $\Theta(y_{12} - t) + 0.15 \times \Theta(t - y_{12})$ $+0.2\Theta(t - y_{30})$ | 1 |
| 5 | 1 | 0.3 | $\theta(y_{12} - t) + \theta(t - y_{20})$ |
| 6 | 1 | $0.3 - 0.2 \times \Theta(t - y_{18})$ | $\theta(y_{12} - t) + \theta(t - y_{20})$ |

reproductive condition, see Table 4.8. As a baseline run, scenario 1, we start with constant levels of fishing rates, loads and reproduction. The annual cycle of the water temperature is chosen as shown in Fig. 2.12. As a typical feature of the model, we show the propagation of the mean individual masses through the different mass classes, Fig. 4.23. While in the lower mass classes the mean individual masses propagates smoothly along the mass axes from one mass class to the next one, there is a gap between the second highest and the uppermost mass class. This is due to the fact that in the uppermost mass-class interval, bigger individuals accumulate, and hence, when new individuals from the lower mass class arrive, the mean individual mass is somewhat reduced—but not to the values near the lower limit of the uppermost mass class.

The results of the first simulation over 40 years in terms of catches and the state variables for cod, herring, sprat and zooplankton are shown in Fig. 4.24. After an adjustment phase of about 10 years, the catches assume an almost steady state for the last 30 years, with slightly decreasing cod and sprat catches, Fig. 4.24(a). The total abundance of the fishes and the zooplankton biomass exhibit an annual cycle. The total biomass and abundance of cod, herring and sprat and the zooplankton biomass remain practically on the same level, after the adjustment phase of about 10 years, Fig. 4.24(b). The state close to an equilibrium can be changed by modifying the nutrient imports, fishery or reproduction conditions. We start with one example of varying loads and consider a reduction of the loads by 30% after 18 years in scenario 2. The scenario is implemented with the aid of the multiplier,

$$Q_N^{\text{import}} \rightarrow Q_N^{\text{import}} \cdot [1 - 0.3\Theta(t - y_{18})],$$

where $y_{18}$ stands for 18 years, i.e. $y_{18} = 18 \times 365$ days, and $Q_N^{\text{import}}$ is the import rate transformed to carbon with the Redfield ratio, $Q_N^{\text{import}} \sim 0.0163 \, \text{mmolC/m}^3/\text{d}$.

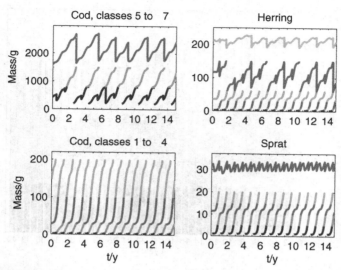

**FIGURE 4.23** The propagation of the mean individual masses through the mass classes of cod, herring and sprat. (See color plate.)

The effects of the reduced loads on the catches and the state variables are shown in Fig. 4.25. For the first 20 years, the results of both simulations are identical, but after 20 years all catches show a clear downward trend. The strongest decrease is found for cod and sprat, while the herring catches vary only weakly. A similar change is found in the fish stocks, Fig. 4.25(b). The total abundances and biomass of the fish and the zooplankton biomass show the same trend. All variables decrease in response to the lower river loads. The two experimental simulations demonstrate that, at least in the box model, increase or decrease of nutrient imports yield ultimately to higher or lower catches. However, this ignores the negative effects of eutrophication in terms of unfavourable oxygen conditions for the reproduction of cod (Matthäus, 1995; Hammer et al., 2008). This implies that stronger nutrient imports may not support, but may even decrease, the cod stock.

In the next experimental simulations, we keep the loads constant but look at the effects of changing fishery mortality. We consider two cases, scenarios 3 and 4, where the fishing mortality of cod is modified by the multipliers,

$$F_{C_i} \rightarrow F_{C_i} \times [0.15 + 0.5 \times \Theta(t - y_{12}) + 0.35\Theta(t - y_{30})],$$

and

$$F_{C_i} \rightarrow F_{C_i} \times [\Theta(y_{12} - t) + 0.15 \times \Theta(t - y_{12}) + 0.2\Theta(t - y_{30})].$$

In the first case (scenario 3), the fishery pressure is small during the first 12 years, but increases after 12 years and again after 30 years. The catches,

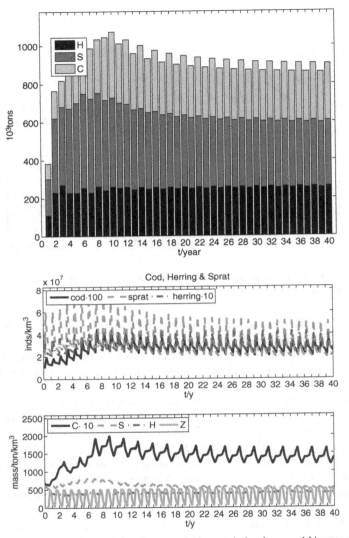

**FIGURE 4.24** Variation of catches and state variables: total abundance and biomass of sprat, herring, cod and biomass of zooplankton for constant rates (fishing mortality, reproduction and loads). (a) Catches for scenario 1 and (b) state variables for scenario 1. (See color plate.)

which are shown in Fig. 4.26(a), increase in response to the enhanced fishery rate, but stabilize towards the end of the simulation period when the cod stock is slightly decreasing by the high fishery rate, Fig. 4.26(b), while the sprat stock is slightly increasing because of a weaker predation. Again, the catches and the stock of herring show only minor effects. In the second case, fishery scenario 4,

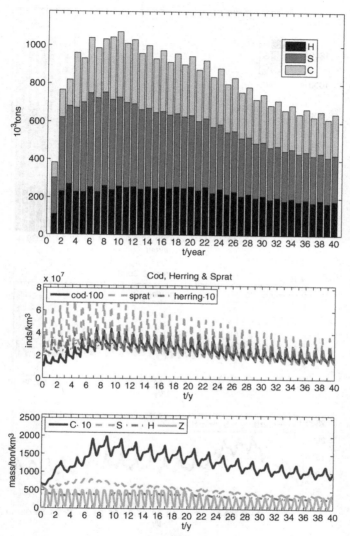

**FIGURE 4.25** Variation of catches and state variables in response to reduced nutrient discharge after 18 years: total abundance and biomass of sprat, herring, cod and biomass of zooplankton. (a) Catches for scenario 2 and (b) state variables for scenario 2. (See color plate.)

we start with high fishery pressure during a period of 12 years and reduce the rate significant for several years. After 30 years, we allow a slight increase of the fishing rate. The high fishing rate of cod gives larger cod catches for a few years, see Fig. 4.27(a), but, at the same time, the cod stock diminishes and the catches decline, Fig. 4.27(b). After moderation of the fishing pressure, the cod

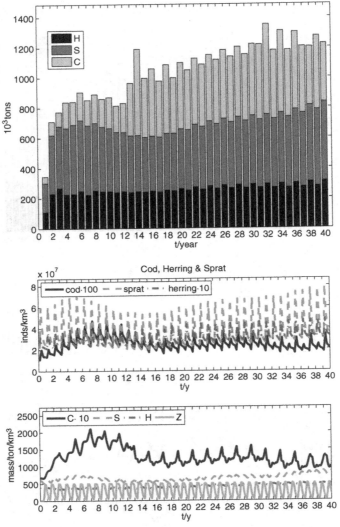

**FIGURE 4.26** Variation of catches and state variables: total abundance and biomass of sprat, herring and cod and the biomass of zooplankton for fishery scenario 3, beginning with low and then increasing fishing mortality. (a) Catches for scenario 3 and (b) state variables for scenario 3. (See color plate.)

stock recovers, and hence, the catches increases again even with a lower fishery intensity. In response to recovered cod stock, the larger consumption of sprat by cod reduced the sprat stock and catches. At the same time, the total zooplankton biomass increases because of the reduced consumption by sprat.

Next, we look at the effects of a failure of reproduction of cod for a certain time span (scenario 5). This could be caused by the anoxic condition of the

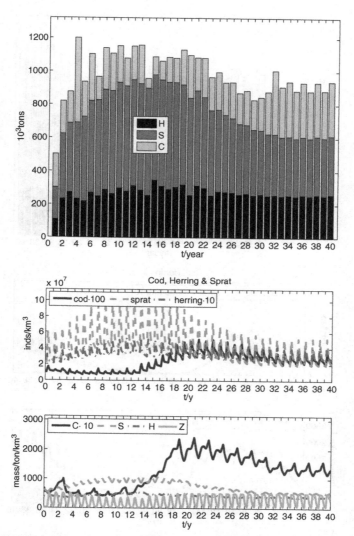

**FIGURE 4.27** Variation of catches and state variables: total abundance and biomass of sprat, herring and cod and the biomass of zooplankton for fishery scenario 4, beginning with high and then decreasing fishing mortality. (a) Catches for fishery scenario 4 and (b) State variables for fishery scenario 4. (See color plate.)

spawning region of cod see, e.g. Hammer et al. (2008). As an example, we chose the following multiplier for the reproduction rates,

$$O_{C_i} \to O_{C_i}[\theta(y_{12} - t) + \theta(t - y_{20})].$$

This means we set the reproduction rate (offspring rate) to zero after 12 years and restore the rate after the year 20, while the fishery multiplier is set to a

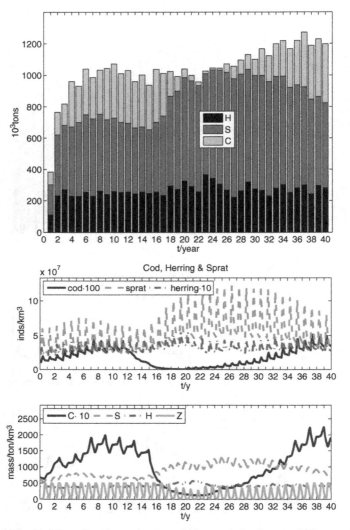

**FIGURE 4.28** Variation of catches and state variables: total abundance and biomass of sprat, herring and cod and the biomass of zooplankton for a low-reproduction scenario (scenario 5); in the time span from 12 to 20 years, there is no cod reproduction. (a) Catches for scenario 5 and (b) State variables for scenario 5. (See color plate.)

relatively small value, $F_{C_i} \rightarrow F_{C_i} \times 0.3$. The results are illustrated in Fig. 4.28. The cod stock and catches start to decrease 2 years after the reproduction of cod was switched off. At the same time period the stock and catches of sprat and, to a lesser extent of herring, increase. With a time lag of about 5–8 years after the restored reproduction, the cod stock and catches recover. This correlates with a reduced sprat stock due to the enhanced predation.

In our last example (scenario 6), we consider the case where the fishery of cod is modified 6 years after the onset of no reproduction, i.e. we invoke a second multiplier,

$$F_{C_i} \rightarrow F_{C_i} \times [0.3 - 0.2 \times \Theta(t - y_{18})].$$

While in the first half of the simulation the results are the same as in the previous case, the second half of the simulation yields smaller cod catches but a somewhat higher cod stock than without a moderation of the fishery, as shown in Fig. 4.29. The sprat catches and stock are somewhat reduced because of a stronger predation by the enhanced cod stock. The catches and stock of herring show some minor changes, apart from some stronger year-to-year variations. The propagation of the mean individual mass of cod differs strongly from Fig. 4.23 because now the lower mass classes no longer exist for the time span of no reproduction. The mean individual cod mass accumulates in the largest mass class, see Fig. 4.30, but at a low abundance level. However, as long as enough fish survive, the cod stock can recover after the onset of favourable reproduction conditions.

All considered scenarios show that significant changes of the catches can be driven by varying fishery, varying the reproduction conditions of cod or changing nutrient inputs, or by a combination of all these factors. The most striking variability is found for cod and sprat, while the changes of the herring catches are relatively small. The strongest predator–prey interaction can be observed between sprat and cod. This is apparently due to the small size and the fast development of sprat, which responds immediately to changed forcing.

The interaction between the lower and upper parts of the model food web can be characterized by the fluxes, $G_F$, which represent the flux of zooplankton biomass to fish, and $L_{FN}$ and $L_{FD}$, which describe the fluxes from fish biomass to nutrients and dead organic matter, the detritus. As an example, we show the fluxes for scenarios 3 and 4 in Fig. 4.31. All signals exhibit a clear annual cycle. The strongest matter flux goes from the zooplankton to fish, followed by the flux from fish to detritus, and the smallest amplitude has the flux from fish to nutrients. In scenario 3, an increase of the fluxes occurs in the second part of the 40-year simulation period; in scenario 1, not shown, the fluxes remain practically on the same level during the simulation. In the other cases, the typical pattern for the scenarios is an decrease of the amplitude of the fluxes during the third and fourth decade of the simulation, which is related by the lower stocks of sprat and herring in the last two decades.

The MATLAB code of this fish model is available on the website of the book. The reader can reproduce the scenarios and can perform is or her own experiments with the model. However, the variation of the parameters and multipliers should be done with some care because there are many possibilities to drive the model in catastrophic states. One obvious example is that of a too strongly reduced cod mortality; the cod stock could then consume the prey

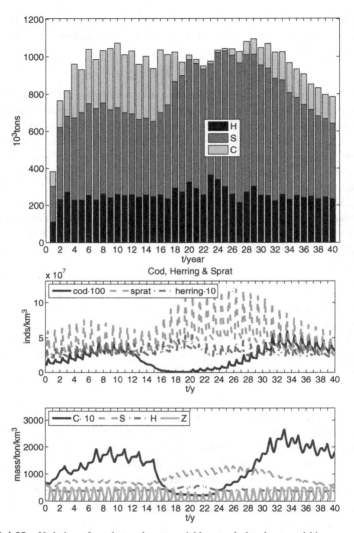

**FIGURE 4.29**   Variation of catches and state variables: total abundance and biomass of sprat, herring and cod and the biomass of zooplankton for a mixed reproduction and fishery scenario (scenario 6); there is a reproduction gap from year 12 to year 20, but cod fishing is moderated after year 18. (See color plate.)

fishes completely, and in the end, the cod stock would be extinguished due to lack of food. However, such strange scenarios are largely caused by the simplified properties of the box model, where a predator can find all food items. In spatially explicit three-dimensional models, the predation processes are more complex because fish have to find food and prey can escape. Moreover, apart

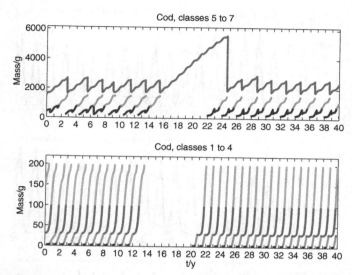

**FIGURE 4.30** The propagation of the mean individual masses through the mass classes of cod for scenario 6. Due to the period of no reproduction, the lower mass classes no longer exist, while the mass accumulates in the largest class. (See color plate.)

from fishing, there are other sources of mortality in the real systems. Sea birds and, in particular, a high seal population might provide some additional top-down control (Hansson et al., 2007).

## 4.4.7 Discussion

The theoretical studies presented in this chapter showed that the description of life cycles of copepods and fish can be included in state-variable models. For coherently developing cohorts of copepods with the same individual growth of all animals, a population model with growth-controlled transfer rates between the stages is sufficient. However, if reproduction is included, the population model approach is no longer sufficient. Then we have to extend the theory by including stage- or mass-class-resolving biomass state variables. The resulting model has more state variables because the dynamic equations for abundance and biomass concentrations of all stages or mass classes must be integrated. The development of the mass of the mean individuals in each class, which can be expressed through the quotient of concentration over abundance, is used to prescribe the transfer from one stage or mass class to the next one.

The models used in this chapter represent combinations of models of individual and state-variable models. The individual-based aspects of the model refer to development and reproduction of individuals that is controlled by behaviour of mean individuals of the different stages. There are several other

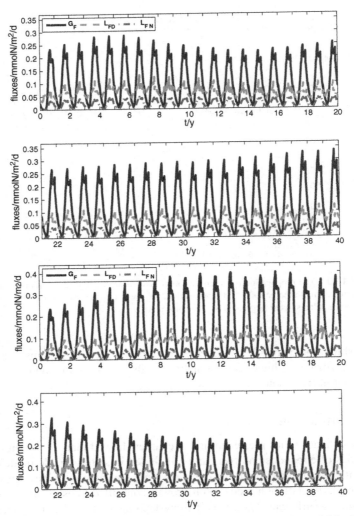

**FIGURE 4.31**   Fluxes of matter between the lower and upper parts of the model food web for scenarios 3 (upper panels) and 4 (lower panels). (See color plate.)

important issues related to individual behaviour of copepods or fish, such as vertical migration and avoidance of predators or active migration to spawning areas and foraging, which cannot be represented in simple box models. To describe behavioural patterns an explicit spatial resolution is required. We will address these issues in Chapter 7, where the stage-dependent zooplankton model and the mass-class structured fish model will be embedded in a fully three-dimensional circulation model.

# Chapter 5

# Physical–Biological Interaction

Modelling marine ecosystems requires an integrated approach, which involves biogeochemical processes, food web interaction and the control of these processes by the physical environment. The role of the physical processes was largely ignored in previous chapters. The important effects of Sun irradiation was parameterized by the control parameter 'length of day'. The vertical mixing was somehow included by the use of stacked boxes with prescribed exchange rates. The important role of physical currents and mixing processes was completely ignored. Thus, it makes sense to start with biological dynamics and take a minimum of physics into account. In this chapter, we discuss some physical properties and processes of the marine systems that affect the ecosystem dynamics. In addition to the discussion of important aspects of physical–biological interactions, we consider related theoretical problems, which concern the theoretical basis of the linkage of individual-based and state-variable models.

We begin with the light, i.e. the Sun's radiation, which controls the starting point of the food web, the primary production.

## 5.1 IRRADIANCE

### 5.1.1 Daily, Seasonal and Annual Variation

The irradiance of the Sun has direct and indirect implications for biological processes in aquatic systems. The direct impact is due to the visible light that occupies the spectral range from 300 to 760 nm and enables photosynthesis. The indirect impact is related to the heat flux imposed by radiation, which affects the water temperature and hence the vertical thermal stratification of the sea.

The irradiance, $I$, received per unit area on the sea's surface is basically determined by the Sun's radiation (solar constant, $I_s$) and depends on the position of the Sun, which is characterized by the zenith angle, $\zeta$. If we ignore, for simplicity, atmospheric effects, it follows that

$$I(\zeta) = I_s \cos \zeta.$$

The zenith angle depends on the latitude, $\vartheta$, and the inclination, $\delta$, which is the seasonally varying angle between the equatorial plane and the Sun. The zenith angle varies with daytime due to Earth rotation as

Introduction to the Modelling of Marine Ecosystems. http://dx.doi.org/10.1016/B978-0-444-63363-7.00005-2

$$\cos \zeta = \sin \vartheta \sin \delta + \cos \vartheta \cos \delta \cos(\omega_h t_h), \tag{5.1}$$

where $\omega_h = 2\pi/24$, and $t_h$ counts the hours from 0 to 24, starting at noon. This implies that the zenith angle at noon is given by

$$\cos \zeta = \cos(\vartheta - \delta).$$

With the help of Equation (5.1), we can express the length of the day, $\Delta d$, by searching for the first zero crossing of $\cos \zeta$,

$$\sin \vartheta \sin \delta + \cos \vartheta \cos \delta \cos \left( \omega_h \frac{\Delta d}{2} \right) = 0.$$

The period from noon to the next zero crossing gives the half length of the day, i.e.

$$\Delta d = \frac{2}{\omega_h} \arccos(-\tan \vartheta \tan \delta). \tag{5.2}$$

The inclination depends on $t_d$, which counts the day of the year from 1 to 365 starting at the first day of January. The inclination can be expressed by a trigonometric polynomial,

$$\delta = \frac{a_0}{2} + \sum_{n=1}^{3} [a_n \cos(n\gamma) + b_n \sin(n\gamma)], \tag{5.3}$$

with $\gamma = \frac{2\pi}{365}(t_d - 1)$. The coefficients $a_n$ and $b_n$ are listed in Table 5.1.

To a good approximation, we can use the simpler relationship,

$$\delta = -0.3999 \cos \gamma + 0.0702 \sin \gamma.$$

With the aid of these formulas, the seasonal and daily variations of the light can be described. Examples for the normalized length of the day, i.e. $\Delta d/24\,\mathrm{h}$, according to Equation (5.2) are shown in Fig. 5.1 for different latitudes.

However, corrections are required to describe the atmospheric effects on the Sun's radiation when penetrating the atmosphere. Energy is lost by scattering and absorption due to water vapour, carbon dioxide, ozone and dust. The

**TABLE 5.1** Coefficients in Equation (5.3)

| Coefficient | Value | Coefficient | Value |
|---|---|---|---|
| $a_0/2$ | 0.006918 | | |
| $a_1$ | −0.399912 | $b_1$ | 0.070257 |
| $a_2$ | −0.006758 | $b_2$ | 0.000907 |
| $a_3$ | −0.002697 | $b_3$ | 0.001480 |

The coefficients of the Fourier series of the astronomical formula (5.3), after Spencer (1971).

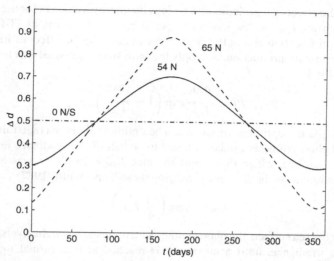

**FIGURE 5.1** Annual cycle of the normalized length of the day at the equator, 54°N, and 65°N.

irradiance available for photosynthesis is also modified by cloud coverage and reflection at the sea surface (albedo), which depends on the Sun's angle,

$$I_0 = I[1 - c_1 C - c_2(\vartheta - \delta)],$$

where $c_1$ and $c_2$ are fitted coefficients, $C$ is the fractional cloud coverage and $\vartheta - \delta$ is the noon zenith angle.

The invisible part of the incoming radiation, i.e. about 50% of $I_0$, is absorbed at the water surface and not used for photosynthesis. The photosynthetically active radiation, $I^{PAR}$, in the water, decreases due to light attenuation of the water and shading effects of plankton and detritus,

$$I^{PAR}(z) = \frac{I_0}{2} \exp\left[ k_w z - k_c \int_z^0 dz'(P + Z + D) \right], \quad (5.4)$$

where $k_w$ is the light attenuation constant of sea water and $k_c$ the attenuation due to shading by the sum of the phytoplankton, $P$, zooplankton, $Z$, and detritus, $D$. Note that the $z$-axis is positive upward.

## 5.1.2 Production–Irradiance Relationship

The relationship between primary production and irradiance are described by so-called PI formulas, which express the intensity of photosynthesis as a function of the photosynthetically active radiation. While irradiation is determined by well-known physical laws, the quantification of the response of cells to the irradiations involves uncertainties, due to biological processes, such as adaption or internal storage of energy. There are several empirical PI relationships, which include the effect of light inhibition by introducing a parameters for the optimal light intensity, $I_{opt}$. A discussion of several proposed choices of PI formulas is, given in Ebenhöh et al. (1997). In order to describe the effect of irradiance upon the primary production, we apply the most frequently used Steele formula (Steele, 1962)

$$PI = \frac{I^{PAR}}{I_{opt}} \exp\left(1 - \frac{I^{PAR}}{I_{opt}}\right), \quad (5.5)$$

where $I_{opt}$ is the optimum irradiance. The optimum daily averaged irradiance for algal photosynthesis can be assumed to be half of the irradiance just below the water surface, where the minimum value $I_{min} = 25\,\text{Wm}^{-2}$ was estimated from measurements in the Baltic Sea (Stigebrandt and Wulff, 1987)

$$I_{opt} = \max\left(\frac{I_0}{4}, I_{min}\right). \quad (5.6)$$

Near the sea surface, the photosynthesis increases at low light levels, linearly with the irradiance, until a maximum is reached at the optimal light level, where $I^{PAR} = I_{opt}$. For higher intensity, photosynthesis decreases in response to the light inhibition of the cells. At even higher light levels, the cells are light adapted and photosynthetic activity remains constant. In deeper parts of the water column, the light level is lower due to the light attenuation. In particular, for $I^{PAR} < I_{opt}$, the photosynthesis-light relationship remains in the linearly increasing part of Equation (5.5). The form of the photosynthesis-light relationship given in Equation (5.5) is sketched in Fig. 5.2 for conditions at the sea surface, $z = 0$.

The euphotic zone is the near-surface layer illuminated by sunlight. The thickness of the euphotic zone, $H_{euph}$, is defined by the depth where the irradiance, $I^{PAR}$, is 1% of the value immediately beneath the surface. In advanced circulation models, Sun's radiation and atmospheric corrections are well described and can be used to force photosynthesis in coupled physical–biological models. However, for simple model studies, we can also use time series of observed irradiance or simple expressions to mimic the daily cycle, e.g.

$$I_0 \sim \sin\left[\frac{\pi}{\Delta d} \text{mod}(t, 24)\right],$$

when the sine function becomes negative $I_0$ is set equal zero. Periodicity is enforced by the modulo, whereby the sine function is reset to zero at the beginning of the day, see Fig. 5.3. If it is not necessary to resolved the daily cycle, we can use the normalized length of the day to prescribe approximately seasonal changes in radiation. Actually, integration over one day gives

$$\frac{1}{24\,\text{h}} \int_0^{24\,\text{h}} dt \sin\left[\frac{\pi}{\Delta d} \text{mod}(t, 24)\right] = \int_0^{\Delta d} dt \sin\left(\frac{\pi}{\Delta d} t\right) = \frac{2}{\pi} \frac{\Delta d}{24\,\text{h}}.$$

Note that in this formula the integration starts at the beginning of the day ($t = 0$).

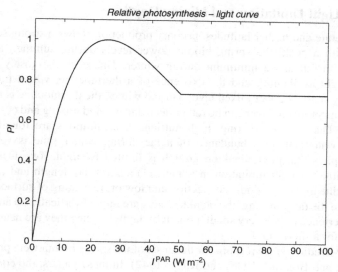

**FIGURE 5.2** Relative photosynthesis–light relationship versus photosynthetically active radiation, $I^{PAR}$, at the sea surface, $z = 0$. Photosynthesis grows linearly at low light intensity, reaches maximum at the optimal light level and decreases for further increase of light (inhibition) until it becomes practically independent of the light intensity. For an irradiance with $I^{PAR} > 2I_{min}$, the light adaptation of the cells is such that an increase in light does not affect the photosynthesis.

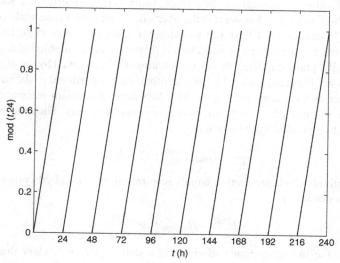

**FIGURE 5.3** Graphical representation of the modulo function mod($t/24$).

### 5.1.3 Light Limitation and Mixing Depth

In temperate and higher latitudes, primary production shows a strong seasonal cycle with a peak, the spring bloom, lower levels during summer, a slight increase in fall and a minimum during winter. This cycle is closely related to the light availability and the existence of a thermocline, which implies a relatively shallow upper mixed layer. During winter, the thermocline is missing and the upper mixed layer can be rather deep due to wind mixing and convective cooling. Due to deep mixing, high nutrient concentrations are found in the surface water, but the abundance of algae during winter time is very low. This implies that phytoplankton growth is limited by light. Two factors are responsible for light limitation in winter: (1) a short day length and (2) deep mixing due to strong winds, convection and low or no heating of surface water. Owing to the deep mixing, the plankton cells are moved vertically up and down and experience only a very small amount of daylight when they are near the sea surface for a short time.

Early studies on the reasons for the onset of the spring bloom were presented by Gran and Braarud (1935) and Riley (1942). In these papers, the concept of a critical depth was introduced and employed. The critical depth is defined as that part of the water column where the vertically integrated daily production due to photosynthesis exceeds the loss terms such as respiration or grazing. An attempt to provide a quantitative analytical formula for the critical depth to understand the necessary conditions for the onset of the spring bloom was given by Sverdrup (1953). This work can perhaps be considered to be the starting point of theoretical work on physical–biological interaction. Let us assume that a thoroughly mixed layer of thickness $H_{mix}$ exists and the plankton cells are evenly distributed throughout this layer. From tracer experiments with (dye tracer Rhodamin), it is known that the overturn is fast with a time scale of about 45 min (Fennel et al., 1986). Let us further assume that there is no limitation in nutrients and the primary production is driven by the available light, or more specifically, photosynthetically active radiation, $I^{PAR}$, received by the algae. The production of organic matter by photosynthesis is proportional to the radiation, which varies with time and depth. As a loss term, a constant respiration rate, $l_{PN}$ (loss of phytoplankton to nutrients), is considered only. Then the equation for the phytoplankton production is,

$$\frac{dP}{dt} = (r_{max} I^{PAR}(z, t) - l_{PN})P.$$

The depth at which production equals respiration, the so-called compensation depth, is defined by

$$I^{PAR}(-H_{comp}, t) = \frac{l_{PN}}{r_{max}}.$$

Because the incoming Sun radiation has a daily cycle, it is clear that $H_{comp}$ depends on time and is not well defined during darkness.

Within the mixed-surface layer, algae moves up and down and experiences relatively high levels of $I^{PAR}$ just below the surface and low levels near the bottom of the mixed layer. Hence, we can assume that the mixing process acts like an average operation, and we may replace $P$ with the daily average of the phytoplankton concentration over the mixed layer. The daily averaged mean value of $I^{PAR}$, sensed by the algae within the mixed layer, is given by

$$\overline{I^{PAR}} = \frac{1}{T} \int_0^T dt' \frac{1}{H_{mix}} \int_{-H_{mix}}^0 I^{PAR}(z', t') \, dz',$$

where $H_{mix}$ is the depth of the mixed layer and $T = 24\,h$. For simplicity, we neglect self-shading effects and use, instead of Equation (5.4) the simpler formula,

$$I^{PAR}(z, t) = I_0^{PAR}(t) \, e^{k_w z}, \tag{5.7}$$

where $k_w$ is an attenuation constant. Integration yields

$$\overline{I^{PAR}} = \frac{1 - e^{-k_w H_{mix}}}{k_w H_{mix}} \frac{1}{T} \int_0^T dt' I_0^{PAR}(t'),$$

or, after introducing the symbol $\widehat{X}$ for the time average over one day,

$$\widehat{X} = \frac{1}{T} \int_0^T dt' X(t'), \tag{5.8}$$

we find

$$\overline{I^{PAR}} = \frac{1 - e^{-k_w H_{mix}}}{k_w H_{mix}} \widehat{I_0^{PAR}} \tag{5.9}$$

Thus, the rate of the mean daily primary production is given by $r_{max} \overline{I^{PAR}}$. Let us further assume that the loss is reasonably well covered by a constant respiration rate, $l_{PN}$. Then the critical depth can be defined as the depth where the mean daily primary production balances the mean daily loss,

$$r_{max} \frac{1 - e^{-k_w H_{crit}}}{k_w H_{crit}} \widehat{I_0^{PAR}} = l_{PN},$$

or

$$\frac{k_w H_{crit}}{1 - e^{-k_w H_{crit}}} = \frac{r_{max}}{l_{PN}} \widehat{I_0^{PAR}}. \tag{5.10}$$

This equation was presented by Sverdrup (1953) and defines the critical depth in relation to physiological rates, $r_{max}$ and $l_{PN}$, and to the daily mean of the available light. It is obvious that production exceeds respiration if $H_{crit}$ is larger than the mixing depth $H_{mix}$, i.e.

$$\frac{1 - e^{-k_w H_{mix}}}{k_w H_{mix}} > \frac{1 - e^{-k_w H_{crit}}}{k_w H_{crit}}, \quad \text{for } H_{crit} > H_{mix}.$$

Although Equation (5.10) represents a transcendent relationship for $H_{crit}$ and $H_{mix}$, it is obvious that the condition applies because $(1 - e^{-x})/x$ decreases monotonously with increasing argument $x$.

Unfortunately, Sverdrup introduced the compensation depth, $H_{comp}$, in his theory. This depth, where local production equals respiration, follows from

$$r_{max} I_0^{PAR}(t) e^{-k_w H_{comp}} = l_{PN}, \quad \text{or } I_c = \frac{l_{PN}}{r_{max}}$$

where $I_c$ was defined by $I_c(t) = I_0^{PAR}(t) e^{-k_w H_{comp}}$. Sverdrup used this relation to eliminated the ratio of the rates $r_{max}/l_{PN}$ in Equation (5.10). However, this is inconsistent because the rates $r_{max}$ and $l_{PN}$ are constants, apart from some dependence on temperature and population structure, while $I_c$, and hence $H_{comp}$, depend on the daily cycle of the Sun's radiation. In particular, $I_c$ assumes zero values during night time. Thus, the practical use of the Sverdrup's critical depth formula is somewhat obscured by this inconsistency. In spite of these problems, it has received an implicit acceptance which is reflected in the way that it is routinely cited. This may be due to the fact that Sverdrup's theory is highly appreciated as a pioneering work in the field of physical biological interaction in the sea. The general idea is sketched in Fig. 5.4 for two cases, $H_{mix} > H_{crit}$ and $H_{mix} < H_{crit}$. The bloom can start only if the depth integral over the production exceeds the integral over respiration. The integrals, which are extended to the mixed-layer depth, $H_{mix}$, correspond to the areas between the curves and the $z$-axis in Fig. 5.4.

A simpler empirical criterion for the onset of the spring bloom in the North Sea and in the Baltic Sea was proposed by Kaiser and Schulz (1976), where

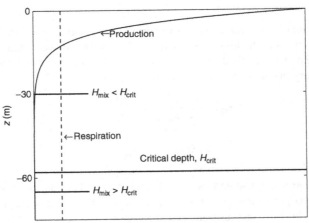

**FIGURE 5.4**  Illustration of the concept of the critical depth according to Sverdrup's theory for $H_{mix} > H_{crit}$, and $H_{mix} < H_{crit}$. The vertical integrals of the production and respiration curves over depths from 0 to $H_{mix}$ correspond to the areas between the curves and the $z$-axis. Only if the integrated production exceeds the integrated respiration can the spring bloom commence.

the critical depth was identified with the depth of the euphotic zone. Then the spring bloom commences if the depth of the euphotic zone exceeds the mixed-layer depth,

$$\frac{H_{\text{mix}}}{H_{\text{euph}}} < 1.$$

This criterion is basically a simpler formulation of Equation (5.10) and is in accordance with series of observational findings (Kaiser and Schulz, 1976). A critical discussion of the Sverdrup formula was given by Smetacek and Passow (1990), where in particular the simple approach of loss in terms of a constant respiration was critically reviewed.

The general importance of the mixed-layer depth for the onset of the spring bloom implies that all factors that control vertical stratification near the surface have to be considered. Apart from heating of the surface layer, a lowering of the salinity of the top layer, related to river discharges or melting of ice, may generate shallow surface layers, which in turn allow the spring bloom to start; see, e.g. Kahru and Nommann (1990). This mechanism, which applies to coastal areas or marginal seas, was mentioned by Sverdrup in his original work.

In complex coupled models with a comprehensive description of physics and biology, the onset of the spring bloom can be adequately simulated provided that the surface layer is vertically sufficiently resolved. The concept of the critical depth is particularly helpful for the set-up of simple box models with appropriate vertical scales.

The discussion of the relation between mixing and illumination indicates that the primary production in a mixed-water column is higher than in calm waters. The vertical mixing also implies that those cells reach the illuminated near surface zone of the water column, which otherwise would be trapped in the dark. In order to illustrate this, we compare the case of calm water with the case of mixed waters. Assume that for calm conditions, the phytoplankton can adapt and still grow at the depth $z = -z_0$, where the daily average of the light level, estimated according to Equation (5.8), is 10% of the surface level, i.e. $\widehat{I^{\text{PAR}}}(z_0) = \widehat{I_0^{\text{PAR}}}/A$, where $A = 10$. The corresponding thickness of the water column within which phytoplankton grows follows from Equation (5.7) as

$$z_0 = \frac{\ln(A)}{k_w}. \tag{5.11}$$

Within a well-mixed layer, we suppose that phytoplankton is adapted to the mean daily radiation, averaged over the depth of the mixed layer, $\overline{I^{\text{PAR}}}$, see Equation (5.9), and we assume that the lower limit where primary production is still possible is 10% of the daily mean of the surface value, $\widehat{I_0^{\text{PAR}}}/A$. Then, from Equation (5.9), it follows a relationship for a maximal mixed-layer depth, $H_{\text{max}}$, which still permits primary production,

$$1 = \frac{k_w H_{\text{max}}}{A} + e^{-k_w H_{\text{max}}}. \tag{5.12}$$

This transcendental equation has a nontrivial solution for $A > 1$. Because of the rapidly decreasing exponential, implying $k_w H_{max}/A \gg e^{-k_w H_{max}}$, we find to a good approximation

$$H_{max} = \frac{A}{k_w}.$$

The comparison of the relevant depth intervals for both cases gives

$$\frac{H_{max}}{z_0} \simeq \frac{A}{\ln(A)} \simeq 4.34, \tag{5.13}$$

i.e. $H_{max}$ can exceed $z_0$ by a factor of 4.3. Hence, mixing enables primary production over a substantially deeper part of the water column than still waters. This fact is employed in biological wastewater cleaning facilities by stirring the wastewater.

## 5.2 COASTAL OCEAN DYNAMICS

Another important class of physical processes relevant for marine ecosystem dynamics are the ocean currents, which transport nutrients, cells, eggs and animals.

Due to the general importance of coastal upwelling we briefly discuss the dynamics of transient upwelling that occurs in shelf and semi-enclosed seas and upwelling systems near ocean eastern boundaries in this section. In particular, we consider the vertical current in upwelling systems, which provides injections of nutrients in the euphotic zone, and the typical currents associated with the upwelling, such as coastal jets and undercurrents near coastal boundaries. We confine the discussion to the relatively simple, idealized case of a stratified, flat-bottomed coastal ocean bounded by straight walls along the $y$-axis. This is one of the rare models that is susceptible to analytical solutions. Although the oceanic circulation can be described by numerical models with a high degree of realism, it is illuminating to go through a few examples that can explicitly be expressed by analytical formulas. The model of the coastal system is sketched in Fig 5.5. The coastline is parallel to the $y$-axis, the $x$-axis is cross shore and the $z$-axis is positive upward with the sea surface at $z = 0$ and the bottom at $z = -H$. For simplicity, we consider the response of a coastal ocean to a sudden onset of alongshore winds, $Y$, and zero cross-shore winds, $X$, acting on an ocean at rest. The wind force is assumed to act as a body force equally distributed over a preexisting upper layer of thickness, $H_{mix}$,

$$Y(t, x, y, z) = \frac{v_*^2}{H_{mix}} \theta(z + H_{mix}) \theta(t) \overline{Y}(x, y), \quad \text{and} \quad X = 0, \tag{5.14}$$

where $v_*^2$ is the friction velocity and $\overline{Y}(x, y)$ prescribes the spatial structure of the along-shore wind. The assumption of a body force acting within an upper layer means that turbulent vertical mixing in the upper layer was implicitly involved

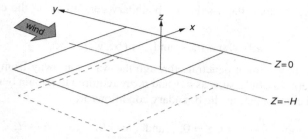

**FIGURE 5.5** Sketch of the geometry of an idealized coastal ocean. The $y$-axis is along shore, the coast stretches along $x = 0$ and the $x$-axis has negative values in offshore direction.

to create the layer of depth $H_{mix}$. In the following, we consider the responses to three examples of wind fields: (1) a spatially constant wind, $\overline{Y} = 1$, (2) a wind band $\overline{Y}(y) = \theta(y - |a|)$ of the width $2a$ and (3) meridional wind with a cosine shape in $x$- and $y$-direction, $\overline{Y}(x, y) = \theta(y - |a|)\cos(\kappa_0 y)\theta(x + L + l)\cos(k_0 x)$, with $\kappa_0 = \pi/(2a)$ and $k_0 = \pi/(2L)$. The parameter, $l$, with $l \leq L$, controls how far the wind patch stretches over the coast.

## 5.2.1 Basic Equations

The governing equations are linearized hydrostatic Boussinesq equations on a $f$ plane, which follow after several traditional approximations:

$$u_t - fv + p_x = X, \tag{5.15}$$

$$v_t + fu + p_y = Y, \tag{5.16}$$

$$p_{zt} + N^2 w = 0, \tag{5.17}$$

$$u_x + v_y + w_z = 0, \tag{5.18}$$

where the inertial frequency, $f$, is defined as a Coriolis parameter at a given latitude, $\vartheta$,

$$f = 2\Omega \sin \vartheta,$$

with $\Omega$ being the magnitude of the angular frequency of the Earth's rotation $\Omega = 2\pi/24$ h. In the southern hemisphere, $\vartheta$ and $f$ become negative. Then we replace $f$ by $-f$, with $f = 2\Omega|\sin \vartheta|$. Further, $N$ is the Brunt–Väisälä-frequency

$$N^2 = -\frac{g}{\rho}\frac{d\rho}{dz},$$

which is a measure of the density stratification. The pressure $p$ in the set (5.15)–(5.18) is the dynamical pressure; i.e. the hydrostatic contribution was subtracted. Partial derivations are denoted by the subscripts $x$, $y$, $z$ or $t$. Because the coastline is along the $x$-axis, the boundary conditions on the cross-shore velocity, $u$, are

$$u(0, y, t) = 0, \quad \text{and} \quad u(\infty, y, t) < \infty, \tag{5.19}$$

the onshore flow cannot penetrate through the wall. Far away from the coast, the dependent variables, such as $u$, $v$ and $p$, are required to remain bounded. For the vertical component, $w$, the boundary conditions are

$$w = \frac{p_t}{g\rho}, \quad \text{for } z = 0, \quad \text{and} \quad w = 0, \quad \text{for } z = -H. \tag{5.20}$$

The vertical variations of the solutions, which depend on $z$, can be separated by expanding the dynamic quantities into a series of vertical eigenfunctions, $F_n(z)$,

$$\Phi(x, y, z, t) = \sum_{n=0}^{\infty} \Phi_n(x, y, t) F_n(z), \tag{5.21}$$

where $\Phi$ stands for $u$, $v$ and $p$. Inserting this into Equations (5.15)–(5.17), it follows that the $F_n(z)$'s, are subject to the vertical eigenvalue problem,

$$\left( \frac{\mathrm{d}}{\mathrm{d}z} \frac{1}{N^2} \frac{\mathrm{d}}{\mathrm{d}z} + \lambda_n^2 \right) F_n(z) = 0, \tag{5.22}$$

with the boundary conditions $F_n'(0) + (N^2(0)/g)F_n(0) = 0$, $F_n'(-H) = 0$. For a constant Brunt–Väisälä-frequency, $N$, we have for the barotropic mode $F_0 = 1/\sqrt{H}$, $\lambda_0 = 1/\sqrt{gH}$, and for the baroclinic modes, $F_n(z) = \sqrt{(2/H)}$ $\cos(\frac{n\pi z}{H})$, $\lambda_n = n\pi/NH (n = 1, 2, \ldots)$. Note that the inverse eigenvalues, $\lambda_n^{-1} = NH/n\pi$, have the dimension of a speed. For simplicity, we will use the case $N = \text{const}$ in the following discussion.

The hydrostatic Boussinesq equations expanded into normal modes are:

$$u_{nt} - fv_n + p_{nx} = X_n, \tag{5.23}$$

$$v_{nt} + fu_n + p_{ny} = Y_n, \tag{5.24}$$

$$u_{nx} + v_{ny} + \lambda_n^2 p_{nt} = 0. \tag{5.25}$$

The expansion of forcing functions of the structure given in Equation (5.14) amounts to a Fourier series of the step function,

$$\frac{\theta(z + H_{\text{mix}})}{H_{\text{mix}}} = \sum_{n=0}^{\infty} \frac{F_n(z)}{h_n}.$$

The coefficients, $h_n$, are

$$\frac{1}{h_0} = \sqrt{\frac{1}{H}} \quad \text{for } n = 0 \quad \text{and} \quad \frac{1}{h_n} = \frac{\sin(\frac{n\pi}{H} H_{\text{mix}})}{\frac{n\pi}{H} H_{\text{mix}}} \quad \text{for } n \geq 1.$$

So far, friction was ignored. In order to include viscous effects in a qualitative manner, we may introduce a linear friction. This amounts to the replacement of the derivative $\partial/\partial t$ by $(\partial/\partial t) + r$, with the friction parameter, $r$, in the set (5.23)–(5.25).

To solve this set of equations for given examples of wind forcing, we first derive an equation for one of the dependent variables alone. Because of the simple boundary for $u$, we choose the $u_n$ component and obtain

$$u_{nttt} + f^2 u_{nt} - \frac{1}{\lambda_n^2}\Delta u_{nt} = X_{ntt} - \frac{1}{\lambda_n^2}X_{nyy} + fY_{nt} + \frac{1}{\lambda_n^2}Y_{nxy}. \tag{5.26}$$

To obtain this equation, we have to manipulate the equations in the set (5.23)–(5.25). First, we differentiate Equation (5.23) with respect to $t$ and use Equation (5.24) to eliminate $v_t$ in the resulting equation,

$$u_{ntt} + f^2 u_{nt} = -fp_{ny} - p_{nxt} + fY_n + X_{nt}. \tag{5.27}$$

We differentiate Equation (5.27) once more with respect to $t$ and find,

$$u_{nttt} + f^2 u_{ntt} = -fp_{nyt} - p_{nxtt} + fY_{nt} + Y_{ntt}. \tag{5.28}$$

With the aid of Equation (5.25), the combination of pressure terms $-fp_{nxy} - p_{nxtt}$, can be expressed through $u$ and $v$ as

$$-fp_{nyt} - p_{nxtt} = \frac{1}{\lambda_n^2}(v_{ntxy} + u_{ntxx} + fu_{nxy} + fv_{nyy}). \tag{5.29}$$

In order to eliminate $u$ in Equation (5.29), we take the $y$-derivative of Equation (5.23) and the $x$-derivative of Equation (5.24), subtract the resulting equations and find

$$u_{nty} - v_{ntx} - f(u_{nx} + v_{ny}) = X_{ny} - Y_{nx}.$$

Differentiating this once more with respect to $y$ and rearranging the terms, we get,

$$v_{ntxy} + f(u_{nxy} + fv_{nyy}) = u_{ntxx} - v_{nxy} - X_{nyy} + Y_{nxy}.$$

After inserting this expression into the right-hand side of Equation (5.29), we arrive at Equation (5.26).

To solve Equation (5.26), a Fourier transformation of this differential equation with respect to $x$ and $t$ according to

$$u_n(x, y, t) = \int_{-\infty}^{\infty} \frac{d\omega}{2\pi} \int_{-\infty}^{\infty} \frac{d\kappa}{2\pi}\, u_n(\kappa, \omega, x)\exp(-i\omega t + i\kappa y).$$

is helpful. Because we consider only along-shore winds, $Y$, we set $X = 0$, and obtain,

$$u_{nxx} - \alpha^2 u_n = -\lambda_n^2 fY_n(\omega, \kappa) + \frac{\kappa}{\omega}Y_{nx}(\omega, \kappa), \tag{5.30}$$

with

$$\alpha_n^2 = \lambda_n^2 f^2 + \kappa^2 - \lambda_n^2 \omega^2. \tag{5.31}$$

Note that with $R_n = (\lambda_n f)^{-1}$, a spatial scale, the Rossby radius, emerges. The Rossby radius depends on the vertical mode number, $n$. Using the eigenvalues $\lambda_n$, for a constant Brunt–Väisälä-frequency, we find for the barotropic mode $n = 0$, the barotropic Rossby radius

$$R_0 = \frac{\sqrt{gH}}{f}. \tag{5.32}$$

For $n = 1$, we obtain the first-mode Rossby radius as

$$R_1 = \frac{NH}{\pi f}, \quad \text{and} \quad R_n = \frac{R_1}{n}. \tag{5.33}$$

The first-mode Rossby radius is usually called the internal or baroclinic Rossby radius. The Rossby radii vary geographically and seasonally. Examples of Rossby radii for the North Pacific and North Atlantic are given in Emory et al. (1984), Chelton et al. (1998) and, for the Baltic Sea, in Fennel et al. (1991).

If $u_n$ is once known, we can use Equations (5.24) and (5.25) to express $v$ and $p$ through $u$ and the forcing. Combining Equations (5.24) and (5.25), we have

$$v_{ntt} - \frac{1}{\lambda_n^2} v_{nyy} = Y_{nt} - f u_{nt} + \frac{1}{\lambda_n^2} u_{nxy}, \tag{5.34}$$

and

$$p_{ntt} - \frac{1}{\lambda_n^2} p_{nyy} = \frac{1}{\lambda_n^2} (-Y_{ny} - f u_{ny} + u_{nxt}), \tag{5.35}$$

or, after Fourier transformation,

$$v_n = \frac{i}{\omega^2 \lambda_n^2 - \kappa^2} (\omega \lambda_n^2 Y - \omega \lambda_n^2 f u_n - \kappa u_{nx}), \tag{5.36}$$

$$p_n = \frac{i}{\omega^2 \lambda_n^2 - \kappa^2} (\kappa Y - \kappa f u_n - \omega u_{nx}). \tag{5.37}$$

Formally, differential equation (5.30), together with the boundary conditions (5.19), represents an inhomogeneous boundary value problem. The appropriate way to solve such a problem is to introduce the corresponding Green's function, $G_n(x, x')$, by means of

$$\frac{d^2}{dx^2} G_n(x, x') - \alpha_n^2 G_n(x, x') = \delta(x - x'), \tag{5.38}$$

with the boundary conditions $G(0, x') = 0$ and $G(\infty, x')$ to be finite. The solution is (Fennel, 1988),

$$G_n(x, x') = \frac{1}{2\alpha_n} \left( e^{\alpha_n(x+x')} - e^{-\alpha_n|x-x'|} \right).$$

With the aid of the Green's function, the solution to Equation (5.30) can be expressed by a source representation as (Green's theorem)

$$u_n(y) = \int_{-\infty}^{0} dx' G_n(x, x') \left[ -\lambda_n^2 f Y_n(\omega, \kappa, x') + \frac{\kappa}{\omega} Y_{nx'}(\omega, \kappa, x') \right].$$

This represents a formal solution for $u_n$. The oceanic response to the wind forcing is given by convolution integrals of the Green's function and the forcing terms Introducing, for the convolution integral, the notation,

$$G_n * Y_n = \int_{-\infty}^{0} dx' G_n(x, x') Y_n(x'), \tag{5.39}$$

the solution reads

$$u_n(y) = -\lambda_n^2 f G_n * Y_n + \frac{\kappa}{\omega} G_n * Y_{nx'}. \tag{5.40}$$

With Equation (5.40), we find from Equations (5.36) and (5.37)

$$v_n = \frac{i}{(\omega^2 \lambda_n^2 - \kappa^2)} \left( \omega \lambda_n^2 Y_n + \omega \lambda_n^4 f^2 G_n * Y_n - \kappa \lambda_n^2 f G_n * Y_{nx'} \right.$$
$$\left. + \kappa \lambda_n^2 f G_{nx} * Y_n - \frac{\kappa^2}{\omega} G_n * Y_{nx'} \right), \tag{5.41}$$

$$p_n = \frac{i}{(\omega^2 \lambda_n^2 - \kappa^2)} \left( \kappa Y_n + \kappa \lambda_n^2 f^2 G_n * Y_n - \frac{\kappa^2 f}{\omega} G_n * Y_{nx'} \right.$$
$$\left. + \omega \lambda_n^2 f G_{nx} * Y_n - \kappa^2 G_{nx} * Y_{nx'} \right). \tag{5.42}$$

Because we consider here only a switch-on forcing, given by a $\theta$-function, see Equation (5.14), with the Fourier transforms,

$$\theta(t) = \int \frac{d\omega}{2\pi} \frac{i e^{-i\omega t}}{\omega + i\epsilon},$$

we have to consider the contributions of two poles: $\omega = 0$, which refers to steady motion, and $\omega^2 \lambda_n^2 = \kappa^2$, which are the Kelvin poles referring to Kelvin waves. According to Equation (5.31), for the Kelvin poles we have $\alpha_n^2 = 1/R_n^2$ and for $\omega = 0$, $\alpha_n^2 = 1/R_n^2(1 + \kappa^2 R_n^2)$. If we apply the long-wave approximation, $\kappa^2 R_n^2 \gg 1$, we can also use $\alpha_n^2 \approx 1/R_n^2$ for this term. After a partial fraction expansion of Equations (5.41) and (5.42), and using the property of the convolution integral, $G_n * Y_{nx'} = -G_{nx'} * Y_n$, we find that terms associated with the pole, $\omega \lambda_n = -\kappa$, are zero and $v_n$ and $p_n$ are given by,

$$v_n = \frac{i}{\omega} G_{nx} * Y_{nx'} + \frac{i \lambda_n}{(\omega \lambda_n - \kappa) R_n} e^{(x+x')/R_n} * Y_n, \tag{5.43}$$

$$p_n = \frac{if}{\omega} G_n * Y_{nx'} + \frac{i}{(\omega \lambda_n - \kappa) R_n} e^{(x+x')/R_n} * Y_n. \tag{5.44}$$

In the following sections we look at subinertial motions, where the time variations are much smaller than the inertial period, i.e. $\omega^2 \ll f^2$. For a discussion of the contributions near the inertial frequency, refer to, e.g. Kundu et al. (1983) and Fennel (1989) for details. Thus, because the method to solve the posed problems is straightforward, we specify the forcing (Equation 5.14), that, after mode expansion and Fourier transformation, has the form,

$$Y_n(\omega, \kappa, x) = \frac{v_*^2}{h_n}\overline{Y}(\kappa, x)\frac{i}{\omega + i\epsilon}, \quad \text{and} \quad X_n = 0 \qquad (5.45)$$

We insert this into the formal solution, Equations (5.40), (5.43), and (5.44), and calculate the corresponding convolution integrals to find the solution.

## 5.2.2 Large-Scale Winds and Coastal Jets

As a first example, we study the oceanic response a large-scale wind in the along-shore direction, which is independent of $x$ and $y$, i.e. $\overline{Y} = -1$ and $Y_x = 0$. In the Fourier space this amounts to $\overline{Y} = -2\pi\delta(\kappa)$, and our forcing terms reads,

$$Y_n(\omega, \kappa) = -\frac{v_*^2}{h_n}2\pi\delta(\kappa)\frac{i}{\omega + i\epsilon}$$

Inserting this into Equations (5.40), (5.43) and (5.44), we see that we have to calculate the convolution integrals, $G*1 = \int_{-\infty}^{0} dx'G$ and $\exp((x+x')/R_n)*1 = \int_{-\infty}^{0} \exp[(x+x')/R_n]dx'$. The second integral is

$$\int_{-\infty}^{0} e^{(x+x')/R_n}dx' = R_n e^x/R_n.$$

For the first integral, we use the equation for the Green's function (5.38) with $\alpha_n = 1/R_n$.

$$\int_{-\infty}^{0} dx' G_n(x, x') = \int_{-\infty}^{0} dx' R_n^2[-\delta(x - x') + G_{nx'x'}(x, x')] =$$

$$= R_n^2[-1 + G_{nx'}(x, 0) - G_{nx'}(x, -\infty)] = -R_n^2(1 - e^{\frac{x}{R_n}}). \qquad (5.46)$$

Then we find,

$$u_n(x, \kappa, \omega) = -\frac{v_*^2}{h_n}2\pi\delta(\kappa)\frac{i}{(\omega + i\epsilon)f}(1 - e^{x/R_n}), \qquad (5.47)$$

$$v_n(x, \kappa, \omega) = -\frac{v_*^2}{h_n}2\pi\delta(\kappa)\frac{-1}{(\omega + i\epsilon)^2}e^{x/R_n}, \qquad (5.48)$$

$$p_n(y, k, \omega) = -\frac{v_*^2}{h_n}\frac{2\pi\delta(\kappa)}{(\omega + i\epsilon)^2\lambda_n}e^{x/R_n}. \qquad (5.49)$$

The inverse Fourier transforms of the expressions in Equations (5.47), (5.48) and (5.49) are trivial with respect to $\kappa$ because of the $\delta$-function, $\delta(\kappa)$. The

Fourier integrals with respect to $\omega$ are easy to calculate from Cauchy's residues theorem,

$$\int \frac{d\omega}{2\pi} \frac{i\,e^{-i\omega t}}{\omega + i\epsilon} = \theta(t), \int \frac{d\omega}{2\pi} \frac{-e^{-i\omega t}}{(\omega + i\epsilon)^2} = t\,\theta(t).$$

After the Fourier transforms, we find the solutions in the physical-space regime,

$$u_n(y, t) = -\frac{v_*^2}{fh_n}(1 - e^{x/R_n}), \tag{5.50}$$

$$v_n(y, t) = -\frac{v_*^2}{h_n} t\, e^{x/R_n}, \tag{5.51}$$

$$p_n(y, t) = -\frac{v_*^2}{\lambda_n h_n} t\, e^{x/R_n}. \tag{5.52}$$

The solutions describe a coastal jet with $p_n$ and $v_n$ and are in geostrophic balance, $p_{nx} = -fv_n$. The vertical component is related to the pressure through

$$w_n(x, t) \sim -p_{nt}(x, t) = \frac{v_*^2}{\lambda_n h_n} e^{x/R_n}.$$

The cross-shore current, $u_n$, consists of two contributions: (1) Ekman off shore transport and (2) a coastal rectification that ensures that the boundary condition at the coast applies. The rectification of the Ekman transport is associated with the divergence, $u_{nx}$, which drives the coastal upwelling, according to

$$u_x + w_z = 0.$$

The final representations for $u$ and $v$ follow after summation over the vertical modes. For $u$, $v$ and $w$, the summation can be performed explicitly and yields a closed-form expressions see Fennel (1988),

$$v(x, z, t) = v_*^2 t \sum_{n=0}^{\infty} \frac{F_n(z)}{h_n} e^{x/R_n} \tag{5.53}$$

$$= v_*^2 t \left[ \frac{e^{x/R_0}}{H} + S^{(1)}(x, H_{\text{mix}} + z) + S^{(1)}(x, H_{\text{mix}} - z) \right].$$

The mode sum, $S^{(1)}$, can directly be performed and is defined by

$$S^{(1)}(x, z) = \sum_{n=1}^{\infty} \frac{F_n(z)}{h_n} e^{x/R_n}$$

$$= \frac{1}{\pi H_{\text{mix}}} \arctan \left\{ \frac{\sin\left[\frac{\pi}{H}(H_{\text{mix}} + z)\right]}{e^{-x/R_1} - \cos\left[\frac{\pi}{H}(H_{\text{mix}} + z)\right]} \right\}.$$

Similarly, we find for the cross-shore component,

$$
\begin{aligned}
u(x, z) &= v_*^2 \sum_{n=0}^{\infty} \frac{F_n(z)}{h_n} (1 - e^{x/R_n}) \\
&= u_*^2 \left[ \frac{\theta(z + H_{\text{mix}})}{H_{\text{mix}}} - \frac{e^{x/R_0}}{H} \right. \\
&\quad \left. - S^{(1)}(x, H_{\text{mix}} + z) - S^{(1)}(x, H_{\text{mix}} - z) \right].
\end{aligned}
\tag{5.54}
$$

In these expressions, the spatial scales are set by the barotropic and baroclinic Rossby radius and by the depth of the mixed layer. In order to calculate the vertical component, we use $w = -(1/N^2)p_{zt}$, and find

$$
\begin{aligned}
w(x, z) &= -\frac{v_*^2}{N^2} \sum_{n=1}^{\infty} \frac{e^{x/R_n}}{\lambda_n h_n} \frac{\mathrm{d}}{\mathrm{d}z} F_n(z) \\
&= \frac{v_*^2}{N H_{\text{mix}}} S^{(2)}(x, z),
\end{aligned}
\tag{5.55}
$$

where

$$
\begin{aligned}
S^{(2)}(x, z) &= \frac{H_{\text{mix}}}{N} \sum_{n=1}^{\infty} \frac{e^{x/R_n}}{\lambda_n h_n} \frac{\mathrm{d}}{\mathrm{d}z} F_n(z) \\
&= \frac{1}{2\pi} \ln \left\{ \frac{\cosh(\frac{x}{R_1}) - \cos\left[\frac{\pi}{H}(H_{\text{mix}} - z)\right]}{\cosh(\frac{x}{R_1}) - \cos\left[\frac{\pi}{H}(H_{\text{mix}} + z)\right]} \right\}.
\end{aligned}
$$

The dynamic balance near the coastal boundary, $x \to 0$, follows as

$$
-fv = p_x,
$$
$$
v_t \approx Y,
$$
$$
u_x + w_z = 0.
$$

The divergence of the cross-shore current, $u_x$, drives upwelling, which builds up a cross-shore pressure gradient, $p_x$, which in turn is in geostrophic balance with the coastal jet. Because the upwelling is steady, the pressure gradient grows with time, and the coastal jet is virtually accelerated by the wind force, $v \approx Yt$.

Far away from the boundary, $x \to -\infty$, or $|x| \gg R_1$, the coastal effects diminish and an Ekman balance remains, where $u$ is determined by the Ekman transport,

$$
u = \frac{Y}{f}.
$$

The cross-shore, along-shore and vertical currents, $u(x, z)$, $v(x, z, t)$, $w(x, z)$, which are given by the analytical expressions in Equations (5.53), (5.54) and (5.55), respectively, are shown in Figs. 5.6, 5.7 and 5.8. The involved parameters

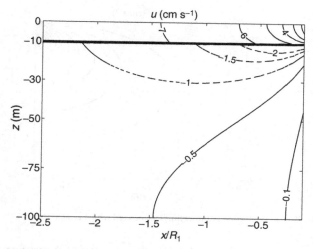

**FIGURE 5.6**   The cross circulation, $u$, associated with the coastal jet. The offshore Ekman transport in the upper layer is rectified by an onshore compensation flow (dashed lines refer to negative flow direction).

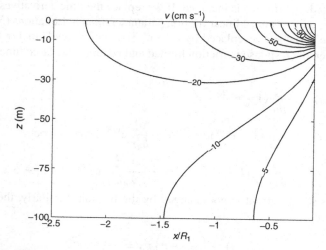

**FIGURE 5.7**   The coastal jet driven by along-shore wind.

are chosen as $R_1 = 5\,\text{km}$, $f = 1.2 \times 10^{-4}\,\text{s}^{-1}$, $H = 100\,\text{m}$, $H_{\text{mix}} = 10\,\text{m}$ and $v_*^2 = 1\,\text{cm}^2\,\text{s}^{-2}$. The choice of $f$ and $R_1$ implies $\lambda_n^{-1} = c_1 = 60\,\text{cm}\,\text{s}^{-1}$. The essential variations in cross-shore direction occur in a coastal strip with the scale of the first-mode Rossby radius, $R_1$. The coastal jet decreases for $|x| > R_1$; the upwelling, characterized by the $w$-component, is confined also to a coastal strip of the width of $R_1$; and the Ekman transport tends towards zero when

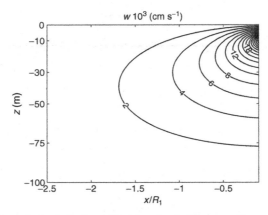

**FIGURE 5.8**    The vertical velocity (upwelling) associated with a coastal jet.

approaching the coast. Obviously, the current systems shows two unrealistic features: (1) the currents do not change in alongshore direction, and (2) the costal jet is proportional to $t$, i.e. the velocity grows unlimited. The latter point can be avoided if friction is included. If we replace the time derivatives $\partial/\partial t$ by $(\partial/\partial t) + r$, with $r$ being a linear friction parameter, then in Equations (5.51) and (5.50), the variable $t$ is replaced by $(1 - e^{-rt})/r$, which tends to $1/r$ for large times. Then the response is friction limited and becomes for large times,

$$u_n(y, t) = -\frac{v_*^2}{fh_n}(1 - e^{x/R_n}), \tag{5.56}$$

$$v_n(x, t) = -\frac{v_*^2}{h_n r}(1 - e^{-rt})e^{x/R_n} \rightarrow -\frac{v_*^2}{h_n r}e^{x/R_n}, \text{ for } t \rightarrow \infty \tag{5.57}$$

$$p_n(x, t) = -\frac{v_*^2}{\lambda_n h_n r}(1 - e^{-rt})e^{x/R_n} \rightarrow -\frac{v_*^2}{\lambda_n h_n r}e^{x/R_n}, \text{ for } t \rightarrow \infty. \tag{5.58}$$

The cross-shore current is not changed by the friction. Similarly, the vertical component,

$$w_n(x, t) \sim \left(\frac{\partial}{\partial t} + r\right)p_n = \frac{v_*^2}{\lambda_n h_n}e^{x/R_n},$$

is not affected because of the relation $((\partial/\partial t) + r)(1 - e^{-rt})/r = 1$.

### 5.2.3  Kelvin Waves and Undercurrents

In the previous subsection, we considered an unbounded wind field. However, in real systems, winds have spatial structures and are of limited extent. In order to illustrate the effects of limited spatial wind distributions, we consider as a

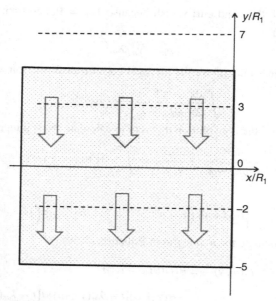

**FIGURE 5.9** A finite band of along-shore wind extending from $-5R_1$ to $5R_1$. Some cross-shore sections (dashed) are indicated for later reference.

second example the coastal response to an along-shore wind restricted to a wind band of the width $2a$, as sketched in Fig. 5.9,

$$Y(t, z) = -\frac{v_*^2}{H_{\text{mix}}}\theta(z + H_{\text{mix}})\theta(t)\theta(a - |y|), \quad \text{and} \quad X = 0.$$

Fourier transformation and expansion into vertical modes gives,

$$Y_n(\omega, \kappa) = -\frac{v_*^2}{h_n}\frac{i}{\omega + i\epsilon}\frac{2\sin a\kappa}{\kappa}.$$

The solution to $u_n$ follows from the formal solution (5.40), i.e. under the conditions of the long wave approximation, as

$$u_n(x, \kappa, \omega) = -\frac{v_*^2}{h_n}2\frac{\sin a\kappa}{\kappa}\frac{i}{(\omega + i\epsilon)}(1 - e^{x/R_1}),$$

which gives in the physical space

$$u_n(x, y, t) = -\frac{v_*^2}{h_n f}\theta(t)\theta(a - |y|)(1 - e^{x/R_1}).$$

For $v_n$ and $p_n$, it follows from Equations (5.43) and (5.44) that

$$v_n(x, \kappa, \omega) = -\frac{v_*^2}{h_n f}\frac{2\sin a\kappa}{\kappa}\frac{i}{(\omega + i\epsilon)}\frac{i\lambda_n}{(\omega\lambda_n - \kappa)}e^{x/R_n}. \tag{5.59}$$

The terms with the wind curl vanish because $Y_{nx} = 0$, and then $p_n$ is simply related to $v_n$ by

$$p_n = v_n / \lambda_n.$$

To solve the $\kappa$ integral for the inverse Fourier transformation, we note that

$$\int \frac{d\kappa}{2\pi} \frac{i e^{i\kappa y}}{\omega \lambda_n - \kappa} = \theta(y) e^{i\omega \lambda_n y}.$$

Then we can use the convolution theorem to calculate the $\kappa$ integral as

$$I_n(\omega, y) = \int \frac{d\kappa}{2\pi} \frac{i e^{i\kappa y}}{\omega \lambda_n - \kappa} \frac{2 \sin(a\kappa)}{\kappa} = \int_{-\infty}^{\infty} dy' \theta(y - y') e^{i\omega \lambda_n (y - y')} \theta(a - |y|).$$

The result is

$$I_n(\omega, y) = \frac{i}{\omega \lambda_n} \left[ \theta(a + y) \left( 1 - e^{i\omega \lambda_n (a + y)} \right) - \theta(y - a) \left( 1 - e^{i\omega \lambda_n (y - a)} \right) \right].$$

Finally, we calculate the $\omega$ integral in Equation (5.59),

$$\int_{-\infty}^{\infty} \frac{d\omega}{2\pi} e^{i\omega t} I_n(\omega, y) \frac{i}{\omega} = \theta(t)(\theta(a - |y|)t$$

$$-\theta(y + a)[t - \lambda_n(y + a)]\theta[t - \lambda_n(y + a)]$$

$$+\theta(y - a)[t - \lambda_n(y - a)]\theta[t - \lambda_n(y - a)]).$$

Then, for the along-shore current in the physical space, we find

$$v_n(x, y, t) = -\frac{v_*^2}{h_n} \theta(t) e^{x/R_n} (\theta(a - |y|)t$$

$$-\theta(y + a)[t - \lambda_n(y + a)]\theta[t - \lambda_n(y + a)]$$

$$+\theta(y - a)[t - \lambda_n(y - a)]\theta[t - \lambda_n(y - a)]). \qquad (5.60)$$

The cross-shore current, $u_n$, is virtually the same for the wind band as for the large-scale wind, except that the signal is confined to the area of the band. It consists of the Ekman transport in the surface layer and a compensation flow, which rectifies the Ekman transport near the coastal boundary. However, the along-shore current in a wind band differs significantly from the response to a large-scale wind. We have both an accelerating jet in the wind band as well as signals that propagate with the phase speeds $c_n = 1/\lambda_n$ along the coast, leaving the coast to the right (on the northern hemisphere). The waves are trapped near the coast. The signals of the coastally trapped waves decrease exponentially in the offshore direction with the baroclinic Rossby radii as trapping scales. These waves are known as Kelvin waves. There is a complete set of waves, one for each mode $n$. The phase speeds, $c_n = 1/\lambda_n = c_1/n$, and Rossby radii, $R_n = R_1/n$, ($n = 1, 2, 3, \ldots$), decrease with increasing mode number, $n$.

For this discussion, we look at the first mode wave, $n = 1$. From Equation (5.60), we see that the Kelvin waves start at the edges of the wind band, at $y = -a$ and $y = a$. The wave starting from $y = -a$ crosses the wind band

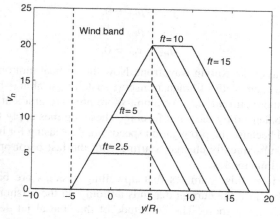

**FIGURE 5.10**   The role of Kelvin wave for the developing and arresting of a coastal jet within the wind band and for the export of signal outside the band. The wind band is indicated by dashed lines.

within a time scale $\tau_1 = 2a/\lambda_1$. After this span, the wave leaves the wind band. The wave starting from $y = a$ propagates outside the wind band and exports the signal of the along-shore currents into the unforced area. An observer at a location $y$ inside the band will see an accelerating jet, increasing proportional to $t$, until after the time span, $t = (y+a)/\lambda_1$, the Kelvin wave arrives and arrests the signal; i.e. the magnitude of the jet is then proportional to $(y+a)/\lambda_1$ and will not change with time. These features are sketched in Fig. 5.10, which indicates how the amplitude of the jet depends on $y$ and $t$. The vertical speed follows as

$$w \sim -p_{nt} = -\frac{v_{nt}}{\lambda_n} = \frac{v_*^2}{h_n} e^{x/R_n} \left(\theta(a - |y|)\right.$$
$$-\theta(y+a)\theta[t - \lambda_n(y+a)] + \theta(y-a)\theta[t - \lambda_n(y-a)]\right),$$

which implies that in band, the upwelling stops after the arrival of the Kelvin waves. At the edge $y = a$, upwelling is exported outside the wind band, but the signal lasts only until the Kelvin wave travelling from the edge $y = -a$ arrives and cancels the vertical motion.

The dynamic balance in the wind band before the Kelvin wave has arrived is

$$-fv_n + p_{nx} = 0,$$
$$v_{nt} + fu_n = Y_n,$$
$$u_{nx} + \lambda_n^2 p_{nt} = 0,$$

where $\lambda_n^2 p_{nt}$ corresponds to the vertical current. This is the coastal jet regime, where the divergence of the Ekman transport, $u_{ny}$, drives a vertical motion, which creates a cross-shore pressure gradient that is in geostrophic balance with the accelerating along-shore flow. This regime changes to

$$-fv_n + p_{nx} = 0,$$
$$fu_n + p_{ny} = Y_n,$$
$$u_{nx} + v_{ny} = 0,$$

after the arrival of the Kelvin wavefront. Now the vertical motion has stopped and the divergence of the Ekman transport sustains an along-shore gradient of the along-shore current, $u_{nx}$. An along-shore pressure gradient, $p_{nx}$, was set up, which is balanced by the wind forcing. These regimes apply to all vertical modes, $n$, and because the propagation speeds are decreasing for higher modes, the balances subsequently change, starting with the fast barotropic mode and then the baroclinic modes.

We learn from this examples that upwelling systems are basically three dimensional because the current depends not only on the offshore coordinate, but also varies along-shore. The amplitude of the coastal jet depends on the travelling time of the Kelvin waves, and the upwelling is only of short duration because it is stopped when the Kelvin waves arrive. A further important aspect emerges if we perform the sums over the vertical modes. The Kelvin waves have introduced along-shore pressure gradients that drive an alongshore undercurrent beneath the jet.

The properties of the current systems are somewhat modified if effects of friction are taken into account. For the sake of simplicity, we consider only a linear friction; i.e. we replace the time derivative $\partial/\partial t$ with $(\partial/\partial t) + r$, which amounts in the Fourier space to the replacement of all $\omega$'s with $\overline{\omega} = \omega + ir$. Thus, in the formal solution (5.43) and (5.44), we replace all $\omega$'s by $\overline{\omega}$, except the $\omega$ in the forcing function. While, in the frame of our approximation, the $u_n$ component is not changed by friction, we obtain for the along-shore current $v_n$

$$v_n(x, y, t) = -\frac{v_*^2}{rfh_n} e^{x/R_n} \theta(t) \left( \theta(a - |y|)(1 - e^{-rt}) \right.$$
$$+ \theta(y + a)\theta[t - \lambda_n(y + a)] \left( e^{-rt} - e^{-r\lambda_n(y+a)} \right) \qquad (5.61)$$
$$\left. - \theta(y - a)\theta[t - \lambda_n(y - a)] \left( e^{-rt} - e^{-r\lambda_n(y-a)} \right) \right).$$

The pressure is given by $v_n/\lambda_n$ and the vertical component follows as

$$w_n(x, y, t) \sim -\left( \frac{\partial}{\partial t} + r \right) p_n(x, y, t) = \frac{v_*^2}{h_n\lambda_n} e^{x/R_n} \theta(t) \left\{ \theta(a - |y|) \right.$$
$$- \theta(y + a)\theta[t - \lambda_n(y + a)]e^{-r\lambda_n(y+a)} \qquad (5.62)$$
$$\left. + \theta(y - a)\theta[t - \lambda_n(y - a)]e^{-r\lambda_n(y-a)} \right\}.$$

In the viscid case, the magnitudes of the Kelvin waves decrease with the travel distances and the damping effect increases with increasing mode number. The trapping scales of the decrease in the along-shore direction is determined by

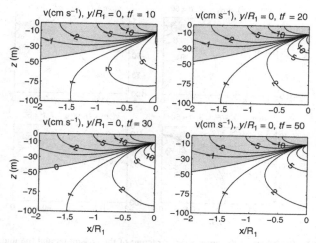

**FIGURE 5.11** Modification of the coastal jet by Kelvin waves in the middle of the forcing area for different times ($ft = 10, ft = 20, ft = 30$ and $ft = 50$). The shaded areas refer to negative flow direction. The coastal jet is increasingly affected by Kelvin waves, which also generate the undercurrent.

the phase speed and the time scale set by the friction parameter, $r$, and depend on the mode number through $r/\lambda_n$.

The upwelling within the wind band is reduced after the arrival of the Kelvin wave, but not completely stopped as in the inviscid case, i.e. $w_n \sim 1 - e^{-r\lambda_n(y+a)}$.

The full solution, which follows from the summation over the vertical eigenfunctions, $F_n(z)$, is shown in Fig. 5.11. The involved parameters are the same as in the previous section, i.e. $R_1 = 5$ km, $f = 1.2 \times 10^{-4}$ s$^{-1}$, $H = 100$ m, $H_{mix} = 10$ m and $v_*^2 = 1$ cm$^2$ s$^{-2}$. But now we have two additional parameters: the width of the wind band, $a = 5R_1$, and the linear friction parameter, $r = 0.05f$. For the graphical representation, it is convenient to scale the argument of the step function, which represents the propagation of the Kelvin waves, by $f$ and $R_n$, i.e. $\theta(t - \lambda_n(y+a)) = \theta(tf - (y+a)/R_n)$. Fig. 5.11 illustrates how the coastal jet and the undercurrent are shaped by the Kelvin waves in the middle of the forcing band. For the selected times, $ft = 10, ft = 20, ft = 30$ and $ft = 50$, the number of modes which have reached the position in the middle of the wind band, $y = 0$, are $n = 2$, $n = 4$, $n = 6$ and $n = 10$, respectively.

The signals exported outside the forcing area consist of two sets of Kelvin waves: those that were excited at the right edge of the forcing area and those that were excited at the left edge of the forcing area. The latter have to cross the wind band before they leave it at the right edge, as indicated in Fig. 5.10. The along-shore and cross-shore structure of the costal jet and the coastal undercurrent are illustrated in Fig. 5.12.

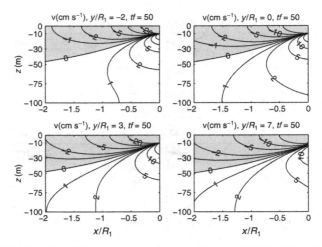

**FIGURE 5.12** Snapshots of the cross-shore structure of the coastal jet and the undercurrent for $ft = 50$ along the four sections: $y = -2R_1$, $y = 0$, $y = 3R_1$ and $y = 7R_1$, as indicated in Fig. 5.9. The shaded areas refer to negative flow direction. The last section, $y = 7R_1$, is outside the wind band and shows the export of the current system.

A further important effect of the friction is that the vertical current is not completely cancelled by the Kelvin waves. The vertical component follows as

$$w_n(x, y, t) \sim -\left(\frac{\partial}{\partial t} + r\right) p_n(x, y, t) = \frac{u_*^2}{h_n \lambda_n} e^{x/R_n} \theta(t) \left[ \theta(a - |x|) \right.$$

$$+ \theta(x + a)\theta\left(tf - \frac{x + a}{R_n}\right) e^{-r\lambda_n(x+a)} \qquad (5.63)$$

$$\left. - \theta(x - a)\theta\left(tf - \frac{x - a}{R_n}\right) e^{-r\lambda_n(x-a)} \right].$$

This implies that the vertical current provided by the higher modes is not completely cancelled through the Kelvin waves if friction is involved.

## 5.2.4 The Role of Wind-Stress Curls

In the typical upwelling areas at the eastern boundaries of the oceans, the large-scale wind patterns are characterized by wind-stress curls, which imply a decrease of the along-shore wind towards the coast (see, e.g. Bakun and Nelson (1991)), and this spatial structure can generate additional flows, which make the current system more complex (see, e.g. McCreary et al. (1987) and Fennel (1999c)). As an example, we consider the Benguela system off southern Africa with the aid of a simple model. A southern along-shore wind with a dome-shaped structure can be used to roughly mimic the mean distribution of the meridional wind stress in the Benguela region (Fennel et al., 2012). The wind is switched on at $t = 0$ and has the analytical shape

$$Y(x, y, z, t) = \frac{v_*^2}{H_{\text{mix}}} \theta(z + H_{\text{mix}}) T(t) Q(y) \Pi(x), \quad X = 0. \quad (5.64)$$

A dome-shaped wind structure can be implemented by the positive half-wave of a cosine profile, i.e.

$$Q(y) = \theta(a - |y|) \cos(\kappa_0 y),$$

with $\kappa_0 = \pi/2a$, where $2a$ is the length of the wind band. Similarly, the cross-shore wind profile, $\Pi(x)$, is chosen as

$$\Pi(x) = \theta[x + (L + l)] \cos[k_o(x + l)],$$

with $k_o = \pi/2L$. The parameter $l$ will be used to shift the maximum of the wind field in zonal direction, i.e, towards or away from the coastal boundary. The spatial patterns of the dome-shaped wind patch are illustrated in Fig. 5.13. For the meridional and zonal extent of the wind patch, we chose 1000 km, i.e. $a = 500$ km and $L = 500$ km.

Because we now consider a region on the southern hemisphere, our formal solution (5.40), (5.43) and (5.44), is somewhat modified: $f$ has changed its sign and the Kelvin wave propagates poleward with the eastern boundary on the left. Then the formal solution for the southern hemisphere reads,

$$u_n(y) = \lambda_n^2 f G_n * Y_n + \frac{\kappa}{\omega} G_n * Y_{nx'}, \quad (5.65)$$

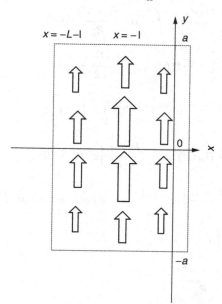

**FIGURE 5.13** Sketch of the wind patch with cosine shape in both meridional and zonal directions. The distance of the wind maximum to the boundary and the width of the band are controlled by the parameters $l$ and $L$.

$$v_n = \frac{i}{\omega} G_{nx} * Y_{nx'} + \frac{i\lambda_n}{(\omega\lambda_n + \kappa)R_n} e^{(x+x')/R_n} * Y_n, \tag{5.66}$$

$$p_n = -\frac{if}{\omega} G_n * Y_{nx'} - \frac{i}{(\omega\lambda_n + \kappa)R_n} e^{(x+x')/R_n} * Y_n. \tag{5.67}$$

The Green's function is the same as for the northern hemisphere. Now we have to take the derivatives of $Y_{nx}$ into account, which will generate new current patterns in comparison with the previously considered cases. One property of the response can directly be seen from the formal solution. The driven signals of the wind-stress curls are not associated with the Kelvin poles and hence not with Kelvin waves. Thus, the wind curl does not generate undercurrents.

After expansion into vertical modes and Fourier transformation with respect to $t$ and $y$, we find for the forcing function,

$$Y_n(x, \omega, \kappa) = -\frac{v_*^2}{h_n} \frac{i}{\omega + i\epsilon} Q(\kappa)\Pi(x),$$

where

$$Q(\kappa) = \frac{\sin[a(\kappa + \kappa_0)]}{\kappa + \kappa_0} + \frac{\sin[a(\kappa - \kappa_0)]}{\kappa - \kappa_0}.$$

Inserting this into our formal solution (5.65), (5.66) and (5.67), we have to calculate several convolution integrals, in particular,

$$A_n(x) = \frac{G(x, x') * \Pi(x')}{R_n^2} = \int_{-\infty}^0 dx' \frac{G(x; x')}{R_n^2} \Pi(x')$$

$$= \frac{1}{1 + k_o^2 R_n^2} \left[ e^{x/R_n} \Pi(0) + \frac{\Pi_x(-L-l)R_n}{2} \right.$$

$$\left. \left( e^{(x-L-l)/R_n} - e^{-(|L+l+x|)/R_n} \right) - \Pi(x) \right],$$

$$B_n(x) = G(x, x') * \Pi_{x'}(x') = \frac{R_n^2}{1 + k_o^2 R_n^2} \left\{ e^{x/R_n} \Pi_x(0) \right.$$

$$- \frac{\Pi_x(-L-l)}{2} \left[ e^{(x-L-l)/R_n} - \text{sign}(L + l + x)e^{-|L+l+x|/R_n} \right]$$

$$\left. - \Pi_x(x) \right\},$$

$$C_n(x) = \Pi(x) * \frac{e^{(x+x')/R_n}}{R_n}$$

$$= \frac{1}{1 + k_o^2 R_n^2} \left[ e^{x/R_n} \left( \Pi(0) - R_n \Pi_x(0) \right) + R_n \Pi_x(-l-L)e^{(x-L-l)/R_n} \right],$$

and

$$D_n(x) = G_{nx} * \Pi_{x'} = \frac{1}{(1 + k_o^2 R_n^2)} \left\{ k_o^2 R_n^2 \Pi(x) \right.$$

$$\left. + R_n e^{x/R_n} \Pi_x(0) - \frac{\Pi_x(-L-l)R_n}{2} \left[ e^{(x-L-l)/R_n} + e^{(-|x+L+l|)/R_n} \right] \right\}.$$

Note that $\Pi(0) = \cos(k_o l)$, $\Pi_x(0) = -k_o \sin(k_o l)$, $\Pi_x(-L-l) = k_o$, $(\partial/\partial x)$
$C_n = C_n/R_n$, $(\partial/\partial x)A_n = (B_n/R_n^2) + C_n/R_n$, $(\partial/\partial x)B_n(x) = D_n(x)$ and $D_n(x)$
$= A_n(x) + \Pi(x) - C_n(x)$. Then the solution is,

$$u_n(x, \kappa, \overline{\omega}) = \frac{v_*^2}{h_n} T(\omega) Q(\kappa) \left[ \frac{\kappa}{\overline{\omega}} B_n(x) + \frac{1}{f} A_n(x) \right], \tag{5.68}$$

$$v_n(x, \kappa, \overline{\omega}) = \frac{v_*^2}{h_n} T(\omega) iQ(\kappa) \left[ \frac{1}{\overline{\omega}} D_n(x) + \frac{i\lambda_n}{\lambda_n \overline{\omega} + \kappa} C_n(x) \right], \tag{5.69}$$

$$p_n(x, \kappa, \overline{\omega}) = -\frac{v_*^2}{h_n} T(\omega) Q(\kappa) \left[ \frac{if}{\overline{\omega}} B_n(x) + \frac{i}{\lambda_n \overline{\omega} + \kappa} C_n(x) \right]. \tag{5.70}$$

The inverse Fourier transform of $u$ with respect to $\omega$ and $\kappa$ is straightforward; note that $i\kappa Q(\kappa)$ gives $(\partial/\partial y)Q(y)$. The transformation with respect to $\omega$ amounts to

$$\int i \frac{T(\omega)}{\overline{\omega}} e^{-i\omega t} \frac{d\omega}{2\pi} = \frac{\theta(t)}{r} (1 - e^{-rt}) \quad \text{and} \quad \int T(\omega) e^{-i\omega t} \frac{d\omega}{2\pi} = \theta(t).$$

Thus, we have

$$u(x, y, z, t) = v_*^2 \sum_n \frac{F_n(z)}{h_n} \theta(t) \left( \underbrace{-\frac{1 - e^{-rt}}{r} \frac{\partial}{\partial y} Q(y) B_n(x)}_{\text{curl}} + \underbrace{\frac{Q(y)}{f} A_n(x)}_{\text{Ekman}} \right),$$

$$\tag{5.71}$$

with $(\partial/\partial y)Q(y) = -\kappa_0 \theta(a - |y|) \sin(\kappa_0 y)$. For the along-shore current, $v_n$, and the pressure, $p$, the relevant $\kappa$ integral is

$$I_n(\kappa, \overline{\omega}) = \int \frac{Q(\kappa)}{\overline{\omega} \lambda_n + \kappa} e^{i\kappa y} \frac{d\kappa}{2\pi},$$

which gives

$$I_n(y, \overline{\omega}) = \frac{1}{\overline{\omega}^2 \lambda_n^2 - \kappa_0^2} (\overline{\omega} \lambda_n Q(y) - i\kappa_0 \theta(a - |y|) \sin(\kappa_0 y)$$

$$+ i\kappa_0 [\theta(a - y) e^{i\overline{\omega}\lambda_n(a-y)} + \theta(-a - y) e^{-i\overline{\omega}\lambda_n(a+y)}]).$$

The transformation with respect to $\omega$, i.e.

$$\int T(\omega) iI_n(y, \overline{\omega}) e^{-i\omega t} \frac{d\omega}{2\pi} = K_n(y, t),$$

can be calculated in a straightforward manner, and we find,

$$K_n(y,t) = \frac{\theta(t)\theta(a-|y|)}{\lambda_n r\left(1+\frac{\kappa_0^2}{r^2\lambda_n^2}\right)}\left[\cos(\kappa_0 y)-\frac{\kappa_0}{r\lambda_n}\sin(\kappa_0 y)-e^{-rt}\Psi_n\left(y+\frac{t}{\lambda_n}\right)\right]$$

$$+\frac{\theta(t-\lambda_n(a-y))\theta(a-y)}{\lambda_n r\left(1+\frac{\kappa_0^2}{r^2\lambda_n^2}\right)}\left[\frac{\kappa_0 e^{-r\lambda_n(a-y)}}{r\lambda_n}+e^{-rt}\Psi_n\left(y+\frac{t}{\lambda_n}\right)\right]$$

$$+\frac{\theta[t+\lambda_n(a+y)]\theta(-a-y)}{\lambda_n r\left(1+\frac{\kappa_0^2}{r^2\lambda_n^2}\right)}\left[\frac{\kappa_0 e^{r\lambda_n(a+y)}}{r\lambda_n}-e^{-rt}\Psi_n\left(y+\frac{t}{\lambda_n}\right)\right],$$

where

$$\Psi_n\left(y+\frac{t}{\lambda_n}\right)=\cos\left[\kappa_0\left(y+\frac{t}{\lambda_n}\right)\right]-\frac{\kappa_0}{r\lambda_n}\sin\left[\kappa_0\left(y+\frac{t}{\lambda_n}\right)\right].$$

Finally, we arrive at the solution for the long-shore current,

$$v(x,y,z,t)=v_*^2\sum_n\frac{F_n(z)}{h_n}\left(\underbrace{\frac{1-e^{-rt}}{r}Q(y)D_n(x)}_{\text{curl}}+\underbrace{\lambda_n K_n(t,y)C_n(x)}_{\text{coastal}}\right),\quad (5.72)$$

and for the pressure,

$$p(x,y,z,t)=-v_*^2\sum_n\frac{F_n(z)}{h_n}\left(\underbrace{f\frac{1-e^{-rt}}{r}Q(y)B_n(x)}_{\text{curl}}+\underbrace{K_n(t,y)C_n(x)}_{\text{coastal}}\right).\quad (5.73)$$

We note that in the frame of the low-frequency and long-wave approximation, Equations (5.40), (5.43) and (5.44) are solutions of the set:

$$fv_n+p_{nx}=0,$$

$$\left(\frac{\partial}{\partial t}+r\right)v_n-fu+p_{ny}=Y_n,$$

$$u_{nx}+v_{ny}+\left(\frac{\partial}{\partial t}+r\right)\lambda_n^2 p_n=0.$$

The vertical velocity component, $w$, is related to the pressure,

$$w(x,y,z,t)=-\frac{1}{N^2}\left(\frac{\partial}{\partial t}+r\right)\frac{\partial}{\partial z}p(x,y,z,t)\qquad (5.74)$$

$$=-\frac{1}{N^2}\sum_n\frac{\partial}{\partial z}F_n(z)\left(\frac{\partial}{\partial t}+r\right)p_n(x,y,t).\qquad (5.75)$$

With

$$-\frac{1}{N^2}\frac{\partial}{\partial z}F_n(z) = \frac{\lambda_n}{N}\sqrt{\frac{2}{H}}\sin\left(\frac{n\pi}{H}z\right),$$

we find for $w$,

$$w(x, y, z, t) = -\frac{v_*^2}{NH_{\text{mix}}}\sum_{n=1}^{\infty}\frac{2}{\pi n}\sin\left(\frac{n\pi}{H}H_{\text{mix}}\right)\sin\left(\frac{n\pi}{H}z\right)$$

$$\times \left(\underbrace{\frac{1}{R_n}Q(y)B_n(x)}_{\text{curl}} + \underbrace{\lambda_n\left(\frac{\partial}{\partial t} + r\right)K_n(y, t)C_n(x)}_{\text{coastal}}\right), \quad (5.76)$$

where

$$\left(\frac{\partial}{\partial t} + r\right)K_n(y, t)$$

$$= \frac{\theta(a - |y|)}{\lambda_n\left(1 + \frac{\kappa_0^2}{r^2\lambda_n^2}\right)}\left[\cos(\kappa_0 y) - \frac{\kappa_0}{\lambda_n r}\sin(\kappa_0 y) + e^{-rt}\frac{\kappa_0}{\lambda_n r}\Phi_n\left(y + \frac{t}{\lambda_n}\right)\right]$$

$$+ \frac{\theta(a - y)\theta[t - \lambda_n(a - y)]}{\lambda_n\left(1 + \frac{\kappa_0^2}{r^2\lambda_n^2}\right)}\frac{\kappa_o}{\lambda_n r}\left[e^{-r\lambda_n(a - y)} - e^{-rt}\Phi_n\left(y + \frac{t}{\lambda_n}\right)\right]$$

$$+ \frac{\theta(-a - y)\theta[t + \lambda_n(a + y)]}{\lambda_n\left(1 + \frac{\kappa_0^2}{r^2\lambda_n^2}\right)}\frac{\kappa_o}{\lambda_n r}\left[e^{r\lambda_n(a + y)} + e^{-rt}\Phi_n\left(y + \frac{t}{\lambda_n}\right)\right],$$

and $\Phi_n$ is defined by

$$\Phi_n\left(y + \frac{t}{\lambda_n}\right) = \sin\left[\kappa_o\left(y + \frac{t}{\lambda_n}\right)\right] + \frac{\kappa_o}{r\lambda_n}\cos\left[\kappa_o\left(y + \frac{t}{\lambda_n}\right)\right].$$

Note that $\Psi_{nt} = -(\kappa_o/\lambda_n)\Phi_n$ and $\Psi_{ny} = -\kappa_o\Phi_n$.

Owing to the wind-stress curl, the cross-shore current is rather complex, with opposite directions at the southern and northern ramp of the along-shore wind patch. The solution for the cross-shore current consists of two contributions. One is proportional to $QA_n/f \sim -Q(y)\Pi(x)/f$ and can be recognized as the Ekman transport, which tends to zero near the meridional borders of the wind patch, at $y = a$ and $y = -a$. The other term, indicated as 'curl' term, is proportional to the divergence of the wind field, $\partial Q/\partial y$, and assumes a maximum at $y = a$, a minimum at $y = -a$ and vanishes in the middle of the wind band, $y = 0$, where in turn the Ekman term reaches its maximum. For the discussion and visualization of the results, we chose the involved parameters as $v_*^2 = 0.8\,\text{cm}^2\,\text{s}^{-2}$, $H_{\text{mix}} = 20\,\text{m}$, $H = 1000\,\text{m}$, $f = 6 \times 10^{-5}\,\text{s}^{-1}$, $r = 0.02f$ and $R_1 = 50\,\text{km}$. In the centner of the wind band, $y = 0$, we have a typical Ekman

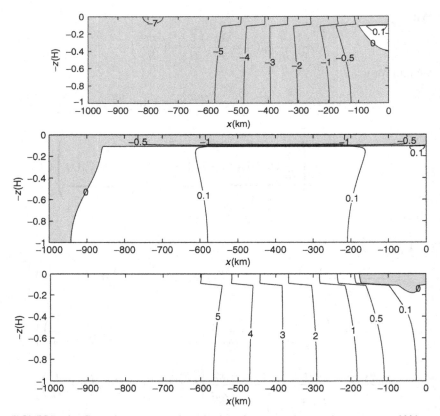

**FIGURE 5.14**   Cross-shore current, $u$ in cm/s, along three cross-shore sections: (a) at $y = 300$ km, (b) $y = 0$ km and (c) $y = -300$ km in the wind band with the choice of $l = 400$ km. In order to show the current in upper part more clearly, $H_{\mathrm{mix}} = 100$ m was chosen (Fennel et al., 2012).

cross-circulation with an offshore transport in the upper layer and an onshore recirculation below (Fig. 5.14(b)). However, in the northern ramp there is a broad offshore transport, which also stretches over the water column below the mixed layer (Fig. 5.14(a)). In the southern ramp, we see only a small part of the section with an offshore Ekman transport, while most of the water moves towards the coast (Fig. 5.14(c)). The offshore motion at the northern ramp and the onshore drift at the southern ramp indicate a circular current around the wind patch.

The assumption of the circular circulation pattern is supported by the fact that, in the middle of the wind patch, the meridional current has a structure similar to the wind pattern,

$$v_n^{\mathrm{curl}} \sim QD_n \sim Q\frac{1}{\left(1 + k_o^2 R_n^2\right)} \left[ k_o^2 R_n^2 \Pi(x) + R_n e^{x/R_n} \Pi_x(0) \right].$$

This is indicated in Fig. 5.16, which only shows a part of the wind patch region. Closer to the coastal boundary, the along-shore currents are largely shaped by the strength of the wind near the coast, which is controlled by the parameter, $l$, and by Kelvin waves. The Kelvin waves propagate the vertical modes along shore with the phase speeds, $c_n$. As previously discussed for the wind-band forcing, we find the near-shore balance is $v_{nt} = Y$ before the arrival of the Kelvin wave front. With the value of $c_1 = 3\,\mathrm{m\,s^{-1}}$, or $c_1 = 285\,\mathrm{km\,d^{-1}}$, the typical propagation time needed by the first mode Kelvin wave to cross the wind band of the width $2a = 1000\,\mathrm{km}$ is $\tau_1 = 3.5\,\mathrm{days}$. Because $c_n = c_1/n$ and $\tau_n = n\tau_1$, it follows that the first five modes have crossed the band after two weeks, time; i.e. the currents are adjusted to the regime behind the Kelvin waves, where, close to the coast, the along-shore pressure gradient balances the forcing $p_{ny} \sim Y_n$. However, the magnitude of the coastal currents depends on the critical parameter $l$. For small values of $l$, the wind is strong near the coast, while for values of $l$ close to $L$, the near shore wind is weak. Near to the coast the leading contributions of the curl term are, with $\Pi_x(0) = -\kappa \sin(k_o l)$,

$$v_n^{\mathrm{curl}} \sim QD_n \sim Q\frac{1}{(1 + k_o^2 R_n^2)}\left[k_o^2 R_n^2 \Pi(x) - R_n e^{x/R_n}\kappa \sin(k_o l)\right].$$

Thus, for $l$ close to $L$, the second term with $\sin(k_o l)$ can exceed the first term, and we find a alongshore current in opposite direction of the wind. How far the countercurrent reaches also depends on the Kelvin waves, which reduce the coastal jet signal near the northern ramp of the wind patch. This poleward countercurrent can merge with the undercurrent.

Examples of the along-shore currents are shown in Fig. 5.15 for different cases of the wind field structure, i.e. for different choices of the distances of the wind maximum from the coast. For $l = 200\,\mathrm{km}$ (Fig. 5.15(a)) the coastal currents have the 'typical' shape that consists of an equator-ward coastal jet flowing above a poleward countercurrent. For $l = 425\,\mathrm{km}$ (Fig. 5.15(b)) the wind maximum is far offshore, the flow reversal due to the curl merges with the undercurrent and the resulting countercurrent reaches the sea surface. The broadening of the undercurrent is due to the wind-stress curl; i.e. the effect of the curl is visible but not dominating. For $l = 450\,\mathrm{km}$, i.e. a small displacement of the curl (Fig. 5.15(c)) the countercurrent strengthens and overcompensates the coastal jet. In a strip of a width of about two times the first-mode Rossby radius, the coastal current flows mostly in the poleward direction, i.e. against the local wind. Only a marginal part of the coastal jet is still visible. Thus, for $l \to L$, the coastal response is dominated by the curl-related contributions, while for $l \to 0$, the coastally trapped terms dominate the flow patterns. For the example of $l = 425\,\mathrm{km}$, the structure of the meridional (along-shore) current near the surface is shown in Fig. 5.16. This horizontal representation corresponds to the section in Fig. 5.15(b). Close to the coast, we see a typical coastal jet that broadens towards the south due

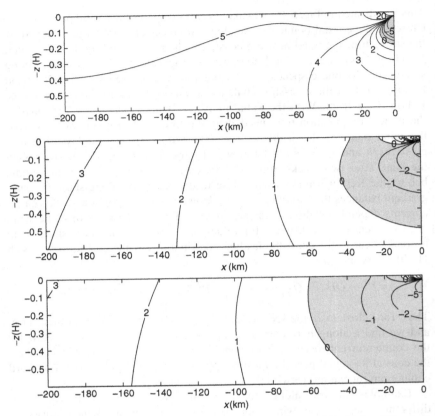

**FIGURE 5.15**     Alongshore, meridional, current, $v$ in cm/s, for (a) $l = 200$, (b) $l = 425$ and (c) $l = 450$ km along the section $y = 0$ in the middle of the wind band. Only the upper 600 m are plotted (Fennel et al., 2012).

to the travel time of the different modes of the Kelvin waves (the later the Kelvin waves arrive, the longer the coastal jet could accelerate). In the northern half of the wind patch, the countercurrent due to the wind-stress curl largely overcompensates for the coastal jet and generates a countercurrent that reaches the sea surface.

The curl contributions, $u_n^{curl}$ and $v_n^{curl}$, are geostrophically balanced by the pressure gradients, $p_{nx}^{curl}$ and $p_{nx}^{curl}$, respectively. The divergence of curl terms vanishes, $u_x^{curl} + v_y^{curl} = 0$. This follows explicitly from Equations (5.71) and (5.72), with $u_x^{curl} \sim -Q_y B_x$, $v_y^{curl} \sim Q_y D$ and $D = B_x$. This means that these terms do not contribute to a vertical motion, but establish a circulation around the wind patch. The relevant terms that generate the upwelling are related to the 'Ekman terms'. The Ekman transport near the coast, $x \rightarrow 0$, is rectified due to

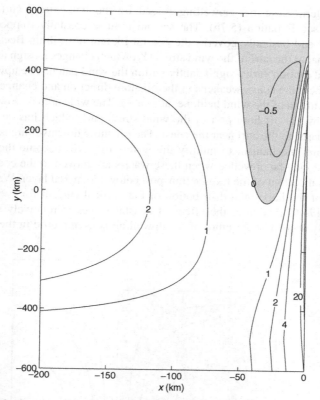

**FIGURE 5.16** Structure of the alongshore current and countercurrent, v in cm/s, near the surface, $z = -10$ m, for $l = 425$ km (Fennel et al., 2012).

the coastal inhibition and amounts to,

$$u_n^{\text{Ekman}} \sim \frac{v_*^2}{h_n} \frac{Q(y)}{f} A_n(x) \sim \frac{v_*^2}{h_n} \frac{Q(y)}{f} \left( -\Pi(x) + \Pi(0) e^{x/R_n} \right),$$

while far offshore, near the western boundary of the wind patch, $x \to -L - l$, the Ekman signal is smoothed out and stretches over the wind edge,

$$u_n^{\text{Ekman}} \sim \frac{v_*^2}{h_n} \frac{Q(y)}{f} A_n(x) \sim \frac{v_*^2}{h_n} \frac{Q(y)}{f} \left[ -\Pi(x) - \frac{\Pi_x(-L-l)R_n}{2} e^{-(|L+l+x|)/R_n} \right]$$

The divergence of the Ekman transport consists of two terms,

$$u_{nx}^{\text{Ekman}} \sim \frac{v_*^2}{h_n} \frac{Q(y)}{f} \frac{\partial}{\partial x} A_n(x) = \frac{v_*^2}{h_n} \frac{Q(y)}{f} \left[ \frac{B_n(x)}{R_n^2} + \frac{C_n(x)}{R_n} \right].$$

The first term describes upwelling forced by the wind-stress curl, Ekman pumping, see Equation (5.76). The second term is coastally trapped by the term $e^{x/R_n}$ and basically governs the coastal Ekman upwelling. Because $w$ is proportional to the curl of the wind stress, $Y_x$, which changes its sign at $x = -l$, the vertical motion varies significantly within the wind patch. The upwelling is strongest near the coast, weakens in the offshore direction and changes its sign at the maximum of the wind field, i.e. at $x = -l$. The wide area of downwelling is forced by the offshore part of the wind stress curl, which has an opposite sign compared to the curl near the coast. The off shore downwelling is possibly an important mechanism to modify the waters that will become thermocline waters and may be upwelled when these waters are moved to the coast by the rectification current of the Ekman transport below the mixed layer. We show an example of the horizontal distribution of the vertical current for $l = 400\,\mathrm{km}$ in Fig. 5.17. In this case, the effect of coastal waves is relatively small but still clearly seen. The structure of the upwelling is asymmetric in the northern

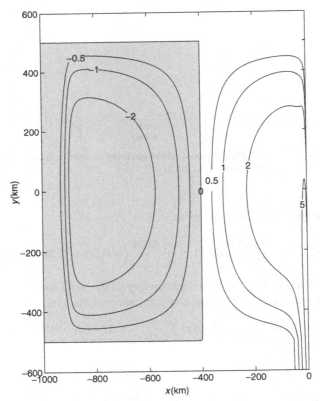

**FIGURE 5.17**   Structure of the vertical current, in $10^{-4}\,\mathrm{cm\,s{-1}}$, for $l = 400\,\mathrm{km}$, $100\,\mathrm{days}$ after the onset of the forcing (Fennel et al., 2012).

and southern ramps of the forcing. The maximum upwelling signal is found near the coast at the latitude of maximum wind stress in the centner of the wind band. But due to downwelling/upwelling and Kelvin waves excited at the northern/southern edge of the wind patch, the coastal upwelling signal has been shifted southward. We note that the patterns shown in Fig. 5.17 correspond to a time period of 100 days after the onset of the wind. This implies that virtually all significant modes of Kelvin waves have crossed the wind band and diminished the coastal upwelling signal. The Kelvin waves generated at the southern edge leave the wind band and export upwelling.

In the real system, the wind fluctuates, implying that upwelling is generated again and again in response to every new wind event and the response to a switch-on wind can be considered as the average pattern. For wind pulses of finite duration, it appears, that during the decaying phase of wind event downwelling/upwelling, Kelvin waves are generated at the northern/southern edge of the wind band, which decrease the upwelling signal (Fennel et al., 2010).

## 5.2.5 Discussion

The dynamics of coastal upwelling are basically three dimensional. The upwelling is driven by the offshore Ekman transport in the surface layer and the onshore compensation current beneath the upper mixed layer. These flows are relatively small, nongeostrophic motions that generate pressure gradients. The pressure gradients are associated with the strong geostrophic coastal jets. The along-shore structure of coastal jets and coastal undercurrents are shaped by the Kelvin waves. The relevant scales are the Rossby radii, the scale of the wind field and the propagation speeds of the different modes of the Kelvin waves.

The small Ekman compensation flow, which drives the upwelling, moves nutrients into the euphotic zone. The vigorous along-shore, near-surface currents and oppositely directed undercurrents are important for the ecosystem response because they can move cells and copepods alongshore. By small vertical excursion, copepods can enter the opposite flows and can move along shore in both directions. The coastal Kelvin waves export the current outside the wind band and can transport nutrients and phytoplankton in areas with no wind forcing. These considerations demonstrate also that a two-dimensional modelling of ecosystems of upwelling areas that takes only the vertical and cross-shore co-ordinates into account is obscured by ignoring the most vigourous, along-shore current signals.

The currents in an upwelling system become more complex if wind-stress curls are taken into account. If an along-shore wind decreases towards the boundary, a coastal countercurrent against the wind can be generated and combined with the coastal jet and undercurrent. The wind-curl-generated upwelling is not affected by the Kelvin waves and is only limited by friction.

Another example of upwelling systems are ice edges, where a pair of jets, currents and countercurrents trapped at the ice edge can be driven by along-edge winds; see, e.g. Sjøberg and Mork (1985) and Fennel and Johannessen (1998).

The Kelvin waves are one important example of waves guided by topography. Here, we considered a very simple topography—namely, a coast idealized as a straight wall. In general, the ocean boundaries are characterized by a shelf with a steep descent in the offshore direction. Those cases are mathematically complex, and the vertical separation in terms of vertical modes no longer applies because the condition $u = 0$ at the coast now assumes the form $uH_x = w$.

Clearly, the analytically solvable problem of the coastal upwelling is greatly simplified. Important nonlinear effects, such as outcropping of isotherms, formation of frontal zones and filaments, as well as shedding of eddies, cannot be not described with linear equations of motion. Involving these processes would require the use of the full nonlinear equations, which are not susceptible to analytical solutions but can be treated in numerical circulation models. The value of the analytical examples lies in the possibility of clear and explicit descriptions of several key processes. This knowledge is helpful to understanding and discussing the results of the numerical models.

## 5.3 ADVECTION–DIFFUSION EQUATION

The interface between the physical transport and mixing processes and biological development is the advection diffusion equation for the state variables. Let $S_i$ be a state variable such as nutrient concentration, phytoplankton biomass or detritus. It is clear that the changes of the state variables can be driven by chemical-biological processes and/or by advection and turbulent mixing. This is governed in the advection-diffusion equation for the state variables written as

$$\frac{\partial}{\partial t} S_i + \nabla \cdot \mathbf{v} S_i - A \Delta S_i = \text{gain}(S_i) - \text{loss}(S_i). \tag{5.77}$$

The advection term, $\mathbf{v}$, may include the active motion of cells or animals described by the state variables. For example, sinking or upward floating $w_i$ is included by $\mathbf{v} = (u, v, w + w_i)$, where $w$ is the vertical motion due to fluid dynamics. The turbulent mixing is controlled by the diffusivity $A$. The distinction of currents (advection) and turbulent exchange governed by the diffusivity implies a filtering or averaging procedure, and it is illuminating to look closer at the derivation of the advection-diffusion equation. The theory described in the following sections applies basically for passive substances, i.e. for substances that are moved by currents but do not affect the currents.

### 5.3.1 Reynolds Rules

Let $c$ be the concentration of a passive substance, which is moved and mixed by a current field $\mathbf{v}$. The concentration does not affect the flow field. Then the conservation of mass requires that the total change of $c$ equals the differences of sources and sinks. Let us start with the case of no external sources or sinks. Then, in a certain moving material element (Lagrangian approach), the total change is zero

$$\frac{d}{dt}c = 0.$$

In a volume element, fixed in space (Eulerian approach), the change in time is balanced by fluxes through the volume element,

$$\frac{\partial}{\partial t}c + \nabla \cdot \mathbf{v}c = 0.$$

This is the equation of continuity for the state variable, $c$, which states the conservation of mass.

In general, spectra of oceanic flows have a broad range of time scales ranging from seconds to months. A current vector can, in principle, be separated into mean or slowly varying motions and fluctuations that may be considered as deviations from the means, $\mathbf{v} = \bar{\mathbf{v}} + \mathbf{v}'$. Similarly, we assume that the concentration, $c$, can be written as $c = \bar{c} + c'$. The mean or slowly varying part of the concentration, $\bar{c}$, can be estimated by averaging a time series, while $c'$ is the deviation from the mean value. Hence, the equation for the change of the mean concentration, $\bar{c}$, follows as

$$\frac{\partial}{\partial t}\bar{c} + \nabla \cdot \overline{c\mathbf{v}} = 0.$$

The average of product, $\overline{\mathbf{v}c}$, emerges as a new quantity, which consists of

$$\overline{\mathbf{v}c} = \bar{\mathbf{v}}\,\bar{c} + \bar{\mathbf{v}}c' + \overline{\mathbf{v}'\bar{c}} + \overline{\mathbf{v}'c'}.$$

Under certain circumstances the so-called Reynolds rules apply,

$$\bar{\bar{\mathbf{v}}} = \bar{\mathbf{v}} \quad \overline{\mathbf{v}'} = 0,$$

and

$$\overline{\bar{\mathbf{v}}\,\bar{c}} = \bar{\mathbf{v}}\,\bar{c}, \qquad \overline{\mathbf{v}c'} = \overline{\mathbf{v}'\bar{c}} = 0.$$

The mean value of the product of the fluctuations, which are generated by turbulence, is in general different from zero, $\overline{\mathbf{v}'c'} \neq 0$. The simplest approach to relate this term to $\bar{c}$ is

$$\overline{\mathbf{v}'c'} = -A\nabla\bar{c},$$

which means that the turbulent flux reduces gradients of the mean distribution. The turbulent diffusivity, $A$, is an empirical quantity that has to be determined by measurements. From observations, it is known that the horizontal mixing, expressed by $A_h$, is significantly larger than the vertical mixing, expressed by $A_v$, implying that

$$\overline{\mathbf{v}'c'} = -A_h\nabla_h\bar{c} - A_v\nabla_z\bar{c},$$

where $\nabla_h = (\mathbf{i}(\partial/\partial x) + \mathbf{j}\partial/\partial y)$, $\nabla_z = \mathbf{k}\partial/\partial z$ and $\mathbf{i}$, $\mathbf{j}$ and $\mathbf{k}$ are the unit vectors. We then arrive at the advection-diffusion equation for the mean concentration, which reads

$$\frac{\partial}{\partial t}\overline{c} + \nabla \cdot \overline{\mathbf{v}c} - A_h\nabla_h^2\overline{c} - A_v\nabla_z^2\overline{c} = 0.$$

In order to solve the advection-diffusion equation, we have to prescribe the current field and the eddy diffusivities. In a numerical model system, this is done in a natural way by the solving the full set of model equations, which computes the velocity field and the turbulent diffusivities at every grid point for every time step and passes the results to the tracer equations. For simpler studies, it is also possible to choose some appropriate examples for $\mathbf{v}$ and $A$.

### 5.3.2 Analytical Examples

There are a few simple examples with prescribed physical properties of the current field that allow an analytical treatment. In particular, studies of the spreading of material from an initial point source amounts to the estimation of the corresponding Green's function, G, which refers to the concentration of a passive tracer,

$$\frac{\partial}{\partial t}G + \overline{\mathbf{v}} \cdot \nabla G - A_h\nabla_h^2 G - A_v\nabla_z^2 G = \delta(t - t_0)\delta(\mathbf{r} - \mathbf{r}_0). \tag{5.78}$$

We have to specify boundary conditions. Let the horizontal boundaries be far away; we then have

$$\nabla_h G \to 0 \text{ for } \mathbf{r} \to \infty.$$

In the vertical direction, we assume a thermocline at a depth $H = H_{mix}$, which is a boundary for the vertical turbulent flux,

$$\frac{\partial}{\partial z}G = 0, \text{ for } z = -H_{mix}.$$

Due to the causality of the response to the pulselike addition of mass, we can try an approach $G(\mathbf{r}, t; \mathbf{r}_0, t_0) = \theta(t - t_0)g(\mathbf{r}; t, \mathbf{r}_0, t_0)$. Inserting this into Equation (5.78) and using

$$\frac{\partial}{\partial t}G = \frac{\partial}{\partial t}\theta(t - t_0)g + \theta(t - t_0)\frac{\partial}{\partial t}g$$

$$= \delta(t - t_0)g(\mathbf{r}; t_0, \mathbf{r}_0, t_0) + \theta(t - t_0)\frac{\partial}{\partial t}g,$$

we find a homogeneous equation for $g$,

$$\frac{\partial}{\partial t}g + \overline{\mathbf{v}} \cdot \nabla g - A_h\nabla_h^2 g - A_v\nabla_z^2 g = 0, \tag{5.79}$$

with the initial condition

$$g(\mathbf{r}; t_0, \mathbf{r}_0, t_0) = \delta(\mathbf{r} - \mathbf{r}_0).$$

Let us assume for simplicity that the horizontal advection is independent of the vertical coordinate, $z$; then we can separate the horizontal and vertical variables,

$$g = C(x - x_0, y - y_0, t)Z(z; z_0). \tag{5.80}$$

Inserting this into Equation (5.79), we obtain two equations

$$\frac{\partial}{\partial t}C + \bar{\mathbf{v}} \cdot \nabla C - A_h \nabla_h^2 C + \lambda^2 C = 0, \tag{5.81}$$

$$A_v \nabla_z^2 Z + \lambda^2 Z = 0, \tag{5.82}$$

where $\lambda$ is a separation constant. The first equation can be simplified with the approach

$$C = C_0 e^{-\lambda^2 t},$$

where $C_0$ obeys

$$\frac{\partial}{\partial t}C_0 + \bar{\mathbf{v}} \cdot \nabla C_0 - A_h \nabla_h^2 C_0 = 0. \tag{5.83}$$

The boundary conditions on $Z$ are

$$\frac{\partial}{\partial z}Z = 0 \quad \text{for } z = 0, \quad \text{and} \quad z = -H_{\text{mix}}.$$

This defines a vertical eigenvalue problem. For constant vertical diffusivity, $A_v$, the normalized eigenfunctions are

$$Z_n(z) = \sqrt{\frac{2}{H_{\text{mix}}}} \cos\left(\frac{n\pi}{H_{\text{mix}}}z\right) \quad \text{for } n = 1, 2, 3, \ldots$$

and

$$Z_0 = \sqrt{\frac{1}{H_{\text{mix}}}}.$$

The corresponding eigenvalues are

$$\lambda_n = \frac{n\pi}{H_{\text{mix}}}\sqrt{A_v}.$$

The completeness theorem for the set of vertical eigenfunctions,

$$\sum_{n=0}^{\infty} Z_n(z)Z_n(z_0) = \delta(z - z_0),$$

implies that Equation (5.80) must be rewritten as

$$g = C_0(x - x_0, y - y_0, t - t_0) \sum_{n=0}^{\infty} e^{-\lambda_n^2(t - t_0)} Z_n(z)Z_n(z_0). \tag{5.84}$$

From this formula, we see that the higher vertical modes, $Z_n(z)$, are rapidly damped by the exponential term, which depends on $n^2$. The smallest damping is provided by the first eigenvalue, $n = 1$. For $\lambda_1^2 t \gg 1$, the vertical distribution is homogeneous, and we can define a vertical mixing time, by

$$\tau_{\text{mix}} = \lambda_1^{-2} = \frac{H_{\text{mix}}^2}{\pi^2 A_v}.$$

For time scales exceeding the vertical mixing time the solution becomes independent of $z$,

$$g = C_0(x - x_0, y - y_0, t - t_0) \frac{1}{H_{\text{mix}}}.$$

With further consideration, we set $x_0 = y_0 = 0$ and $t_0 = 0$ without loss of generality.

### 5.3.3  Turbulent Diffusion in Collinear Flows

We briefly describe and discuss solutions of the advection-diffusion equation for a linearly varying horizontal current field, where

$$u = u_0 + u_x x + u_y y, \tag{5.85}$$

$$v = v_0 + v_x x + v_y y. \tag{5.86}$$

The derivatives of $u$ and $v$ are constants. Such a current field is called 'collinear flow'. We start the discussion with the consideration of nondivergent flow, $u_x + v_y = 0$. For a nondivergent field, we can define a stream function, $\psi(x, y)$, by

$$u = -\frac{\partial}{\partial y} \psi, \text{ and } v = \frac{\partial}{\partial x} \psi.$$

To calculate the stream function, we use Equations (5.85) and (5.86) as

$$-\frac{\partial}{\partial y} \psi = u_0 + u_x x + u_y y,$$

$$\frac{\partial}{\partial x} \psi = v_0 + v_x x - u_x y,$$

and find, after integration, that

$$\psi = v_0 x - u_0 y + v_x \frac{x^2}{2} - u_y \frac{y^2}{2} - u_x xy.$$

Thus, the streamlines are described by curves of second order, i.e. circles, ellipses and hyperbolas. While cycles and ellipses correspond to eddylike patterns, the hyperbolas refer to deformation fields. The distortion of the flow field can be related to deformation, $v_x + u_y$, and vorticity, $v_x - u_y$. The structure of the streamline patterns are roughly characterized by the invariant, $\delta = -u_y v_x - u_x^2$, e.g. Okubo (1978). Let us assume that $v_x$ is positive while $u_y$ may be positive or negative in different cases. For negative values, $u_y = -|u_y|$, the $\delta$ can be positive, $\delta = |u_y| v_x - u_x^2 > 0$. Then the curves are circles or ellipses. In this case, $v_x - |u_y| < v_x + |u_y|$, and hence, vorticity exceeds deformation. For positive values of $u_y = |u_y|$, the $\delta$ is negative and the streamlines are governed by hyperbola. In this case, $v_x + |u_y| > v_x - |u_y|$, and hence, deformation exceeds vorticity. In particular, for $u_y = v_x = 0$ and $u_x \neq 0$ the flow pattern amounts to simple hyperbolas, $\psi \sim u_x xy$. The structure of the current patterns for a shear flow, eddy flow and deformation flow are illustrated in Figs. 5.18–5.20.

Regarding the diffusion problem we have to solve Equation (5.83). This can be done by Fourier transforms

$$C_0(x, y, t) = \int \int \frac{dk}{2\pi} \frac{d\kappa}{2\pi} e^{ikx + i\kappa y} C_0(k, \kappa, t). \qquad (5.87)$$

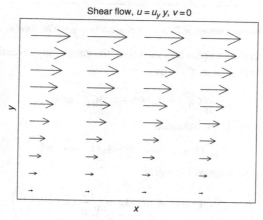

**FIGURE 5.18** Sketch of a linear shear flow, $u(x) = u_y y$, $v = 0$. The shear $u_y$ is constant.

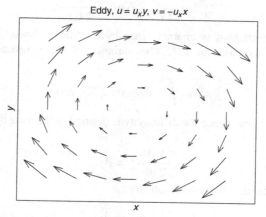

**FIGURE 5.19** Currents in a linear eddy flow, $u(x) = u_x y$, $v = -u_x x$. The shear $u_x$ is constant.

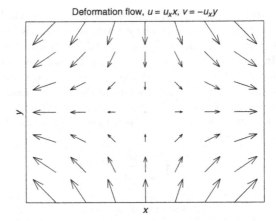

**FIGURE 5.20**   A linear deformation flow, $u(x) = u_x x$, $v = -u_x y$. The shear $u_x$ is constant.

We consider several examples and start with the case of no advection. We denote the solution for this case as $C_0^{(0)}$, then

$$\frac{\partial}{\partial t} C_0^{(0)} + A_h(k^2 + \kappa^2)C_0^{(0)} = 0.$$

The solution can easily be obtained

$$C_0^{(0)}(k, \kappa, t) = \exp[-A_h(k^2 + \kappa^2)t].$$

With the help of the standard integral,

$$\int_{-\infty}^{+\infty} \exp(-ax^2 + ibx)\mathrm{d}x = \sqrt{\frac{\pi}{a}} \exp\left(-\frac{b^2}{4a}\right), \tag{5.88}$$

we find the solution as Gaussian distribution

$$C_0^{(0)} = \frac{1}{4\pi A_h t} \exp(-\frac{x^2 + y^2}{4A_h t}). \tag{5.89}$$

Thus the concentration is characterized by circular isolines. The maximum concentration decreases as $t^{-1}$. The variance of the Gaussian distribution, defined as

$$\sigma_x^2 = \int_{-\infty}^{\infty} \mathrm{d}x\mathrm{d}y x^2 C_0^{(0)}(x, y, t),$$

is equal to $2A_h t$, and hence the diffusivity is defined as the time derivative of the variance

$$\frac{1}{2}\frac{\mathrm{d}}{\mathrm{d}t}\sigma_x^2 = A_h,$$

and, because the distribution is isotropic,

$$\frac{1}{2}\frac{\mathrm{d}}{\mathrm{d}t}\sigma_y^2 = A_h.$$

As a second example, we consider the effect of a constant flow and denote the solution for this case as $C_0^{(1)}$; then we have to solve the equation

$$\frac{\partial}{\partial t}C_0^{(1)} + \left(u_0\frac{\partial}{\partial x} + v_0\frac{\partial}{\partial y}\right)C_0^{(1)} - A_h\nabla_h^2 C_0^{(1)} = 0.$$

This case can be solved in a similar way to the previous example. The solution is

$$C_0^{(1)} = \frac{1}{4\pi A_h t}\exp\left[-\frac{(x - u_0t)^2 + (y - v_0t)^2}{4A_h t}\right],$$

i.e. the whole distribution moves with the constant flow field. The maximum concentration decays algebraically like $t^{-1}$.

### 5.3.3.1 Turbulent Diffusion in a Shear Flow

As the third example, we consider a shear flow, with $u = u_0 + u_y y$ and $v = 0$; see Okubo (1967). We denote the solution for this case as $C_0^{(2)}$. The principal structure of the flow field is sketched in Fig. 5.18. This amounts to the equation

$$\frac{\partial}{\partial t}C_0^{(2)} + \left(u_0 + u_y y\frac{\partial}{\partial x}\right)C_0^{(2)} - A_h\nabla_h^2 C_0^{(2)} = 0.$$

Fourier transformation of the equation gives

$$\left[\frac{\partial}{\partial t} + iku_0 - u_y k\frac{\partial}{\partial \kappa} + A_h(k^2 + \kappa^2)\right]C_0^{(2)} = 0. \qquad (5.90)$$

We expect a structure similar to the preceding cases, and try the approach

$$C_0^{(2)} = \exp[-a(t)k^2 - b(t)\kappa^2 - ck\kappa - ikd - i\kappa e], \qquad (5.91)$$

where $a$, $b$, $c$, $d$ and $e$ are test functions that need to be estimated. The initial condition

$$C_0^{(2)}(k, \kappa, 0) = 1$$

implies

$$a(0) = b(0) = c(0) = d(0) = e(0) = 0.$$

Inserting the approach (5.91) into Equation (5.90) gives

$$-\frac{da}{dt}k^2 - \frac{db}{dt}\kappa^2 - \frac{dc}{dt}k\kappa - \frac{de}{dt}ik - \frac{dd}{dt}i\kappa$$
$$-u_y(-2b\kappa - ck - ie) + A_h(k^2 + \kappa^2) = 0.$$

Comparing the coefficient of the different combinations of $k$ and $\kappa$ gives a set of simple equations for the test functions:

$$\frac{da}{dt} = A_h + u_y c, \qquad \frac{d}{dt}d = u_0 + u_y e,$$

$$\frac{db}{dt} = A_h, \quad \frac{de}{dt} = 0,$$

$$\frac{dc}{dt} = 2u_y b.$$

The equations are easy to integrate, and we find

$$a(t) = A_h t \left(1 + u_y \frac{t^3}{3}\right), \quad b(t) = A_h t,$$

$$c(t) = u_y A_h t^2, \quad d = u_0 t, \quad e = 0.$$

Because $a$ and $b$ are related to the variance of the distribution, we can define effective diffusivities as

$$A_x^{\text{eff}} = \frac{d}{dt} a = \frac{1}{2} \frac{d}{dt} \sigma_x^2 = A_h (1 + u_y t^2),$$

and

$$A_y^{\text{eff}} = \frac{d}{dt} b = \frac{1}{2} \frac{d}{dt} \sigma_y^2 = A_h.$$

This means that after a certain period of time, which is set by the current shear (vorticity), $u_y$, the local diffusivities increase with time due to the presence of spatial variations in the current field.

The solution in the physical space can be obtained by inversion of the Fourier transforms (5.87). Repeated application of the standard integral (5.88) gives

$$C_0^{(2)}(x, y, t) = \frac{\exp\left[\frac{-b(x-d^2)-a(y-e)^2+c(x-d)(y-e)}{4(ab-c^2)}\right]}{4\pi\sqrt{ab - \frac{c^2}{4}}},$$

or, explicitly,

$$C_0^{(2)}(x, y, t) = \frac{\exp\left[\frac{-(x-u_0 t)^2 - y^2\left(1+u_y^2\frac{t^2}{3}\right)+(x-u_0 t)y u_y t}{4A_h t\left(1+\frac{u_y^2 t^2}{12}\right)}\right]}{4\pi A_h t \sqrt{1 + \frac{u_y^2 t^2}{12}}}. \tag{5.92}$$

Contrary to the case of a constant flow field, the maximum concentration decreases only for small time scales like $t^{-1}$ but decays like $t^{-2}$ for large times, $t \gg 1/u_y$. The decrease of the maximum concentrations for the examples of different flow patterns are shown in Fig. 5.21. The distribution is only circular in the initial phase and then becomes elliptic. The angle, $\varphi$, of the elliptic patch to the flow direction, i.e. the $x$-axis, is initially $\pi/4$ but approaches zero for large times,

$$\tan 2\varphi = -\frac{c}{a - b} = -\frac{3}{u_y t}.$$

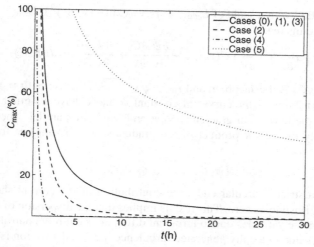

**FIGURE 5.21** Illustration of the different time behaviour of the maximum concentration for the different cases of current regimes. Cases (0), (1) and (3) correspond to no advection, constant advection and an eddy pattern, respectively. Case (2) corresponds to a shear flow, case (4) to a deformation field and case (5) to a convergent flow.

Thus, for $t \to 0$ we have $\tan 2\varphi = -\infty$, implying $\phi = \pi/4$, and for $t \to \infty$ we have $\tan 2\varphi = 0$, implying $\varphi = 0$.

### 5.3.3.2 Turbulent Diffusion in Eddies

The enhanced diffusion in a current field with shear (vorticity) may suggest that the turbulent diffusion in an eddy may be higher than in other flow fields. However, it appears that within an eddy, the turbulent diffusion is weak. In order to show this, we consider a pure eddy, which can be characterized in a collinear flow field by the shear, $u_y = -v_x = \zeta$. The flow structure is shown in Fig. 5.19. We denote this case as $C_0^{(3)}$ and have to solve the advection diffusion equation

$$\frac{\partial}{\partial t} C_0^{(3)} + \left( \zeta y \frac{\partial}{\partial x} - \zeta x \frac{\partial}{\partial y} \right) C_0^{(3)} - A_h \Delta C_0^{(3)} = 0.$$

It can easily be shown that

$$\left( \zeta y \frac{\partial}{\partial x} - \zeta x \frac{\partial}{\partial y} \right) f(x^2 + y^2) = 0,$$

and, therefore, the solution without advection terms (5.89) is also the solution for a circular eddy, i.e. $C_0^{(0)} = C_0^{(3)}$. Thus, the shear in a circular eddy field does not enhance the local diffusivities. This result applies only if the point source of the spreading material is located in the centner of the eddy. We can easily show that the result also holds true for more complex eddy structures, such as a localized eddy with a scale of $L_{eddy}$, $\psi = \psi_0 \exp(-(x^2 + y^2)/L_{eddy})$. Then the advection terms reads

$$\left( u\frac{\partial}{\partial x} + v\frac{\partial}{\partial y} \right) C(\mathbf{r}_h) = -\frac{\partial \psi}{\partial y}\frac{\partial C}{\partial x} + \frac{\partial \psi}{\partial x}\frac{\partial C}{\partial y} = J(\psi, C),$$

where $J(\psi, C)$ is the Jacobian and $r_h = \sqrt{x^2 + y^2}$. Because both $\psi$ and $C$ are circular distributions, the curves of constant $\psi$ and $C$ have a similar shape. In particular, the horizontal gradients, $\nabla_h \psi$ and $\nabla_h C$, are parallel. The Jacobian is obviously the vector product of the gradients, which vanishes for parallel vectors,

$$J(\psi, C) = \nabla_h \psi \otimes \nabla_h C = 0.$$

Thus, for a strictly circular eddy, the contribution of the current shear to the effective diffusivity is zero. This results implies that the dispersion of chemical-biological state variables due to turbulent diffusion is entirely controlled by the diffusivities due to locally generated turbulence; i.e. the dispersion is relatively weak.

### 5.3.3.3 Turbulent Diffusion in Deformation Fields

A further example of a collinear flow is the deformation field, with $u_x = -v_y = h$. The structure of the flow field is illustrated in Fig. 5.20. We denote the solution for this case as $C_0^{(4)}$, and the corresponding advection diffusion equation reads

$$\frac{\partial}{\partial t}C_0^{(4)} + \left( hx\frac{\partial}{\partial x} - hy\frac{\partial}{\partial y} \right) C_0^{(4)} - A_h \Delta C_0^{(4)} = 0.$$

Repeating the procedure used to obtain the solution for the case of $C_0^{(2)}$, see Equations (5.90) and (5.91), we find the equations

$$\frac{d}{dt}a = 2ha - A \quad \text{and} \quad \frac{d}{dt}b = -2hb + A.$$

The solutions are obviously

$$a(t) = \frac{A_h}{2h}(e^{2ht} - 1), \tag{5.93}$$

$$b(t) = \frac{A_h}{2h}(1 - e^{-2ht}). \tag{5.94}$$

Using this we obtain, after inverse Fourier transformation, the solution for $C_0^{(4)}$ in the physical space as

$$C_0^{(4)} = \frac{\exp(-\frac{x^2}{4a} - \frac{y^2}{4b})}{4\pi\sqrt{ab}},$$

or, explicitly

$$C_0^{(4)}(x, y, t) = h \frac{\exp\left[-\frac{x^2 h}{2A(e^{2ht}-1)} - \frac{y^2}{2A(1-e^{-2ht})}\right]}{4\pi A\sqrt{2[\cosh(2ht) - 1]}}.$$

For large times, $t \to \infty$, or more specifically, $t \gg h^{-1}$, it follows that

$$C_0^{(4)}(x, y, t) = h \frac{\exp\left(-\frac{x^2 h e^{-2ht}}{2A} - \frac{y^2 h}{2A}\right)}{4\pi A e^{ht}}.$$

The maximum concentration decays rapidly like $e^{-ht}$. The variances of the distribution are given $2a$ and $b$, implying that the concentration spreads rapidly in the $x$-direction, but remains relatively narrow in the $y$-direction.

### 5.3.3.4 Aggregation at Convergence Lines

As final example we consider the case of a convergence line. We choose $u = -hx$ and $v = 0$; i.e. the convergence line stretches along the $y$-axis. The corresponding horizontal flow field is illustrated in Fig. 5.22.

In this case, the vertical component $w$ is nonzero because the water must sink near the convergence line, $u_x = -w_z$. However, the vertical motion is irrelevant if we consider buoyant cells like cyanobacteria, which float at the sea surface. The distribution pattern is governed by a two-dimensional problem for the concentration, which we denote as $C_0^{(5)}$,

$$\frac{\partial}{\partial t}C_0^{(5)} - hx\frac{\partial}{\partial x}C_0^{(5)} - A_h\Delta C_0^{(5)} = 0,$$

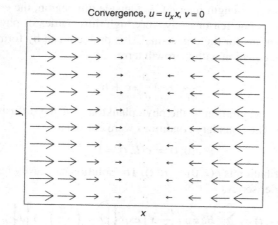

Convergence, $u = u_x x, v = 0$

**FIGURE 5.22**  A convergent flow pattern, $u(x) = u_x x$, $v = 0$. The shear $u_x$ is constant.

and we find

$$a(t) = \frac{A_h}{2h}(1 - e^{-2ht}),$$    (5.95)

$$b(t) = A_h t.$$    (5.96)

Initially, the variance in the $x$- and $y$- directions increases linearly with time, but for larger times, $ht \gg 1$, the distribution will no longer spread into the $x$-direction.

The solution in the physical space reads

$$C_0^{(5)}(x, y, t) = \frac{1}{4\pi A_h}\sqrt{\frac{2h}{(1 - e^{-2ht})t}} \exp\left[-\frac{(xe^{-ht})^2}{\frac{2A_h}{h}(1 - e^{-2ht})} - \frac{y^2}{4A_h t}\right].$$

The maximum concentration decreases like $t^{-1/2}$. Thus, the concentration aggregates along the convergence line and decays even slower, as in an eddy.

In summary, from the consideration of some examples of current patterns, we saw that the spreading of passive substances in the sea can show remarkable differences, which are related to the spatial structure of the flow, which in turn is governed by the gradients of the current components. Because many chemical and biological state variables behave like passive substances, the evolution of distribution patterns is a relevant example of physical–biological interactions.

### 5.3.4 Patchiness and Critical Scales

From satellite imagery, it is well known that biological quantities such as chlorophyll occur as patchy distributions. The phenomenon of 'patchiness' has stimulated attempts to quantify the competition of physical mixing against growth. Long before satellite images became available, a first step was made by Kierstead and Slobodkin (1953) in their classical work on critical-length scales. They started with the simplest case of a one-dimensional distribution of phytoplankton over a length interval, $L$. Outside this region, the concentration drops to zero for some reasons, say, very high diffusivities or physiologically unsuitable conditions for the organisms. This problem can by formulated by a simple diffusion equation with a growth term

$$\frac{\partial}{\partial t}c = A\frac{\partial^2}{\partial x^2}c + rc,$$

where $c$ is the concentration of the phytoplankton, $A$ is the diffusivity and $r$ is the growth rate. The boundary conditions were chosen as

$$c(0, t) = c(L, t) = 0.$$

The initial distribution is $c(x, 0) = B(x)$. The solution can easily be obtained in terms of a Fourier series,

$$c(x, t) = \sum_{n=1}^{\infty} B_n \sin\left(\frac{n\pi}{L}x\right) \exp\left\{\left[r - \left(\frac{n\pi}{L}\right)^2 A\right]t\right\},$$    (5.97)

where the $B_n$'s are the Fourier coefficients of the initial distribution,

$$B_n = \frac{2}{L} \int_0^L B(x) \sin\left(\frac{n\pi}{L}x\right) dx.$$

The time evolution in Equation (5.97) is controlled by the e-function. If $(r - (n\pi/L)^2 A) > 0$, then the population can grow. This depends on the square of the mode number, $n$. Because the smallest mode number is $n = 1$, it follows as a necessary condition for the maintenance or growth of the population, $r \geq (\pi/L)^2 A$. We can define a critical-length scale,

$$L_c = \pi\sqrt{A/r}, \tag{5.98}$$

such that a population can increase if $L > L_c$, decrease if $L < L_c$ and remain static if $L = L_c$. Only if the condition $L \geq L_c$ is fulfilled the amount of phytoplankton lost by fluxes through the boundaries, i.e.

$$-A\frac{\partial}{\partial x}c|_{x=0}, \quad \text{and} \quad -A\frac{\partial}{\partial x}c|_{x=L},$$

can be balanced by phytoplankton growth inside the region. Although this approach is somewhat oversimplified, it is useful to illuminate the interaction of the physical and biological processes in the upper ocean, and it has generated further work towards improvements through more realistic formulations of the problem. Among the most obvious biological processes missing in the Kierstead-Slobodkin approach are the variations of the growth rate due to nutrient limitation. It is implicitly assumed that there is no limitation of nutrients inside the patch, although the phytoplankton continues to grow. Top-down control, e.g. grazing pressure on the population, was not included. Moreover, the role of current patterns, which can provide retention mechanisms in eddies or can generate rapidly growing effective diffusivities due to shear, is ignored.

Another example to address the question of the stability of patches is to ask whether or not the response of phytoplankton to a suddenly generated nutrient patch will generate a stable patch or not. The nutrient patch may be created by upwelling or entrainment of nutrients into the euphotic layer. A well-defined plankton patch can only develop in response to a nutrient injection if turbulent diffusion will not disperse nutrients too much within the duration of one generation. The relevant equations are, see Equation (5.83),

$$\frac{\partial}{\partial t}P + \bar{\mathbf{v}} \cdot \nabla P - A_h\nabla_h^2 P = rP, \tag{5.99}$$

$$\frac{\partial}{\partial t}N + \bar{\mathbf{v}} \cdot \nabla N - A_h\nabla_h^2 N = -rP,$$

with an initial condition $N(x,y,0) = N_0(x,y)$. Because $P$ can only develop where the nutrients are available, it follows that the rate $r$ depends, through $N$, also on $x, y$ and $t$, i.e. $r = r(x,y,t)$. Although this is a vastly simplified problem, we cannot solve it analytically for varying advection, e.g. collinear flows. However, we can use the examples of the previous section to apply some scaling arguments. Let the initial nutrient patch, $N_0$, have a Gaussian shape, which implies in the Fourier space

$$N(k, \kappa, 0) = \exp\left(-\frac{n_x^2}{2}k^2 - \frac{n_y^2}{2}\kappa^2\right).$$

Then the evolution of the nutrient patch, without loss through uptake by phytoplankton, is governed by

$$N(k, \kappa, t) = \exp\left[-\left(\frac{n_x^2}{2} + \frac{\sigma_x^2}{2}\right)k^2 - \left(\frac{n_y^2}{2} + \frac{\sigma_y}{2}\right)\kappa^2 - \sigma_{xy}k\kappa\right].$$

The initial scale of the nutrient patch can be defined as

$$L_0 = \sqrt{\frac{n_x^2}{2} + \frac{n_y^2}{2}},$$

while the dispersion scale is defined by

$$L(t) = \sqrt{\frac{\sigma_x(t)^2}{2} + \frac{\sigma_y(t)^2}{2}}.$$

Then we can expect that a patch can develop if for $t = r^{-1}$, $L_0 \gg L(r^{-1})$, while a patch cannot develop if $L_0 \ll L(r^{-1})$. A critical scale can be defined as

$$L_c = L(r^{-1}).$$

Using examples considered in the previous section, we find, for an eddy field,

$$L_c = \sqrt{\frac{2A}{r}}.$$

This expression is, apart from a numerical factor, quite similar to the Kierstead-Slobodkin critical length (5.98). For a deformation field, it follows that

$$L_c \sim \sqrt{\frac{A_h}{2u_x}} \exp\left(\frac{u_x}{r}\right).$$

Due to the exponential, the critical scale in this case is much larger, and a plankton patch cannot develop in a rapidly dispersing deformation field. For the second example, we also have to consider that the concentration of the nutrient decreases like $\exp(-u_x t)$. Thus, we quickly reach nutrient-limiting conditions, and therefore, the generation time becomes significantly smaller than in an eddy.

The analytical examples given so far are helpful to understanding a few of the processes that generate phytoplankton patchiness. However, these examples are greatly simplified. A more realistic description can be achieved with more complex models that must be solved numerically. In coupled-circulation models and biological models, a suite of important additional factors, such as sinking, grazing and more complex circulation patterns, can be incorporated. These types of models are described in Chapter 6.

## 5.4  UPSCALING AND DOWNSCALING

Upscaling or downscaling of models corresponds to choosing a coarser or higher resolution of models. Starting with the advection-diffusion equation, we discuss how box models and one-dimensional water-column models are related to three-dimensional models. For a spatially constant state, the derivations with respect to $x$, $y$ and $z$ vanish and we have

$$\frac{\partial}{\partial t} S = \text{biological dynamics}.$$

This amounts to a box model. However, real marine systems generally show spatial variations. The transition to a box model implies a spatial-average process, which can be achieved by integration of the advection-diffusion equation over the box volume, $V$. The size of the box can be chosen to study a problem at hand. It can be an enclosed lagoon, a region of the ocean or a marginal sea. Assuming for simplicity no flux through the boundaries, we have

$$\frac{\partial}{\partial t} \langle S \rangle = \frac{1}{V} \int_V d\mathbf{r} \frac{\partial}{\partial t} S = \langle \text{biological dynamics} \rangle.$$

For a simple linear biological process, this procedure is straightforward. For example, the change of the phytoplankton concentration, $P$, due to nutrient uptake under nutrient-rich conditions is governed by the relationship

$$\frac{\partial}{\partial t} \langle P \rangle \sim r_{\max} \langle P \rangle.$$

However, if we consider phytoplankton development under limiting conditions of the nutrient, $N$, and assume Michaelis-Menten uptake kinetics, see Equation (2.10), then we have, with $N \ll k_N$,

$$\frac{\partial}{\partial t} \langle P \rangle \sim \frac{r_{\max}}{k_N} \langle NP \rangle.$$

The temporal changes of the average of the phytoplankton concentration, $\langle P \rangle$ are related to a new function, defined by the average of the product, $\langle PN \rangle$. This problem is similar to the one discussed earlier in Section 5.3.1. Traditionally the term $\langle NP \rangle$ is replaced by the product of mean values, $\langle P \rangle \langle N \rangle$. We may repeat a formal decomposition in mean values and fluctuations, $P = \langle P \rangle + P'$ and $N = \langle N \rangle + N'$, and assuming that relationships similar to the Reynolds rules apply, it follows that

$$\langle NP \rangle = \langle N \rangle \langle P \rangle + \langle N'P' \rangle.$$

In order to close the model equations, we have to express the average of the product of the fluctuations, $\langle N'P' \rangle$, somehow in terms of the mean values.

For physical quantities, such as velocity and tracers, we can argue that the turbulent fluctuations produce additional effective diffusion, which acts to decrease gradients of the slowly varying mean fields. However, the turbulent effects on the chemical-biological variables are already taken into account by the advection-diffusion equation. Because the biological rates are set by the intrinsic properties of the cells, we can expect that the fluctuation are basically due to physical effects. An example of biological fluctuations might be transient deviations of the internal nutrient pools in cells from the Redfield ratio (Geider et al., 1998). In a turbulent regime, we could argue that the turbulent fluctuations increase the contact rates, i.e. the probability for a cell to hit a nutrient molecule. Obviously, there is no *a priori* biological principle that helps to approximate the term $\langle N'P' \rangle$.

It can be assumed that for unevenly distributed nutrients and phytoplankton, their mean values are more weakly correlated than their local concentrations. The local concentration of phytoplankton, $P$, reacts directly to the local concentration of nutrients, $N$. The mean value $\langle N \rangle$ can comprise a patchy distribution that may not match a patchy distribution of the mean value of phytoplankton, $\langle P \rangle$. In a first approach, this can be reflected by a nonlinear uptake rate at small concentrations, e.g.

$$\frac{\partial}{\partial t}\langle P \rangle \sim \frac{r}{k_N}\langle NP \rangle \sim r\frac{\langle N \rangle^2}{k_N^2}\langle P \rangle.$$

Because we must require the uptake to be independent of the nutrients at high nutrient concentration, this amounts to formulas like

$$\frac{\langle N \rangle^2}{k_N^2 + \langle N \rangle^2},$$

instead of the Michaelis-Menten formula (2.10), which gives a linear dependence on $N$ for small nutrient concentrations.

With respect to physics, box models are the simplest approach because physical processes are largely removed by spatial averaging. In order to include the important vertical processes, e.g. light penetration and mixed-layer dynamics, box models can be expanded to a series of vertical stacked boxes like those considered in Chapter 3.

A better description of vertical variations can be achieved with so-called water-column models. This type of model follows from the integration of Equation (5.77) with respect to the horizontal coordinates. Let $S$ be one of the state variables, then the horizontally averaged equation follows as

$$\frac{\partial}{\partial t}\overline{S} + \int dxdy \nabla_h \cdot (\mathbf{v}_h S - A\nabla_h S) + \frac{\partial}{\partial z}\overline{wS} \qquad (5.100)$$

$$-A\frac{\partial^2}{\partial z^2}\overline{S} = \int dxdy \quad \text{(biological dynamics)},$$

where $\bar{S}(z, t) = \int dx dy S(x, y, z, t)$. For a basin with straight vertical walls, the horizontal fluxes $[\mathbf{v}_h S - A \nabla_h S]_{\text{boundaries}} = 0$ vanish, and we find

$$\frac{\partial}{\partial t}\bar{S} + + \frac{\partial}{\partial z}w\bar{S} - A\frac{\partial^2}{\partial z^2}\bar{S} = \overline{\text{biological dynamics}}. \qquad (5.101)$$

Generally, the volume of the deeper water layers decreases due to sloping bottoms; the horizontal area, $F(z)$, is a function of the depth, $z$, described by a hypsographic function. Then the equation reads

$$\frac{\partial}{\partial t}\bar{S} + \frac{1}{F}\frac{\partial}{\partial z}\left(w F \bar{S} - F A \frac{\partial}{\partial z}\bar{S}\right) = \overline{\text{biological dynamics}}. \qquad (5.102)$$

This relation is not as obvious as for straight walls, and we illustrate this case for a basin with an arbitrary bathymetry, see Fig. 5.23.

We start with the equation of conservation of mass for $S$ before we perform the Reynolds decomposition, and ignore for a moment the biological dynamics

$$\frac{\partial}{\partial t}S + \nabla \cdot \mathbf{v}S = 0.$$

Thus, in a volume between to areas $F(z + dz)$ and $F(z)$, the decrease of mass is equal to the mass flow through the upper and lower areas of the volume, see Fig. 5.23.

$$\frac{\partial}{\partial t}\int dV S = -\int_F \int dz d\mathbf{f} \mathbf{v}S = -dz \int_{F(z+dz)} d\mathbf{f}\, w\, S,$$

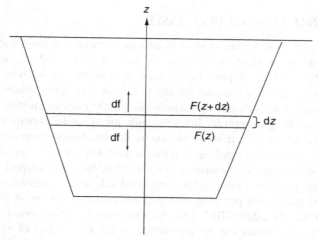

**FIGURE 5.23** Example topography to illustrate the horizontal integration of the advection-diffusion equation.

or, if we assume $dz$ to be small,

$$\frac{\partial}{\partial t}\int dVS = -dz\left[\int_{F(z+dz)} dfw(z+dz)S(z+dz) - \int_{F(z)} dfw(z)\,S(z)\right].$$

The volume integral is approximately $dV = dz[F(z+dz)+F(z)]/2 = dz(\partial/\partial z)$ $F(z)$, and we find

$$\frac{\partial}{\partial t}\int dVS = dz\frac{\partial}{\partial t}S\frac{\partial}{\partial z}F(z)$$

$$= -dz\left[w(z)S(z)\frac{\partial}{\partial z}F(z) + F(z)S(z)\frac{\partial}{\partial z}w(z)\right] + w(z)F(z)\frac{\partial}{\partial z}S(z)$$

$$= dz\frac{\partial}{\partial z}[F(z)w(z)S(z)].$$

Thus,

$$\frac{\partial}{\partial t}S + \frac{1}{F(z)}\frac{\partial}{\partial z}[F(z)w(z)S(z)] = 0.$$

If we now make the Reynolds decomposition of $w$ and $S$ and add the biological dynamics on the right-hand side, we arrive at Equation (5.102). This one-dimensional model might be helpful for some considerations; however, for many studies full three-dimensional marine ecosystem models with mesoscale resolution are required to resolve coastal shelf processes, effects of river plumes, fronts and eddies. In these examples advection of chemical and biological tracer variables is an important issue.

## 5.5 RESOLUTION OF PROCESSES

Upscaling and downscaling of models also involves process resolution, which can differ from problem to problem. In general, it is neither possible nor desirable to consider all parts of the food web in one model. For example, throughout the book we consider models that do not explicitly resolve bacteria. Their dynamic role is implicitly included in several rates to provide recycling processes and biogeochemical fluxes through, for example, mineralization and denitrification. Similarly, in some models we do not explicitly consider fish, but include the top-down control through fish in mortality rates of zooplankton.

We mainly consider biomass state variables, but with respect to stage-resolving zooplankton or mass-class structured fish, we also include abundance (number of individuals per unity volume) and properties of mean individuals. Individual-based models (IBMs), are not explicitly being discussed. However, because the individuals can be considered as the basic units of ecosystems, we wish to show the relationship among the different types of models, state variables and individuals. As mentioned in Chapter 1, the individuals have biomass and form populations that interact up, down and across the food web. This complex network can be represented by an abstract phase space; compare Fig. 1.1. Different cuts through the multidimensional parameter space can look at a set of individuals or a population or may represent functional groups by their biomass. Consequently, there are model classes that focus on individuals, populations, biomass or combinations thereof. In this section, we expand the discussion in Chapter 1 to show the common biological roots of the different model classes and their linkage to physical processes.

## 5.5.1 State Densities and Their Dynamics

All individuals in a marine ecosystem, e.g. phytoplankton cells, copepods, fish, can be represented as points in the phase space. Let us choose the mass, $m$, to characterize the biological states of the individuals, and other parameters may be added if required. The state density then is the sum of the state densities of the different species, e.g.

$$\varrho(\mathbf{r}, t, m) = \varrho^{\text{phyto}}(\mathbf{r}, t, m) + \varrho^{\text{cop}}(\mathbf{r}, t, m) + \varrho^{\text{fish}}(\mathbf{r}, t, m),$$

where

$$\varrho^{\text{phyto}}(\mathbf{r}, t, m) = \sum_{i=1}^{N^{\text{phyto}}} \delta\left[\mathbf{r} - \mathbf{r}_i(t)\right] \delta(m - m_i^{\text{phyto}}(t)), \tag{5.103}$$

$$\varrho^{\text{cop}}(\mathbf{r}, t, m) = \sum_{j=1}^{N^{\text{cop}}} \delta[\mathbf{r} - \mathbf{r}_j(t)] \delta[m - m_j(t)], \tag{5.104}$$

and a similar expression for fish. The cloud of points in the phase space changes with time in response to physical motion and biological processes such as cell division, egg laying, ingestion of cells by copepods, mortality, etc.

Assuming that the velocity $\mathbf{v}(\mathbf{r}, t)$ is the resulting vector of water motion and the individual's motion relative to the water, then the location of an individual is specified by

$$\mathbf{r}_i(t) = \int_0^t dt' \left[\mathbf{v}(\mathbf{r}_i, t') + \xi(t')\right],$$

where $\xi(t)$ represents motions due to small-scale turbulence.

The 'mass-coordinate' for a phytoplankton cell varies by a factor of about 2. After the cell division, the two new cells start from the initial mass until they reach the level of the division mass and divide again. In other words, a cell cycles along the $m$-axis between the minimum and maximum mass back and forth. For copepods and fish, the individual mass can increase by two orders of magnitude or even more. This amounts to a propagation of the individuals along the mass axis, from the eggs to the adults. The growth of the individuals is governed by an equation of the type

$$\frac{d}{dt}m_i = (g - l)m_i.$$

Underneath the point cloud, which represents all individuals of a given system, there are continuous fields of physical quantities, such as currents, temperature, salinity and chemical quantities such as nutrients. These fields control the dynamics of the individuals.

By integration of the state densities over small volumes, $\Delta V$, which can represent either an Eulerian volume element or a Lagrangian parcel of water, along mass intervals, we obtain abundance or biomass state variables, see Equations (1.5) and (1.6),

$$n(\mathbf{r}, t) = \int_{m_{\min}}^{m_{\max}} dm \, \frac{1}{\Delta V} \int_{\Delta V} d\mathbf{r} \varrho(\mathbf{r}, m, t),$$

and

$$B(\mathbf{r}, t) = \int_{m_{\min}}^{m_{\max}} dm\, m \frac{1}{\Delta V} \int_{\Delta V} d\mathbf{r} \varrho(\mathbf{r}, m, t).$$

Thus, starting with the state densities, there is a choice of whether we consider the dynamics of individuals, abundance or biomass or combinations thereof. For example, we can look at individual copepods interacting with a continuous field of phytoplankton biomass.

The dynamics of the state density involves the movement of individuals in the phase space, which is governed by the velocity fields and growth equations and creation and annihilation of individuals. This complex behaviour can be described by operators acting on state densities. One possible way to find the mathematical representation for these operators is to start with the total number of individuals,

$$N^{\text{tot}} = \int dm d\mathbf{r} \varrho,$$

where $\varrho$ is a sum of several state densities of cells, copepods and fish as given by Equation (5.103). The change of the total number over a certain time interval, $\Delta t$, can be written as

$$\frac{\Delta N}{\Delta t} = \frac{\Delta N^{\text{phyto}}}{\Delta t} + \frac{\Delta N^{\text{cop}}}{\Delta t} + \frac{\Delta N^{\text{fish}}}{\Delta t}, \tag{5.105}$$

where, for simplicity, only three large groups of individuals are considered. In the following subsections, we outline how a theoretical description of the associated processes can be developed.

## 5.5.2 Primary Production Operator

The change of the number of phytoplankton cells is controlled by primary production, i.e. creation of new cells by cell division, and by mortality and grazing. First we look at the cell division and construct a primary production operator $\Pi$. The growth of a cell is controlled by a maximum rate that depends on intrinsic cell properties and external factors like light intensity, $I^{PAR}$, temperature, $T$, and nutrients. The external factors are represented by functions of space and time. As a limiting function, we choose nutrient uptake and assume that nitrate $NO_3$ is the limiting nutrient. To avoid confusion with the numbers of individuals, we denote the nutrient distribution by $NO_3(\mathbf{r}, t)$. Then the growth rate is

$$R(\mathbf{r}, t) = r_{\max}[I^{PAR}(\mathbf{r}, t), T(\mathbf{r}, t)] \frac{NO_3(\mathbf{r}, t)}{k_{NO_3} + NO_3(\mathbf{r}, t)}, \tag{5.106}$$

and the growth equation for a single cell is

$$\frac{d}{dt} m^{phyto} = (R - l_0) m^{phyto} \quad \text{for } m^{phyto} \leq m_{division}, \tag{5.107}$$

where $l_0$ stands for loss through respiration. When the division mass is reached the cell divides and the growth equation applies to each of two new cells with the initial mass, $m_0$. In Fig. 5.24, the increase of mass is sketched for a single cell, where the mass cycles between the initial mass, $m_0$, and $m_{division}$, and for the sum of all cells.

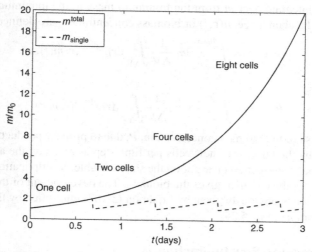

**FIGURE 5.24** The increase of mass starting with one cell (solid) and the cycling of mass of a single cell between the initial mass and the division mass (dashed). If we multiply the mass of the single cell by the number of cells after the divisions, we obtain the solid curve from the dashed curve.

The primary production operator can be written as

$$\Pi(m) = \frac{1}{\Delta t}\theta(m - m_{\text{division}}).$$

The step function is one if the mass of any individual cell reaches the division mass and zero otherwise. The application of the operator, $\Pi$, which acts only on phytoplankton, onto the state density gives

$$\Pi\varrho = \Pi\varrho^{\text{phyto}}(\mathbf{r}, t, m) = \sum_{i=1}^{N^{\text{phyto}}} \Pi(m)\delta[\mathbf{r} - \mathbf{r}_i(t)]\delta(m - m_i^{\text{phyto}})$$

$$= \frac{1}{\Delta t}\sum_{l=1}^{\Delta N^{\text{phyto}}} \delta[\mathbf{r} - \mathbf{r}_l(t)]\delta(m - m_l^{\text{phyto}}),$$

where $\Delta N^{\text{phyto}}$ is the number of cells with a mass close to the division mass, $m \simeq m_{\text{division}}$. As a result, we get the state density of all new cells generated by cell division during the time interval, $\Delta t$. The integral over $\Pi\varrho$ yields

$$\int dm d\mathbf{r}\,\Pi\varrho = \frac{\Delta N^{\text{phyto}}}{\Delta t},$$

i.e. the number of new cells created during the time interval, $\Delta t$. The creation of new cells affects the nutrient background field because, in each volume element where cells are generated, a defined amount of nutrients was removed.

If we are not interested in the specific properties of different individual cells but wish to describe the evolution of a population or its biomass of very many similar individuals, then we can introduce the population density in a parcel of water and integrate over $m$ from the minimum mass to the maximum mass of the cells. The abundance, $n(\mathbf{r}, t)$ or biomass concentration, are defined by

$$n(\mathbf{r}, t) = \int_{m_{\text{min}}}^{m_{\text{max}}} dm\,\frac{1}{\Delta V}\int_{\Delta V} d\mathbf{r}\varrho^{\text{phyto}}(\mathbf{r}, m, t),$$

and

$$P(\mathbf{r}, t) = \int_{m_{\text{min}}}^{m_{\text{max}}} dm\,m\frac{1}{\Delta V}\int_{\Delta V} d\mathbf{r}\varrho^{\text{phyto}}(\mathbf{r}, m, t).$$

The increase of the biomass concentration, $P$, due to primary production is then described by the number of new cells per time step as given by the application of the primary production operator on the state variable. Multiplication with the mean mass of the cell then gives the biomass. The development of the biomass concentration is obviously given by Equation (5.107) multiplied by the number of cells; compare Fig. 5.24.

### 5.5.3 Predator–Prey Interaction

The primary production operator is associated with continuous functions of physical and chemical quantities. Next we deal with processes where two individuals, e.g. a copepod and a food item, such as a plankton cell, interact. As an example case, we introduce a grazing operator, which removes plankton cells from the state density and enters the growth equation by increasing the mass of the copepods due to the ingested cells. A simple way of defining the grazing operator is to consider two necessary conditions: (1) The food must be available in a finite search volume around each copepod, which allows them to capture the food item, and (2) there must be a sufficient difference in size between prey and predator so that the food item can be eaten by the copepod. Because we considered only the mass coordinate in the phase space to characterize biological properties, we use the mass of the individuals as a rough measure of their size.

The grazing operator can be written as

$$\gamma = -\sum_{k=1}^{N^{\mathrm{cop}}} \gamma_k,$$

where

$$\gamma_k(\mathbf{r}, m) = -\frac{1}{\Delta t}\theta(|\mathbf{r} - \mathbf{r}_k| - \epsilon_k)\theta[(m_k - \Delta m_k) - m].$$

Here, $\mathbf{r}$ is the position of the potential food item, $\mathbf{r}_k$ is the position of a copepod and $\epsilon_k$ is the radius of the search volume, which may depend on the properties of the specific copepod, $k$, and on the level of turbulence in the volume element. The turbulence controls the contact rates, which may be increased up to a certain level of turbulence (see, e.g. Rothschild and Osborn, 1988), but for too-high turbulence, the grazing of the copepods is reduced. The difference in size of prey and predator is characterized by $\Delta m_k$. Only a food item with $m < (m_k - \Delta m_k)$ can be ingested. These properties are sketched in Fig. 5.25. Applying this operator, which acts only onto the state densities of phytoplankton, gives

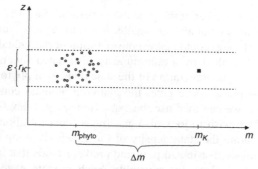

**FIGURE 5.25** Illustration of the concept of search volume and size differences relevant for the predator–prey interaction for a two-dimensional phase space. The predator has the mass, $m_k$, and the food items have a mass smaller than $(m_k - \Delta m)$.

$$\gamma \varrho = \gamma(\mathbf{r}, m) \varrho^{\text{phyto}}(\mathbf{r}, t, m)$$

$$= -\sum_{k=1}^{N^{\text{cop}}} \sum_{i=1}^{N^{\text{phyto}}} \gamma_k(\mathbf{r}_i, m^{\text{phyto}}) \delta[\mathbf{r} - \mathbf{r}_i(t)] \delta(m - m_i^{\text{phyto}})$$

$$= -\frac{1}{\Delta t} \sum_{l=1}^{\Delta N^{\text{phyto}}} \delta[\mathbf{r} - \mathbf{r}_l(t)] \delta(m - m_l^{\text{phyto}}).$$

As a result, we obtain the state density of the cells that are removed from the phase space. The integral over $\gamma \varrho$ yields

$$\int \mathrm{d}m \mathrm{d}\mathbf{r} \gamma \varrho = -\frac{\Delta N^{\text{phyto}}}{\Delta t},$$

i.e. the number cells that are removed due to grazing. The ingested food drives the growth of the copepods. Let the ingested mass of the food items per time unit, $\Delta t$, be $m^{\text{ingested}} = v m^{\text{phyto}}$, where $v$ is the number of cells. Then, we can write the grazing rate, $g$, in the growth equation (4.4) for a copepod of the mass, $m_k^{\text{cop}}$, as

$$g = \frac{v m^{\text{phyto}}}{m_k^{\text{cop}} \Delta t}.$$

Thus, the predator–prey interaction involves ingestion of prey items, growth of predators and mortality of prey. If food items are abundant, it is clear that only a certain number, $v^{\text{max}}$, of cells are ingested in the time interval, $\Delta t$. This can be related to the grazing function (2.22) for continuously distributed food as

$$g_{\text{max}} \approx \frac{1}{\Delta t} \frac{v^{\text{max}} m^{\text{phyto}}}{m_k^{\text{cop}}},$$

i.e.

$$v^{\text{max}} \approx g^{\text{max}} \Delta t \frac{m_k^{\text{cop}}}{m^{\text{phyto}}}.$$

In a similar manner, we can write predation operators for fish, which involves feeding on zooplankton but also on smaller fish (multispecies interaction).

Another model setup could be the consideration of individual copepods that ingest the food described by a continuously distributed function rather than discrete food items. The integration of the state density of phytoplankton along the $m$-axis of the phase space gives the plankton biomass concentration. For the ingestion rate, we can then use the expressions, e.g. Ivlev functions, given in Chapter 4. If we wish to look also at biomass concentrations of copepods, then we can integrate the state density of the copepods along the $m$-axis and arrive at continuously distributed prey and predator fields that interact through ingestion of phytoplankton by copepods. Such models were considered in previous chapters.

## 5.5.4 Mortality Operators

Natural mortality can be prescribed by mortality operators that act on the corresponding state densities and that can be controlled by food history (starvation) or age and may involve random-number generators to select which individual is going to die (see, e.g. Batchelder and Williams, 1995; DeAngelis and Rose, 1992). For example, we consider the operator, $\Lambda^{\text{cop}}$, which removes $M$ individuals within the time interval $\Delta t$,

$$\Lambda^{\text{cop}} \varrho^{\text{cop}} = -\sum_{k=1}^{N^{\text{cop}}} \sum_{l=1}^{M} \delta_{kl} \delta(\mathbf{r} - \mathbf{r}_k(t))\delta(m - m_k^{\text{cop}}). \tag{5.108}$$

Here, $\delta_{kl}$ is the Kronecker delta, with $\delta_{kl} = 1$ for $k = l$ and $\delta_{kl} = 0$ for $k \neq l$, which is generated with a random-number generator. The integral over the phase space gives

$$\int dm d\mathbf{r} \Lambda \varrho^{\text{cop}} = -\frac{N^{\text{cop}}}{\Delta t},$$

i.e. the number of removed copepods. If we wish to consider continuously distributed state variables, there is no longer the need to state which of the individuals will die at a certain time step. We only have to specify the percentage of the population that will be removed.

## 5.5.5 Model Classes

In the previous section, we showed in terms of the state density how, in principle, the dynamic processes and interaction among individuals can be expressed by creation or annihilation operators acting on the state densities and by adding or removing individuals. By different choices of integrals over sections of the phase space, we can relate these operations to changes of state variables. There are many different choices of state variables, as discussed in the previous chapters, ranging from bulk phytoplankton concentration to functional groups or from bulk zooplankton to stage resolved biomass concentration or abundance.

The specific setup of a model is guided by the problem at hand. Obviously, individual-based models are restricted to a manageable number of individuals. Although individuals are basic units of the ecosystem, there is only a small subset of problems that require resolving individual properties. For example, individual-based models can help to explore how genetic variations among individuals provide advantages in competition for resources. Simple individual-based models are used for particle tracking to study the trajectories of a set of cells or eggs. Individual-based models of marine systems usually look at cells or animals of one trophic levels (see, e.g. Caswell and John, 1992; Woods, 2002). It was proposed to introduce so-called super-individuals to reduce the number of individuals and, hence, to reduce the computational costs (Scheffer et al., 1995). This concept requires decisions of which part of a population should be represented by a super-individual and how the internal number of individuals within a super-individual should change over time. This implies that technical rules are needed to deal with the properties of super-individuals. If interactions of individuals of different trophic levels will be considered, the predator–prey interaction must be included explicitly. A certain number of prey individuals vanishes, while predator individuals grow, reproduce and maintain their metabolism by means of the ingested prey mass. An attempt to describe the interaction of two kinds of super-individuals, which represent copepods, *Calanus finmarchicus* and planktivorous fish, can be found in Utne et al. (2012).

We consider throughout this book state-variable models, which use biomass concentrations or numbers of very many individuals per unit of volume. In the previous subsections, we showed how state variables emerge from the individuals. An attractive feature of biomass models is that they are directly constrained by conservation of mass. This does not apply to the number of cells or animals. Thus, by looking at the total mass in the model system, we have a simple way to check the consistency of our model codes. This is important and convenient for calculations of budgets and fluxes of matter, in particular, for model integrations over many years.

# Chapter 6

# Coupled Models

## 6.1 INTRODUCTION

As mentioned in Chapter 5, there are a variety of models with different spatial resolutions ranging from box models to full coupled three-dimensional models. A box model follows from integrating the advection-diffusion equations for state variables over the spatial coordinates, while a one-dimensional water-column model follows from integrating over the horizontal dimensions. Box models are also the appropriate means of simulating tank experiments. Models with simplified physics, such as box models, are useful as workbenches for developing and testing chemical-biological model components. They are also helpful for theoretical studies with very complex chemical-biological models to explore complex relationships and to provide a ranking of the most important processes. Models with very complex biology are easily manageable as box or one-dimensional models, which apply to those cases where the physical control can be simplified. For example, the long-term behaviour of a lagoon can be studied with a box model (Fennel et al., 2002; Humborg et al., 2000; Savchuk, 2002). Basinwide budget models for the Baltic Sea can be represented by several contiguous boxes that exchange matter through their common borders, although the exchange may be difficult to parameterize in a proper way (e.g. Wulff and Stigebrandt, 1989).

Important processes are associated with clear gradients perpendicular to the coastal boundaries. An example is the process of denitrification, which is stronger in shallow areas with sandy bottoms than in deeper waters (e.g. Shum and Sundby, 1996; Voss et al., 2005). If we wish to study how mesoscale currents affect the distribution patterns of state variables, coupling to circulation models is indispensable for realistic simulations. In oceanic systems, mesoscale eddies, which are associated with some upward tilt of isopycnals, contribute significantly to the upward nutrient transports, and hence to the fluxes of material in the ocean; see McGillicuddy et al. (1998) and Oschlies and Garzon (1998). The typical scales of mesoscale current patterns such as eddies, river plumes, coastal jets and associated upwelling fronts are related to the first-mode baroclinic Rossby radius, which varies between 2 and 8 km in different parts of the Baltic Sea. In order to reduce the computational demands of coupled models, the biology may be simplified as far as reasonable. An example of coupled modelling with mesoscale resolution is given in Fennel and Neumann (1996).

Introduction to the Modelling of Marine Ecosystems. http://dx.doi.org/10.1016/B978-0-444-63363-7.00006-4

To illustrate how the model resolution affects the results of the simulation, we consider the retention of matter discharged by a river within a mesoscale buoyant plume. In the model, the discharged matter is distributed over the grid box next to the river mouth. If the resolution is reduced by, say, a factor of 3, $\Delta x \rightarrow 3\Delta x$, the discharged matter is distributed over a larger grid cell and the concentration is reduced by a factor of $1/9$. A dilution of the nutrients implies that the uptake rate, e.g. Equation (2.17), may become small and, therefore, the biological time scale increases. This may support an offshore transport of nutrients in the coarser model grid, while in nature the nutrients are consumed near the coast. Clearly, such problems are even more critical in box models, which cover whole basins by one box and ignore the near-shore retention of matter.

Ecosystem modelling requires the combination of both the biological and the physical system. The physical features of a system shape the potential range of the chemical–biological developments. Obvious examples of physical determinants are the yearly cycle and range of temperatures, typical mixing depths and vertical stratification, as well as transports and distribution of nutrients due to advection and turbulent diffusion. Inhomogeneous physical forcing provides scales that need to be resolved in models for many applications. Strong spatial gradients in the biogeochemical variables also require adequate spatial resolution. This applies to semi-enclosed systems with estuarine circulation, areas near river outlets and many coastal systems with high nutrient concentrations near the coastal boundary.

## 6.2 REGIONAL TO GLOBAL MODELS

In the recent scientific literature, there are many studies based on coupled models ranging from regional to global systems. For marginal and semi-enclosed systems, such as the Baltic Sea, the Black Sea, the Mediterranean Sea or the North Sea, we refer, e.g. to Aksnes et al. (1995), Neumann and Schernewski (2008), Eilola et al. (2011), Gregoire and Lacroix (2001) and Lacroix and Gregoire (2002). For open regional systems, such as George bank, we refer to Miller et al. (1998) and McGillicuddy et al. (2001). Oceanwide and global-scale models are described in Sarmiento et al. (1993), Slater et al. (1993), Oschlies et al. (2008) and Six and Maier-Reimer (1996).

The step from a qualitative understanding of a system to quantitative estimates and predictive capabilities can be substantially supported by the application of three-dimensional ecosystem models. A limiting factor for studies of global scale processes on longer time scales, such as effects of climate changes or anthropogenic impacts on ecosystems, is the achievable spatial resolution required for an adequate description of the relevant physical processes. For example, many global models neglect the continental shelves, which play an important role in the physical–biological interactions and many biogeochemical processes. However, new modelling techniques such as adaptable grid geometries and the breathtaking progress in computer technology will provide new opportunities.

Large international programs and projects tackle the questions of matter fluxes and modification: JGOFS (Joint Global Flux Study) and the impact of the physical environment on the higher trophic levels, GLOBEC (Global Ocean Ecosystem Dynamics). The GEOHAB program (Global Ecology and Oceanography of Harmful Algal Blooms) concentrates specifically on groups of species that are harmful from the viewpoint of man, e.g. toxic cells or noxious mass accumulations of algae. Models are integral components of these programs—in particular, three-dimensional ecosystem models, which are used to synthesize the observational findings, understand the functioning of the systems and develop predictive capability. An integration of the aims of JGOFS and GLOBEC was pursued in the international project IMBER (Integrated Marine Biogeochemistry and Ecosystem Research).

The goal of the JGOFS programme is to understand and quantify the fluxes of $CO_2$ that are related to the carbon bound in plankton and sinking to the bottom. The global synthesis of carbon fluxes can be achieved through application of coupled biological models and circulation models and requires the combination of measurements and models, calibrated by field observations. The biological model components in JGOFS developed in the 1990s can be typified by the model of Fasham et al. (1990), i.e. food web models with highly integrated state variables and process rates which describe the biogeochemical cycles under the assumption of a generally constant Redfield elementary ratio. Three-dimensional basin and global scale biogeochemical simulations with the Fasham-type model are described by Sarmiento et al. (1993), Slater et al. (1993) and Six and Maier-Reimer (1996). A critical discussion of the state of the art is given in Doney (1999).

An important aim of the modelling in GLOBEC is to understand and quantify how mesoscale advection acts on development and distributions of zooplankton. One obvious advantage of coupling biological model components and circulation models lies in the potential to estimate advective transports of cells or animals. In several situations, advection may cause changes which can be more significant than the biological changes. This allows simplified biological model components by considering eggs or larvae as passively drifting particles and simulating their trajectories. Such an approach can be very useful in recruitment studies, e.g. Hinrichsen et al. (2001). Similarly, in the context of GEOHAB studies, the trajectories of dormant cells can be a very important issue. Dormant cells are known as stages of the life cycle of a variety of phytoplankton species. Nonmotile dormant cells are transported by advection and may sink to the bottom, where they may be subject to resuspension and deposition like fine sedimentary material, until they arrive in areas of settlement, where they rest at the seabed. Such processes can be studied with the help of circulation models, which, however, should be linked to wave models in order to include the role of wind waves for the resuspension.

The modelling in IMBER continues the development of the families of models used in JGOFS and GLOBEC. In particular, development and application of models with a specific end-to-end approach that encompasses larger parts of the marine food web are the focus of the IMBER Science Plan.[1]

As a further important issue, we consider eutrophication in marginal and semi-enclosed seas, where river loads can be transported and biologically transformed near the coastal boundaries before they may enter the open sea. An important application of three-dimensional models concerns the mitigation of eutrophication by reduction of river loads. Such efforts need scientific substantiation in order to understand and quantify the possible effects of load reductions and to quantify the time scales of the reactions of ecosystems to decreased nutrient loads. This is an obvious case for experiments with numerical models.

## 6.3 CIRCULATION MODELS

Circulation models are based on the equations of motion of the geophysical fluid dynamics and the thermodynamics of sea water. The state of an oceanic system is described by the velocity, temperature and salinity at each point within it. To compute the time evolution of such a system, the full three-dimensional model equations are solved, usually in spherical coordinates, to provide a high degree of flexibility regarding regional to global applications.

In most ocean models, three important approximations are made to reduce the computational demands. The model ocean is incompressible; vertical acceleration is negligible. In the horizontal equations, small density changes can be neglected, except in the horizontal pressure gradient terms. The resulting equations, often called primitive equations (Bryan, 1969), are the Boussinesq approximations of

- *the horizontal momentum equation*

$$\frac{\partial \mathbf{u}}{\partial t} + \mathbf{u} \cdot \nabla_h \mathbf{u} + w \frac{\partial \mathbf{u}}{\partial z} + f \times \mathbf{u} = -\frac{1}{\rho_0} \nabla_h p + D_\mathbf{u} + F_\mathbf{u},$$

- *the pressure (vertical momentum) equation*

$$\rho g = \frac{\partial p}{\partial z},$$

- *the incompressibility equation*

$$\nabla_h \cdot \mathbf{u} + \frac{\partial w}{\partial z} = 0,$$

- *and the density equation*

$$\rho = \rho(T, S, p).$$

---

1. http://www.imber.info/.

The variables are $\mathbf{u}$, the vector of the horizontal velocity, $\nabla_h$, the horizontal nabla operator, $w$, the vertical velocity, $T$, the temperature, $S$, the salinity, $p$, the pressure, and $\rho$, the density. Moreover, $f$ is the Coriolis term $f = 2\Omega \sin \vartheta$, where $\Omega = 2\pi/24h$ is the Earth rotation rate and $\vartheta$ is the latitude. The diffusion and forcing terms are represented by the terms $D_\mathbf{u}$ and $F_\mathbf{u}$. An $f$-plane approximation of the linearized equations in Cartesian coordinates was used in the discussion of the coastal ocean in Chapter 5.

The three-dimensional advection-diffusion equations for tracers, $\chi$, e.g. temperature and salinity, have the form

$$\frac{\partial \chi}{\partial t} + \mathbf{u} \cdot \nabla_h \chi + w \frac{\partial \chi}{\partial z} = D_\chi + F_\chi. \qquad (6.1)$$

This equation represents the interface between the chemical-biological model components and the physical part of the model system. We note that (5.77) is a simplified version of Equation (6.1), written in Cartesian coordinates and with constant diffusivity.

Several families of ocean circulation models were developed in the past decades, and model development and refinements are ongoing. The models are usually available and are subject to regular upgrades. Upgrades are required both to improve parameterizations and to ensure that the codes can be run on advanced computer facilities. Technical differences between models are related to the vertical resolution and types of numerical grids. We mention only a few examples of models and refer to the literature for a more complete list, e.g. Haidvogel and Beckmann (1998). The $z$-coordinate models use the simplest and most intuitive vertical arrangement of grid points along the $z$-axis. Examples are the GFDL model and its descendants known as Modular Ocean Model (MOM), see (Bryan, 1969; Pacanowski and Griffies, 2000; Pacanowski et al., 1990), or the Hamburg Ocean Shelf Model (HAMSOM), see (Backhaus, 1985). The so-called $\sigma$-coordinates adjust the vertical spacing of grid points to the depths, e.g. the Princeton Ocean Model (POM) see Blumberg and Mellor (1987),

$$\sigma = \frac{z + \eta}{H(x, y) + \eta}.$$

This approach implies a high vertical resolution in shallow areas and a low resolution of deeper parts of the model ocean. Similarly, the $S$-Coordinate Primitive Equation Model (SPEM) solves the hydrostatic primitive equations using a generalized bottom-following coordinate in the vertical, which allows an effective incorporation of strong variations in bathymetry; see (Haidvogel et al., 1991). The $S$-coordinate is defined as a function of the $\sigma$-coordinate,

$$S = S\left(\frac{z + \eta}{H(x, y) + \eta}\right).$$

Such a model is the $S$-Coordinate Rutgers University Model (SCRUM), which solves the free surface, hydrostatic, primitive equations over variable topography using stretched terrain-following coordinates in the vertical and orthogonal curvilinear coordinates in the horizontal, as described in Song and Haidvogel (1994). An upgraded version is the Regional Ocean Model System (ROMS), see (Haidvogel et al., 2000).

Another type of model includes the isopycnic coordinate models, e.g. the Miami Isopycnic Coordinate Model (MICOM) (Bleck et al., 1991), where density surfaces are used as coordinates. It is not easy to say which model is the best one because all models have advantages and disadvantages and it depends on the specific task to which the model should be applied. However, rapid development in computer technologies allows better spatial resolution and longer simulation periods and improves the realism of simulations for all types of models.

Because the models are based on first principles, which are formulated in the primitive equations, the main problems in ocean modelling are the parameterization of the unresolved subgrid processes, as well as the quality of data sets needed for the initialization and forcing of the models.

Among the choices of circulation models that can be used for a coupled physical-biological model system, we use the MOM throughout this book. Governing physical principles for MOM are introduced in detail by Griffies (2004). In particular, we use the version MOM 3.1 (Griffies et al., 2001; Pacanowski and Griffies, 2000; Pacanowski et al., 1990). However, MOM has recently become available as an upgraded version 5 from www.gfdl.noaa. gov/mom-ocean-model. The model includes an explicit free surface, which is important for a semi-enclosed sea like the Baltic, and several other components, such as open boundary conditions to the North Sea, a consistent treatment of freshwater fluxes due to river discharge and precipitation and evaporation, as well as a thermodynamic ice model. In addition to the traditional UNESCO equation of state, the new standard for sea water, TEOS-10 (IOC et al., 2010), is available in MOM.

## 6.4 BALTIC SEA

In this and the next chapter, we use the Baltic Sea as an example system to apply and discuss coupled models. The Baltic Sea is a particularly well-observed marine environment. Due to many field studies and a well-designed, regular, long-term monitoring programme, there are unique data sets, which are invaluable for initialization and skill assessment of the models. The Baltic Sea is one of the largest semi-enclosed brackish water systems. The exchange with the North Sea is restricted by shallow and narrow channels, the Danish straits. The relatively shallow Baltic Sea has a mean depth of about 50 m and consists of several basins separated by sills. In its central part, the Gotland basin, the depth is about 270 m. The bathymetry of the Baltic Sea is characterized by these basins of different sizes and depths and the connecting sill areas and channels, see Fig. 6.1.

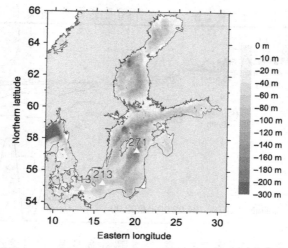

**FIGURE 6.1**   Bathymetry of the Baltic Sea. The triangles refer to monitoring stations: station 113 in the central Arkona Sea, station 213 in the central Bornholm Sea and station 271 in the central Gotland Sea.

Located between 54 and 66°N, the Baltic stretches from subarctic to temperate climatic conditions and is exposed to highly variable winds and many storm events. The northern parts of the Baltic are regularly ice covered during the winter. The Baltic Sea has a positive water balance. While the annual average of precipitation slightly exceeds the evaporation, there is a substantial freshwater input by many rivers distributed more or less evenly around the Baltic.

The freshwater surplus drives a general estuarine circulation with outflowing brackish surface water and an inflow of saline water near the bottom. Consequently, the vertical stratification of the Baltic Sea is characterized by a strong permanent halocline that separates the bottom water from the upper layers (Fig. 6.2, upper panel).

The associated high vertical stability implies that the waters below the halocline can only be ventilated horizontally by near-bottom flows of dense waters formed in the transition area between Kattegatt and Baltic Proper (the central Baltic Sea). There is also a pronounced horizontal gradient, with decreasing salinity for increasing distance from the entrance to the Baltic. The halocline is closely connected with gradients in the oxygen distribution. Sinking organic material accumulates below the halocline, and the subsequent microbial decomposition, which consumes oxygen in the near-bottom layer, causes depletion of oxygen. Thus, the occurrence of hydrogen sulphide is a common feature in the deep waters (Fig. 6.2, lower panel).

The permanent halocline sets a physical condition for the Baltic Sea ecosystem on long time scales. Above the halocline, a seasonal thermocline

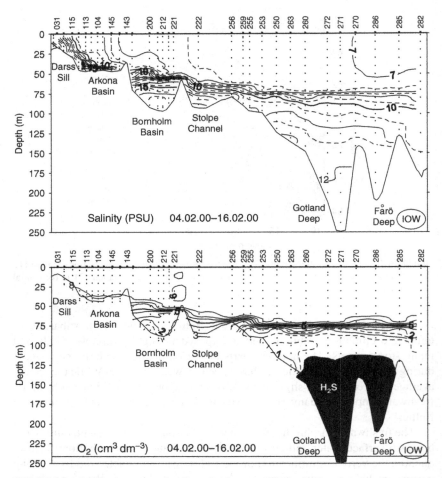

**FIGURE 6.2** Salinity (upper panel) and oxygen (lower panel) distribution along a section through the Baltic Sea.

develops in spring and vanishes in the autumn (Figs. 6.3 and 6.4). The vertical stratification and the different depths of the basins imply different first-mode Rossby radii, which are for constant Brunt-Väisälä frequency defined by (5.33). For the strongly varying Brunt-Väisälä frequencies, with local maxima at the seasonally thermocline and the permanent halocline, the eigenvalue problem (5.22) must be solved numerically in order to determine the Rossby radii. This was done in Fennel et al. (1991), where it was shown that typical values of $R_1$ vary from 5 km (winter) to 8 km (summer) in the central Baltic (Baltic Proper) and from 2 km (winter) to 4 km (summer) in the western Baltic. General overviews of the oceanography of Baltic Sea are given, e.g. in Kullenberg (1981) and Stigebrandt (2001). The dynamics of the thermocline

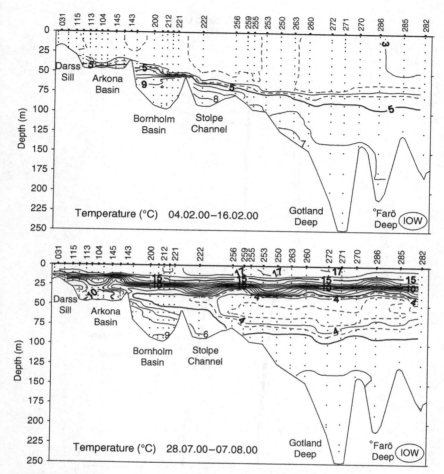

**FIGURE 6.3** Winter temperature (upper panel) and summer temperature (lower panel) distribution along a section through the Baltic Sea.

control the annual biological cycles of the ecosystem. The physical signatures are basically established as responses of the Baltic Sea to atmospheric forcing. The ecosystem is controlled by physical processes and forced by external input of nutrients due to river loads and atmospheric deposition. Interannual changes in the forcing cause variations in the physical responses and affect, therefore, the chemical-biological features of the system. The space scales of the thermocline and halocline are basinwide, while the time scales are seasonally and longer. At shorter time periods, say, a few days, which correspond to the typical time scales of weather patterns, there are a variety of mesoscale currents, such as eddies, coastal jets and associated up- and downwelling patterns. The

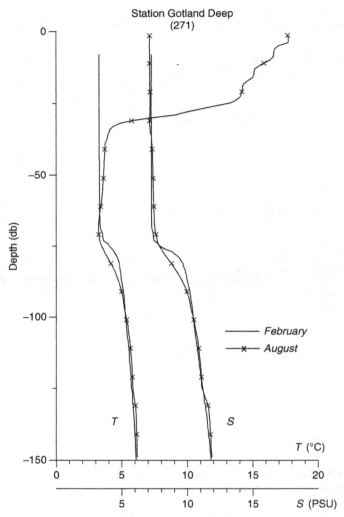

**FIGURE 6.4**    Vertical profiles of temperature and salinity at the central station of the eastern Gotland sea.

scales of the mesoscale structures are typically associated with the baroclinic Rossby radius, $R_1$. Further important mesoscale flow patterns are river plumes, which are strongly affected by alongshore winds and which carry the nutrient loads of the watersheds. The mesoscale features affect the distribution patterns of nutrients and plankton as chemical and biological variables are dispersed or concentrated by the combined effect of advection and turbulent diffusion.

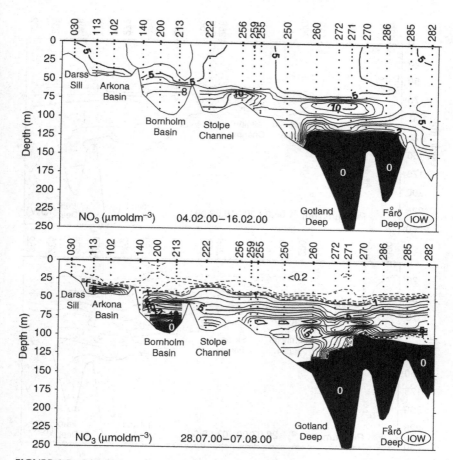

**FIGURE 6.5** Distributions of winter (upper panel) and summer (lower panel) concentrations of nitrate along a section through the Baltic Sea.

The distributions of nitrate and phosphate are depicted in Figs. 6.5 and 6.6 along the same section as the temperature and salinity in Figs. 6.2 and 6.3. Owing to denitrification, the nitrate concentration in the deeper waters are zero in the areas where hydrogen sulphide occurs. In the near-surface waters, we find the typical high winter concentrations, while in the summertime nitrogen is depleted. Similarly, higher winter concentrations of phosphate are found in the surface waters, while in summer the phosphate is at the detection limit. In the deeper part, the phosphate concentration is higher, in particular in the anaerobic parts, due to the release of phosphate from iron-phosphate complexes in the presence of hydrogen sulphide. This implies that substantial inputs of phosphate into the upper layer are possible through mixing events.

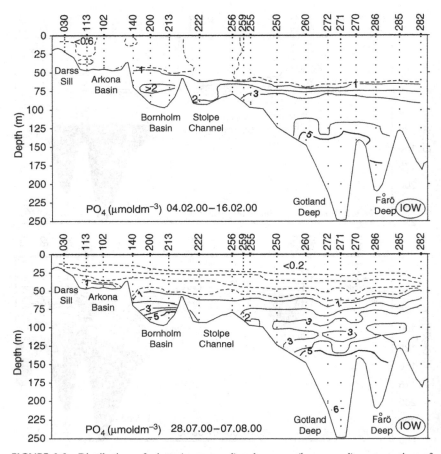

**FIGURE 6.6** Distributions of winter (upper panel) and summer (lower panel) concentrations of phosphate along a section through the Baltic Sea.

Between 1970 and 1985, a significant increase of nutrients in the Baltic Sea was observed. Time series of nitrate and phosphorous are shown in Figs. 6.7 and 6.8. The surface concentrations of nitrate and phosphate in the winter increased threefold; see Nehring and Matthäus (1991) and Matthäus (1995). This eutrophication was obviously the response of the marine system to enhanced river loads, which resulted from an excessive usage of fertilizers in agriculture within the drainage area of the Baltic Sea. For example, the nitrogen loads increased by a factor of 2 loads by factor of 4 between 1950 and 1980 (Larsson et al., 1985; Nausch et al., 2008).

An important and scientifically interesting set of questions concerns the response of the system to these loads. Is it possible to reverse the effects of the

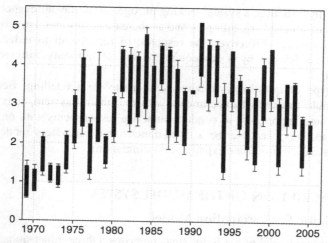

**FIGURE 6.7** Observed concentrations of nitrate in the winter surface water of the central Baltic, at station 271, indicated in Fig. 6.1. (Source: Data bank, IOW.) The filled bars show plus/minus one standard deviation; the thin error bars indicate maximum and minimum values.

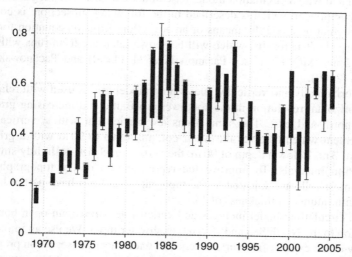

**FIGURE 6.8** Observed concentrations of phosphate in the winter surface water of the central Baltic, at station 271, indicated in Fig. 6.1. (Source: Data bank IOW.) The filled bars show plus/minus one standard deviation; the thin error bars indicate maximum and minimum values.

eutrophication by a reduction of river loads? What are the time scales at which the system reacts? Owing to the nonlinearity and complexity of the food web, and with several pathways of nutrients, we cannot expect a linear response of the marine system to the reduced inputs. Some processes can even counteract load reduction. In particular, a system lacking nitrogen could favour cyanobacteria, which can fix atmospheric nitrogen, and an increase of unwanted cyanobacteria blooms may occur. Moreover, the response of the system to reduced river loads can be delayed by a subsequent release of nutrients buffered in the sediments.

This type of question is obviously an issue for modelling. Because it is impossible to conduct experiments with the natural system, it is a great challenge and opportunity to conduct numerical experiments with models. In the next sections, we describe a three-dimensional ecosystem model of the Baltic Sea and discuss some experimental simulations.

## 6.5  DESCRIPTION OF THE MODEL SYSTEM

### 6.5.1  Baltic Sea Circulation Model

The ecosystem model of the Baltic Sea comprises a three-dimensional circulation model and an embedded biogeochemical model. The evolution of currents, temperature and salinity in response to atmospheric forcing and river discharges is calculated with a circulation model based on the MOM. The biogeochemical model component, which is described in the following subsection, is coupled to the circulation model by means of an advection-diffusion equation for each state variable. The results, which will be shown later, are from runs with both MOM 2 and MOM 3.1; see Pacanowski et al. (1990) and Pacanowski and Griffies (2000).

A horizontally and vertically telescoping model grid is used with relatively high horizontal resolution in the southwestern Baltic and increasing grid size to the north and east. The simulations are carried out with a vertical grid arrangement which resolves the water column down to 90 m with a grid box height of 3 m. Below a depth of 90 m, the vertical grid size is slightly stretched up to 6-m intervals. To improve the representation of the topography, the bottommost cells are adjusted to the topography and can have partial heights with a minimum cell thickness of 1 m.

The circulation model includes an explicit free surface, an open boundary condition to the North Sea and riverine freshwater input. We use an implementation with an explicit free-surface scheme, with tracer conservation properties as described in Griffies et al. (2001). Horizontal velocity components, tracers and sea-level elevation are prognostic variables, while the vertical velocity is diagnosed from the divergence of the horizontal flow. The timesteps are 300 s for baroclinic currents and tracers and 15 s for the barotropic mode. The pressure is calculated from the density distribution. The UNESCO equation of state is used to calculate the densities from temperature and salinity. A thermodynamic ice model simulates the ice cover.

Horizontal subgrid processes are parameterized by the Smagorinsky scheme (Smagorinsky, 1993) where the horizontal eddy viscosity is related to the deformation rate of the flow field and variations in horizontal grid spacing are taken into account.

For the vertical subgrid mixing, the Pacanowski–Philander scheme is used, where vertical mixing is amplified by shear and reduced by stratification (Pacanowski and Philander, 1981). In the two uppermost layers, a constant viscosity is prescribed to simulate wind-induced mixing. The viscosities chosen are as small as possible, but large enough to maintain numerical stability of the model. The numerical values of the parameter choices are listed in Table 6.1. The topography of the model system, driving atmospheric forces and initialization data to provide initial conditions are required in order to run the model. The model topography is adapted from the bathymetry of the whole Baltic Sea compiled by Seifert and Kayser (1995), see Fig. 6.1. As the external force, we can use available products such as from the ERA project.[2] Data sets of river runoff, nutrient loads and atmospheric nutrient deposition are provided by the Baltic Environmental Database[3] of the University of Stockholm. The initial distribution of temperature and salinity fields can be taken from climatological data, e.g. Janssen et al. (1999).

## 6.5.2 The Biogeochemical Model ERGOM

The biogeochemical model ERGOM (Ecological Regional Ocean Model) used in this chapter has several features similar to the model described in Chapter 3. This a biogeochemical model treats aspects of sedimentary material explicitly.

**TABLE 6.1** Viscosity and Diffusivity

| Variable | Symbol | Value |
|---|---|---|
| Background horizontal viscosity | $A_0^m$ | $10^8 \, cm^2 \, s^{-1}$ |
| Background horizontal diffusivity | $A_0^t$ | $10^7 \, cm^2 \, s^{-1}$ |
| Background vertical viscosity | $\kappa_0^m$ | $5.0 \, cm^2 \, s^{-1}$ |
| Background vertical diffusivity | $\kappa_0^t$ | $0.5 \, cm^2 \, s^{-1}$ |
| Maximum diffusivity | $\kappa_{max}^t$ | $20 \, cm^2 \, s^{-1}$ |
| Vertical viscosity scale | $\kappa_{max}^u$ | $50 \, cm^2 \, s^{-1}$ |
| Wind mixing (upper layers) | $\kappa^u$ | $\kappa_{max}^u$ |

Choice of parameters for the Smagorinsky and Pacanowski–Philander subgrid parameterization.

---

2. http://www.ecmwf.int/research/era.
3. http://nest.su.se/models/bed.htm.

A first description of the model was given in Neumann (2000). The most recent version of the model can be find at the website, www.ergom.net. We briefly summarize the model equations and discuss the involved process descriptions before we consider the simulations with the model. The model includes nine state variables and generally describes the nitrogen cycle. The nutrient state variables are dissolved ammonium (A), nitrate (N) and phosphate (PO), (Table 6.2).

Primary production is provided by three functional phytoplankton groups: diatoms, flagellates and cyanobacteria. The state variables of three functional groups of phytoplankton, or autotrophs, are $P_1$ for diatoms, $P_2$ for flagellates and $P_3$ for cyanobacteria. The autotrophs take up the nutrients, i.e. ammonium, nitrate and phosphate, and are grazed by the model zooplankton. Diatoms represent larger cells that grow quickly in nutrient-rich conditions and sink to the bottom. The functional group of model flagellates represents smaller, neutrally buoyant cells, which have an advantage at lower nutrient concentrations during summer conditions. The cyanobacteria are able to fix atmospheric nitrogen and are, therefore, only limited by the phosphate model. Due to their ability to fix nitrogen, the cyanobacteria constitute a nitrogen source for the system.

In the state variable zooplankton, $Z$, all species and stages are lumped together into one bulk heterotroph formulation. The dynamically developing bulk zooplankton variable provides grazing pressure on the phytoplankton

**TABLE 6.2** State Variables of the Biogeochemical Model

| Symbol | Variable | Dimension |
|--------|----------|-----------|
| $P_1$ | Diatoms | – |
| $P_2$ | Flagellates | – |
| $P_3$ | Blue-greens | – |
| $Z$ | Zooplankton | – |
| $D$ | Detritus | – |
| $S$ | Sediment-detritus | m |
| $A$ | Ammonium, $NH_4$ | – |
| $N$ | Nitrate, $NO_3$ | – |
| PO | Phosphate, $PO_4$ | – |
| Ox | Oxygen | – |

All variables are normalized by reference concentrations and are, therefore nondimensional, except the sediment-detritus, $S$. Normalization constant for the nitrogen and phosphorus variables is $4.5\,\mathrm{mmol\,m^{-3}}$ and $375\,\mathrm{mmol\,m^{-3}}$ for oxygen.

model. The model description of zooplankton dynamics is not designed for zooplankton studies. Coupled three-dimensional models that focus explicitly on copepods are considered in Chapter 7. Dead organic material accumulates in the detritus state variables, $D$ and $S$, where $D$ refers to material in the water column while $S$ describes sedimentary material that accumulates at the seabed. The detritus sinks and a substantial amount will reach the bottom, where it is passed to the sediment-detritus, which can be buried in the sediment or resuspended, depending on near-bottom currents. The detritus will be mineralized into dissolved ammonium and phosphate.

The consumption and production of oxygen is linked to biogeochemical processes. At low oxygen levels, nitrate will be denitrified in the water column and sediments. We also take into account the state variable oxygen, which includes hydrogen sulphate as negative oxygen equivalent. The system model is sketched in Fig. 6.9 as a conceptual diagram.

As in Chapter 3, we use nondimensional state variables; i.e. the concentrations are divided by $N_{norm} = 4.5\,\text{mmol}\,\text{m}^{-3}$. Only the sediment detritus, $S$, which is given in mass per area, carries (after normalization) the dimension 'length' in 'm'. With the exception of the uptake or growth rates, $R_n$, and the ingestion or grazing rates, $G_n$, we describe the processes by terms like $L_{XY}$ or $l_{XY}$, which are read as 'loss of X to Y'. The lowercase $l$ refers to constant rates, while the uppercase $L$ involves functions of parameters or state variables.

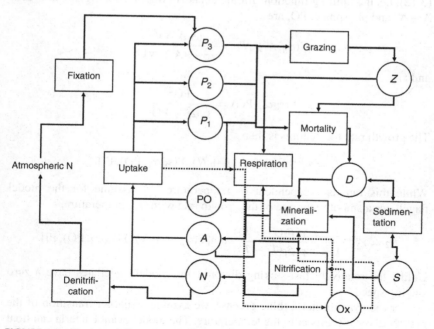

**FIGURE 6.9**  Conceptual diagram of the biogeochemical model.

We denote the sum of the concentration of all phytoplankton groups as $P_{\text{sum}} = \sum_{n=1}^{3} P_n$. For simplicity, we assume that the same primary production-light relationship, expressed by Steele's formula (Steele, 1962), applies to all functional groups,

$$\text{PI} = \frac{I^{\text{PAR}}}{I_{\text{opt}}} \exp\left(1 - \frac{I^{\text{PAR}}}{I_{\text{opt}}}\right).$$

The shape of the formula is illustrated in Fig. 5.2. The photosynthetically active radiation, $I^{\text{PAR}}$, is given by

$$I^{\text{PAR}} = \frac{I_0}{2} \exp\left[k_{\text{w}}z - k_{\text{c}} \int_z^0 \mathrm{d}z' N_{\text{norm}}(P_{\text{sum}} + Z + D)\right],$$

with $I_0$ being the solar radiation at the sea surface, which depends on the latitude, time and cloudiness and which enters the coupled model system as a prescribed forcing field. The light optimum, $I_{\text{opt}}$ is given by (5.6) and $k_{\text{w}}$ and $k_{\text{c}}$ are light attenuation constants of water and planktonic particles (self-shading), respectively. Moreover, we assume a constant respiration rate, $l_{\text{PA}}$, i.e. loss of phytoplankton to ammonium, and a constant mortality rate, $l_{\text{PD}}$, i.e. loss of plankton to detritus, for all groups.

Differences in the physiological rates of the functional groups occur in the uptake efficiencies, sinking speeds and the grazing preferences of the zooplankton. In order to prescribe the limiting properties, we use the notation (3.12); i.e. the limiting functions for the uptake of dissolved inorganic nitrogen, $A + N$, and phosphate, PO, are

$$Y(\alpha_n, A + N) = \frac{(A + N)^2}{\alpha_n^2 + (A + N)^2},$$

and

$$Y(s_{\text{R}}\alpha_n, \text{PO}) = \frac{\text{PO}^2}{s_{\text{R}}^2 \alpha_n^2 + \text{PO}^2}.$$

The growth rate for diatoms is then

$$R_1 = r_1^{\max} \min[Y(\alpha_1, A + N), Y(s_{\text{R}}\alpha_1, \text{PO}), \text{PI}].$$

While this rate is independent of temperature, we assume for the model flagellates a smooth increase of the growth rate with the temperature,

$$R_2 = r_2^{\max} \frac{2 + \exp[\beta_{\text{f}}(T_{\text{f}} - T)]}{1 + \exp[\beta_{\text{f}}(T_{\text{f}} - T)]} \min[Y(\alpha_2, A + N), Y(s_{\text{R}}\alpha_2, \text{PO}), \text{PI}].$$

Unlike the relatively fast-sinking diatoms, the model flagellates have a zero sinking speed.

For the cyanobacteria (blue-greens), we assume a sharper transition of the growth rate with respect to the temperature. The model cyanobacteria can float upward towards the sea surface at a small vertical velocity, and we assume that the cyanobacteria are not limited by dissolved inorganic nitrogen,

$$R_3 = r_3^{\max} \frac{1}{1 + \exp[\beta_{bg}(T_{bg} - T)]} \min[Y(s_R \alpha_3, PO), PI].$$

The structure of this expression was motivated by the observational findings summarized in Wasmund (1997). The numerical values of the involved parameters are listed in Table 6.3.

The grazing by bulk zooplankton is assumed to depend on temperature and has a lower preference for the ingestion of cyanobacteria,

$$G_n = g_n^{\max} \left[ 1 + \frac{T^2}{T_{opt}^2} \exp\left( 1 - \frac{2T}{T_{opt}} \right) \right] [1 - \exp(-I_{Ivlev} P_{sum}^2)],$$

where $g_1^{\max} = g_2^{\max}$, $g_3^{\max} = \frac{1}{2} g_1^{\max}$, $T_{opt}$ is an optimum temperature and $I_{Ivlev}$ is a modified Ivlev constant.

The phytoplankton equations are explicitly:

$$\frac{d}{dt} P_1 = R_1 P_1 - l_{PA} P_1 - l_{PD} P_1 - G_1 \frac{P_1}{P_{sum}} Z - w_1^{sink} \frac{\partial}{\partial z} P_1, \qquad (6.2)$$

$$\frac{d}{dt} P_2 = R_2 P_2 - l_{PA} P_2 - l_{PD} P_2 - G_2 \frac{P_2}{P_{sum}} Z, \qquad (6.3)$$

**TABLE 6.3** Phytoplankton and Zooplankton Rates

| Parameter | Diatoms | Flagellates | Blue-greens |
|---|---|---|---|
| Growth rate | $r_1^{\max} = 1\,d^{-1}$ | $r_2^{\max} = 0.7\,d^{-1}$ | $r_3^{\max} = 0.5\,d^{-1}$ |
| Half-saturation | $\alpha_1 = 0.3$ | $\alpha_2 = 0.15$ | $\alpha_3 = 0.5$ |
| Sinking speed | $w_1 = -50\,cm\,d^{-1}$ | $w_2 = 0$ | $w_3 = 10\,cm\,d^{-1}$ |
| Respiration | $l_{PA} = 0.01\,d^{-1}$ | $l_{PA} = 0.01\,d^{-1}$ | $l_{PA} = 0.01\,d^{-1}$ |
| Mortality | $l_{PD} = 0.02\,d^{-1}$ | $l_{PD} = 0.02\,d^{-1}$ | $l_{PD} = 0.02\,d^{-1}$ |
| Grazing | $g_1^{\max} = 0.5\,d^{-1}$ | $g_2^{\max} = 0.5\,d^{-1}$ | $g_3^{\max} = 0.25\,d^{-1}$ |
| Temperature | – | $T_f = 10\,°C$ | $T_{bg} = 16\,°C$ |
| Parameters | | $\beta_f = 1\,°C^{-1}$ | $\beta_{bg} = 1\,°C^{-1}$ |
| Zooplankton Parameters | | | |
| Exudation | $l_{ZA} = 0.3\,d^{-1}$ | | |
| Mortality | $l_{ZD} = 0.6\,d^{-1}$ | | |
| Ivlev constant | $I_{Ivlev} = 1.2$ | | |
| T-optimum | $T_{opt} = 20\,°C$ | | |

$$\frac{d}{dt}P_3 = R_3 P_3 - l_{PA} P_3 - l_{PD} P_3 - G_3 \frac{P_3}{P_{sum}} Z - w_3^{sink} \frac{\partial}{\partial z} P_3. \qquad (6.4)$$

For the zooplankton model, we have the equation

$$\frac{d}{dt}Z = \frac{G_1 P_1 + G_2 P_2 + G_3 P_3}{P_{sum}} Z - l_{ZA} Z^2 - l_{ZD} Z^2,$$

where constant rates for the mortality, $l_{ZD}$ (loss of zooplankton to detritus), and excretion, $l_{ZA}$ (loss of zooplankton to ammonium), has been assumed. Note that quadratic expressions for the zooplankton are introduced here (Neumann, 2000). All phytoplankton and zooplankton loss terms implicitly involve a background value, $P_0$, $Z_0$, as introduced in Chapter 3, to prevent the stationary state $P, Z = 0$. We skipped it here to keep the equations more clearly represented.

The dead algae and zooplankton are sources of detritus, which is dynamically governed by the equation

$$\frac{d}{dt}D = l_{PD} P_{sum} + l_{ZD} Z^2 - L_{DA} D - w_D^{sink} \frac{\partial}{\partial z} D$$

$$+ (L_{SD} S - L_{DS} D)\delta_{k,k_{bot}},$$

where the mineralization of detritus depends on the temperature as $L_{DA} = l_{DA}[1 + \beta_{DA} Y(T_{DA}, T)]$, with $l_{DA}$ being the mineralization constant at zero degrees Celsius and $T_{DA}$ a temperature parameter. The sinking speed in the water column is $w_D$, $L_{DS}$ is the sedimentation rate (loss of detritus to sediment) and $L_{SD}$ prescribes the rate of resuspension (loss of sediment to detritus). The Kronecker delta is defined by

$$\delta_{i,j} = \begin{cases} 1 & i = j \\ 0 & i \neq j, \end{cases}$$

Thus, $\delta_{k,k_{bot}}$ indicates that this term exists only in the bottom layer of the model with index $k = k_{bot}$ and thickness $\Delta H_{bot}$. The processes of sedimentation and resuspension depend on the near-bottom stress, $\tau$, as

$$L_{SD} = L_{SD}\theta(\tau - \tau^{crit}),$$

and

$$L_{DS} = L_{DS}\theta(\tau^{crit} - \tau),$$

where $\tau^{crit}$ is the critical stress. For the sediment-detritus, $S$, we assume the dynamical equation

$$\frac{d}{dt}S = l_{DS} D \delta_{k,k_{bot}} - l_{SD} S - L_{SA} S - L_S^{denit} S, \qquad (6.5)$$

where

$$L_{SA} = l_{SA} \exp(\beta_{SA} T)[\theta(Ox) + 0.2\theta(-Ox)]$$

prescribes the rate at which ammonium is released from the sediment through mineralization under aerobic and anaerobic conditions, see Stigebrandt and Wulff (1987), while $L_S^{denit} = \theta(Ox)L_{SA}$ prescribes the rate at which nitrogen is lost by subsequent nitrification and denitrification. Note that the oxygen variable refers to the model layer nearest to the bottom.

The dynamics of the nutrients, driven by metabolic losses and biogeochemical processes, are governed by separate equations for ammonium, $A$, nitrate, $N$, and phosphate, PO,

$$\frac{d}{dt}A = -\frac{A}{A+N}(R_1P_1 + R_2P_2) + l_{PA}P_{sum} + l_{ZA}Z^2 \tag{6.6}$$

$$+L_{DA}D - L_{AN}A + L_{SA}\frac{S}{\Delta H_{bot}}\delta_{k,k_{bot}}$$

$$+\frac{A^{flux}}{H_0}\delta_{k,k_0},$$

$$\frac{d}{dt}N = -\frac{N}{A+N}(R_1P_1 + R_2P_2) + L_{AN}A + \frac{N^{flux}}{H_0}\delta_{k,k_0} \tag{6.7}$$

$$-s_1\left(L_{DA}D + L_{SA}\frac{S}{\Delta H_{bot}}\delta_{k,k_{bot}}\right)\theta(-Ox)\theta(N),$$

$$\frac{d}{dt}PO = s_R\left\{-\sum_{n=1}^{3}R_nP_n + l_{PA}P_{sum} + l_{ZA}Z^2 + L_{DA}D \right. \tag{6.8}$$

$$\left. +L_{SA}[1 - p_1\theta(Ox)Y(p_2, Ox)]\frac{S}{\Delta H_{bot}}\delta_{k,k_{bot}}\right\}$$

$$+\frac{PO^{flux}}{H_0}\delta_{k,k_0}.$$

The nutrient uptake of diatoms and flagellates involves a preference for ammonium by means of the ratios $\frac{A}{A+N}$ and $\frac{N}{A+N}$. The uptake of phosphate is scaled by the stoichiometric rate, $s_R$, which is the Redfield ratio of phosphorus to nitrogen. All nutrient equations contain prescribed atmospheric fluxes as source terms, which are acting in the uppermost model layer with the index, $k_0$. This is indicated by the Kronecker delta, $\delta_{k,k_0}$. The recycling of nitrogen due to respiration and mineralization of detritus follows the pathway of ammonium, which is transferred to nitrate through the process of nitrification. The nitrification rate, $L_{AN}$, i.e. loss of ammonium to nitrate, is controlled by the concentration of oxygen and depends on the temperature according to

$$L_{AN} = \theta(Ox)\frac{Ox}{O_{AN} + Ox}l_{AN}\exp(\beta_{AN}T), \tag{6.9}$$

where $l_{AN}$, $O_{AN}$ and $\beta_{AN}$ are constant parameters. This approach is a combination of (3.31) and (3.13).

The process of denitrification, means that under anaerobic conditions nitrate is reduced to molecular nitrogen that can leave the system. The released oxygen is used to oxidize detritus. This is taken into account by the last term in the equation for N, i.e. (6.7). The stoichiometric ratio, $s_1$, is related to the process of nitrate reduction and detritus oxidation. Moreover, the release of phosphate from the sediment under aerobic and anaerobic conditions is described in the equation for PO (6.8), with the aid of the limiting function $Y$.

Finally, we need a dynamic equation for the state variable of oxygen, Ox, which requires several stoichiometric ratios.

$$\frac{d}{dt}Ox = \frac{N_{norm}}{O_{norm}}\left\{ \frac{s_2 A + s_3 N}{A + N}\sum_{n=1}^{3} R_n P_n - s_2(l_{PA}P_{sum} + l_{ZA}Z^2) \right.$$

$$-s_4 L_{AN}A - s_2[\theta(Ox) + \theta(-Ox)\theta(-N)]$$

$$\left( L_{DA}D + L_{SA}\frac{S}{\Delta H_{bot}}\delta_{k,k_{bot}} \right)$$

$$\left. -s_4 L_{SA}\frac{S}{\Delta H_{bot}}\delta_{k,k_{bot}} \right\} + \frac{Ox^{flux}}{H_0}\delta_{k,k_0}.$$

The stoichiometric ratios are listed in Table 3.3. The sources of oxygen are the primary production, here measured as growth through nutrient uptake, and gas exchange through the sea surface. The gas exchange is described by

$$Ox^{flux} = p_{vel}(O_{sat} - Ox),$$

where $p_{vel}$ is a piston velocity. The saturation concentration depends on the temperature according to

$$O_{sat} = \frac{a_0}{O_{norm}}(a_1 - a_2 T),$$

with $a_0 = 31.25\,\text{mmol m}^{-3}$, $a_1 = 14.603$ and $a_2 = 0.4025\,(°C)^{-1}$ (Livingstone, 1993). Oxygen is consumed through respiration, nitrification and mineralization. The term with the switches $\theta(-Ox)\theta(-N)$ takes into account the fact that, under anaerobic conditions and depleted nitrate, hydrogen sulphide is produced. Due to microbial activity, sulphate is reduced to hydrogen sulphide while the released oxygen is used to oxidize detritus. The hydrogen sulphide is counted as negative oxygen, the involved rates are listed in Table 6.4.

After the detailed description of the model equation and the involved biogeochemical processes, we can connect this model component to the circulation model. With the exception of the sediment detritus, i.e. (6.5), we replace the total derivative, $\frac{d}{dt}$, in the equations of the state variables by the partial derivative, $\frac{\partial}{\partial t}$, and the advection-diffusion terms of the tracer equation (6.1), where the forcing term, $F_\chi$, on the right-hand side of (6.1) corresponds to the right-hand side of the equations of the state variables. For the simple case of

**TABLE 6.4** Biogeochemical Process Rates

| Nitrification | |
|---|---|
| Nitrification constant | $l_{AN} = 0.1\,\mathrm{d}^{-1}$ |
| Oxygen parameter | $O_{AN} = 0.01$ |
| Temperature control | $\beta_{AN} = 0.11\,^{\circ}\mathrm{C}^{-1}$ |
| Mineralization | |
| Detritus mineralization rate | $l_{DA} = 0.003\,\mathrm{d}^{-1}$ |
| Temperature control | $T_{DA} = 13\,^{\circ}\mathrm{C},\ \beta_{DA} = 20\,(^{\circ}\mathrm{C})^{-1}$ |
| Sediment mineralization rate | $l_{SA} = 0.001\,\mathrm{d}^{-1}$ |
| Temperature control | $\beta_{SA} = 0.15\,(^{\circ}\mathrm{C})^{-1}$ |
| Release of phosphate | $p_1 = 0.15,\ p_2 = 0.1$ |
| Other Rates | |
| Detritus sinking | $w_D^{sink} = 3\,\mathrm{m\,d}^{-1}$ |
| Sedimentation rate | $l_{DS} = w_D^{sink} \Delta H_{bot}^{-1}$ |
| Resuspension rate | $l_{SD} = 25\,\mathrm{d}^{-1}$ |
| Critical stress | $\tau^{crit} = 0.7\,\mathrm{dyn\,cm}^{-2}$ |
| Nitrogen normalization | $N_{norm} = 4.5\,\mathrm{mmol\,m}^{-3}$ |
| Oxygen normalization | $O_{norm} = 375\,\mathrm{mmol\,m}^{-3}$ |
| Piston velocity | $p_{vel} = 5\mathrm{m\,d}^{-1}$ |

constant diffusivity, and in Cartesian coordinates, the equations assume the form of the advection-diffusion equation (5.77).

## 6.6 SIMULATION OF THE ANNUAL CYCLE

We now describe the results of a series of model runs. We start with the simulation of an annual cycle to show how the model estimates the seasonal variations of the state variables. To this end, we first display the annual cycle in terms of monthly means of several chemical and biological state variables, averaged over the upper 10 m of the water column. The typical course of phytoplankton in the Baltic Sea is shown in Figs. 6.10 and 6.11. The spring bloom starts in the southwestern Baltic and in shallow coastal areas in March. In April and May, the bloom occurs in the central and northern parts.

Thus, the onset of the spring bloom in the central Baltic can lag behind that of the southwestern Baltic. The reason for this lag is a delayed formation of

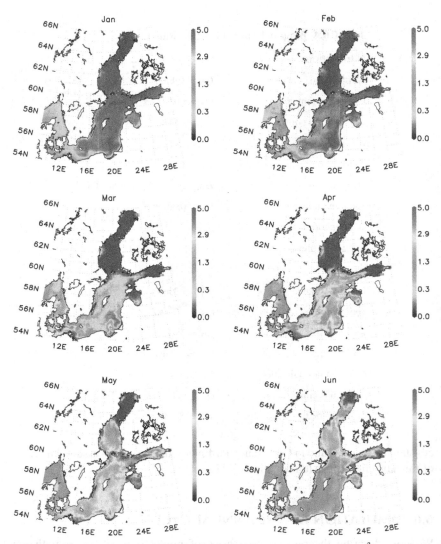

**FIGURE 6.10**   Monthly means of simulated chlorophyll *a* concentration in mg m$^{-3}$ averaged over the upper 10 m for January to June of the model year 1984. (See color plate.)

the thermocline in the central Baltic, in particular after cold winters when the surface water temperature was below the temperature of the density maximum (Fennel, 1999a) and less solar radiation in higher latitudes. However, this is not resolved in the monthly mean values. In the northern Baltic, which is regularly ice covered during the winter, the bloom cannot start before the melting of the ice.

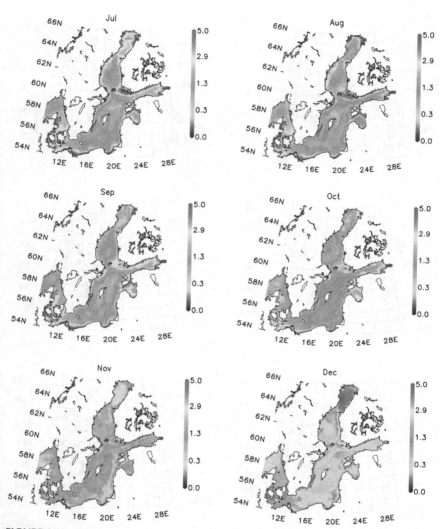

**FIGURE 6.11** Monthly means of simulated chlorophyll $a$ concentration in mg m$^{-3}$ averaged over the upper 10 m for July to December of the model year 1984. (See color plate.)

The nutrients, i.e. nitrogen and phosphorus, show an inverse behaviour compared to the phytoplankton. This is shown in the Figs. 6.12–6.15. During the increase of phytoplankton biomass, the nutrients are depleted. In summertime, there are some provinces where nitrogen is exhausted, e.g. western Baltic Proper in June, while in other regions the phosphate is completely used up, e.g. eastern Baltic Proper in July. After the vegetation period, in late autumn, frequent occurrence of strong winds and decreasing temperatures enforce a

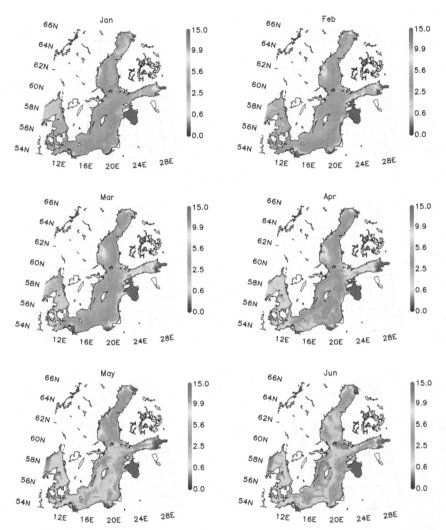

**FIGURE 6.12** Monthly means of simulated DIN (ammonium and nitrate) concentration in mmol m$^{-3}$ averaged over the upper 10 m for January to June of the model year 1984. (See color plate.)

deepening and dissolution of the thermocline and, in turn, an enrichment of the upper waters with nutrients. Thus, the annual variations of the meteorological conditions control the yearly cycle of the biological system.

The monthly mean values show several spatial structures in the phytoplankton and nutrient distributions. In particular, near the outlets of the rivers, elevated concentrations are found due to the input of the riverine nutrient loads. However, the monthly averages smooth out the mesoscale variation.

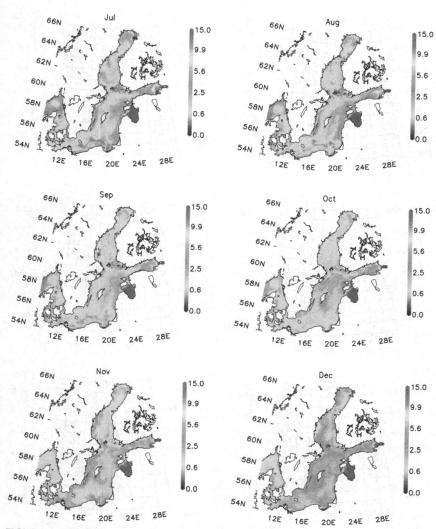

**FIGURE 6.13** Monthly means of simulated DIN (ammonium and nitrate) concentration in mmol m$^{-3}$ averaged over the upper 10 m for July to December of the model year 1984. (See color plate.)

To visualize the spatial patterns (patchiness) of state variables and their variation at a time scale of a few days, we show a series of snapshots of surface model chlorophyll distributions averaged over 5 days in Fig. 6.16.

The Baltic Sea is forced by varying winds with a time scale of a few days. Coastal upwelling events, which are driven by along-shore winds with the coast to the left, transport nutrients from below into the upper layer. These nutrients

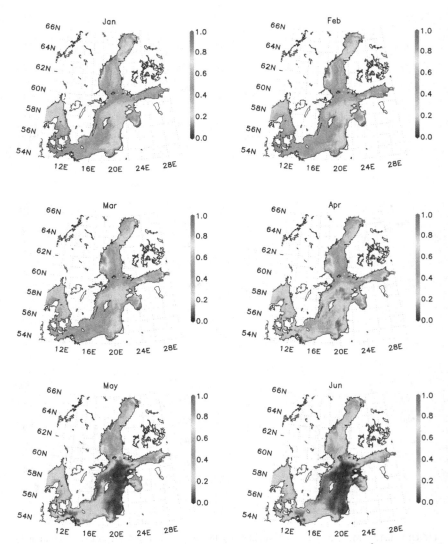

**FIGURE 6.14** Monthly means of simulated phosphate concentration in mmol m$^{-3}$ averaged over the upper 10 m for January to June of the model year 1984. (See color plate.)

are rapidly consumed by phytoplankton. The eastern coastal boundary is also strongly influenced by the nutrient loads of the river discharges.

It appears that the coastal jets, which are driven by alongshore winds, are perhaps more important for the redistribution of nutrients than the upward transport of nutrients due to episodic upwelling events. Along-shore winds with the coast to the right are associated with onshore Ekman transports. Hence, the

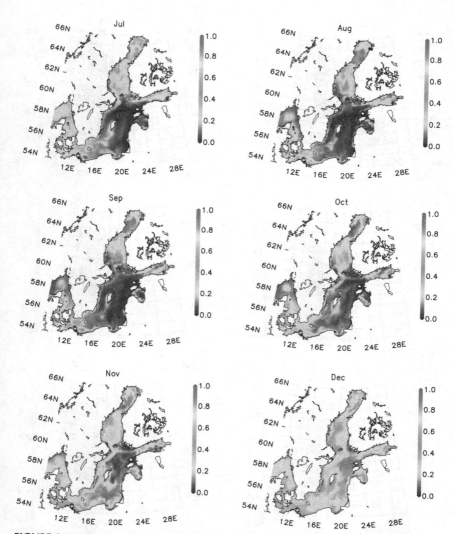

**FIGURE 6.15** Monthly means of simulated phosphate concentration in mmol m$^{-3}$ averaged over the upper 10 m for July to December of the model year 1984. (See color plate.)

downwind coastal jets can transport river plumes in a narrow band along the coast. If the wind turns to the opposite direction, the offshore Ekman transport will move the river plume offshore. Bands of high chlorophyll, which had developed in response to the riverine nutrients, can be detected, as shown in the eastern part of the central Baltic in Fig. 6.16. In the southwestern part of the Baltic, there are several eddies and filaments, which are visible in the model chlorophyll distributions. These patterns, which cannot be explained in terms

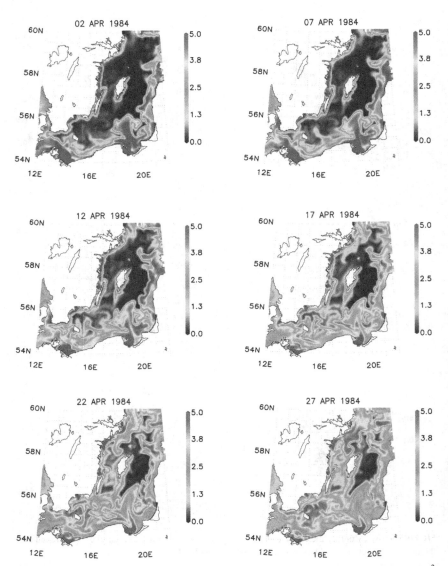

**FIGURE 6.16**   Selected snapshots of 5-day means of the simulated model chlorophyll in mg m$^{-3}$ averaged over the upper 10 m. (See color plate.)

of a linear theory as coastal jets, reflect nonlinear hydrodynamic processes in conjunction with the biological dynamics, and their theoretical description requires numerical models.

The series of 5-day averages of chlorophyll distributions show also how the spring bloom propagates from the southwestern parts of the Baltic towards the

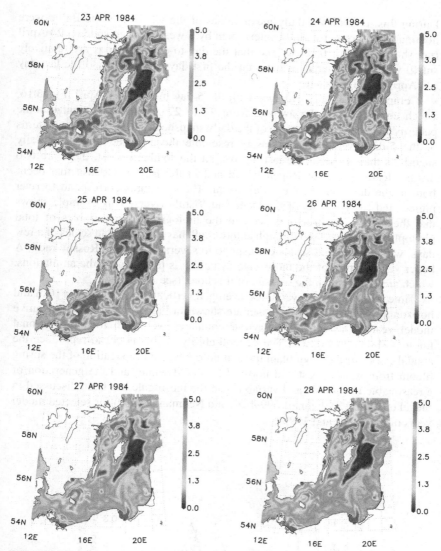

**FIGURE 6.17** Daily snapshots of surface chlorophyll $a$ concentration in $mg\,m^{-3}$ averaged over the upper 10 m. (See color plate.)

north and from the coastal region towards the central Baltic. The richness of the mesoscale patterns, which is still somewhat smoothed by the 5-day averaging in Fig. 6.16, becomes more complex in daily or hourly means of the state variables. In order to illustrate this, we select a series of daily averaged surface chlorophyll distributions for the period from 23 to 28 April, which is shown in Fig. 6.17.

During this interval, the daily mean values of the winds over the Baltic were dominated by northern winds, interrupted by a west-wind episode on 24 April (not shown). In Fig. 6.17, we see that the day-to-day variations are relatively small, while the differences between the first day (23 April) and the last day (28 April) are significant.

Comparing the daily averages with the 5-day mean of 27 April, Fig. 6.16, which correspond to the average from 22 to 27 April, we recognize some patterns that occur only in some of the daily snapshots. The distribution patterns of the 5-day average more closely resemble those of the first three daily snapshots than those of the last 3 days. At the southern coast, these are two distinct features east of Rügen Island and in the Bay of Gdansk that occur both in the daily and in the 5-day mean. These features correspond to river plumes fed by the loads of the Oder and Vistula rivers. This example shows that the mesoscale structures seen in the biological state variables of total phytoplankton, expressed through chlorophyll, have typical time scales of a few days, while the spatial scales correspond to several baroclinic Rossby radii. A better visualization of the mesoscale dynamics is provided by the animations, which are available on the website of this book (see Appendix).

Time series of several variables, averaged vertically over the upper 10 m and horizontally over the Arkona basin are shown in Fig. 6.18 for two consecutive model years. The general behaviour compares well with the observational findings shown in Fig. 2.8. The general ability of the model to reproduce the annual cycles of the phytoplankton and nutrients, the propagation of the spring bloom from west to east and north after a cold winter and the generation of a reasonable magnitude and variability at the mesoscale has been discussed in Siegel et al. (1998), Fennel (1999b,a), and Neumann (2000) for selected model years of the 1980s and 1990s.

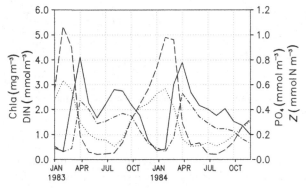

**FIGURE 6.18**   The cycling of several state variables (phytoplankton – solid, inorganic nitrogen – dashed, phosphate – dots, and zooplankton – dash-dotted) for the years 1983 and 1984 at a station in the Arkona sea. The figure corresponds to the observed cycle in Fig. 2.8.

## 6.7 SIMULATION OF THE DECADE 1980–1990

Next we discuss a decadal simulation in order to show the responses of the Baltic Sea system to interannual variations in meteorological forcing and nutrient loads. This simulation serves as a control run for the load reduction studies, which are described in the following section. The discussion is largely based on the paper of Neumann et al. (2002), except that we show here results of simulations with higher vertical resolution of the deeper parts of the water column. The model was driven by meteorological data from the ERA-40[4] project and by data sets of river runoff, nutrient loads and atmospheric nutrient deposition provided by the Baltic Environmental Database[5] of the University of Stockholm. A comprehensive data set, which comprises meteorological forcing, river loads and atmospheric deposition, was available in particular for the period from 1980 to 1990. The initial distributions of temperature and salinity were taken from climatological data (Janssen et al., 1999). Unfortunately, there is only a sparse database for the biogeochemical variables. In order to overcome this problem, we started with the available data and used a model run for a spinup period of 4 years to 'compute' an initialization data set.

In order to test the model performance, we compare model results with observations. First we consider the observed and modelled water levels at a Landsort gauge, which is known to represent the changes of the volume of the Baltic Sea, e.g. Lisitzin (1974). The low-pass filtered water levels, coherence spectra of observed and modelled hourly sea level values and a scatterplot of observations and modelled data are shown in Fig. 6.19. The model reproduces the gross features of observed water levels; however, the peak values and the high-frequency variability are slightly underestimated. We assume that the main reason for these differences is due to the meteorological forcing data, which tend to underestimate the wind peaks. The reanalysis data are averaged over a period of six hours and have a spatial scale of 125 km. This implies a temporal and spatial smoothing of the wind data.

Next we compare several parameters of the model with observations of the Baltic Monitoring Programme. In particular, we use the monitoring Station 113 in the Arkona sea, Station 213 in the Bornholm sea and Station 271 located in the Gotland sea, see Fig. 6.1 with indications of the stations. A comparison of the modelled and observed sea surface temperature at these stations is shown in Fig. 6.20. Taking the general problem of observational undersampling into account (which occurs if point measurements of a patchy distributed quantity are compared with the model results), we find that the model reproduces the observed data reasonably well. The interannual trends of the minimum temperatures are also seen in the model. However, the maxima of the summer temperatures are less well reproduced. This can be attributed to the much higher

---

4. http://www.ecmwf.int/research/era/.
5. http://nest.su.se/models/bed.htm.

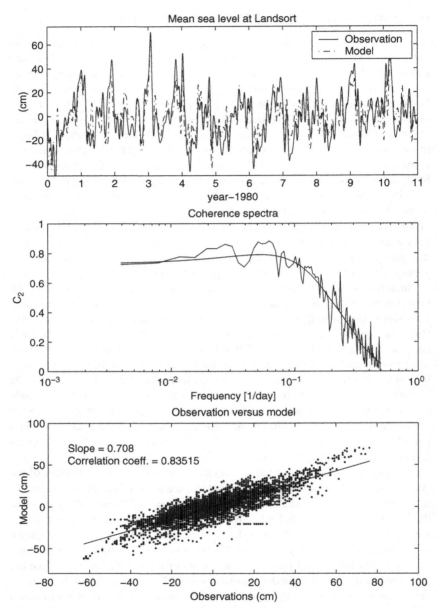

**FIGURE 6.19**    Modelled and observed sea levels at the Landsort gauge. Shown are a time series (top), coherence spectra (middle) and scatterplot (bottom) of observed and modelled sea levels.

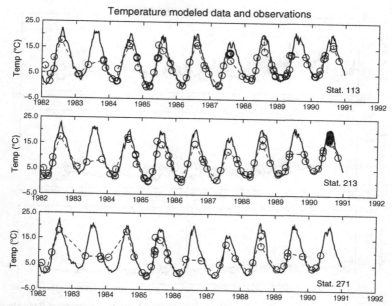

**FIGURE 6.20** Sea surface temperature modelled (solid line) and observed (cycles) at three stations. The locations of the stations are indicated in Fig. 6.1.

temporal and spatial variations of the sea surface temperatures in summertime, which are not adequately sampled at the monitoring stations.

While the model sea surface temperature is strongly influenced by the surface boundary conditions, the salinity signatures in the model develop in response to river runoff, salt exchange at the open boundary to the North Sea and internal water mass transformations. A comparison of observed and simulated surface salinity is shown in Fig. 6.21. Most of the observed variations in the salinity are displayed in the model data, except the strong salinity increase in autumn 1989, which was observed at all stations but not reproduced by the model. The salinity in the bottom layers of the model is not well reproduced. The modelled values are, in many cases, smaller than the observed data. In particular, for the bottom layer of the Bornholm Sea, the model shows an offset, while in the central Baltic the deviations are relatively small, see Fig. 6.22. This shows that the model needs further refinement regarding the treatment of vertical mixing near the bottom, especially in sill regions with strongly varying topography.

To discuss the chemical and biological variables, we start with the model phytoplankton and compare the sum of the three functional groups expressed in bulk-chlorophyll units near the sea surface with observed values, see Fig. 6.23. The model chlorophyll is generally in the range of the observation, but there is no one-to-one correspondence of observed and modelled values. However, the

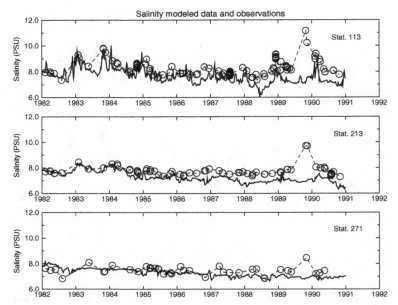

**FIGURE 6.21**   Sea surface salinity modelled (solid line) and observed (circles) at three stations. The locations of the stations are indicated in Fig. 6.1.

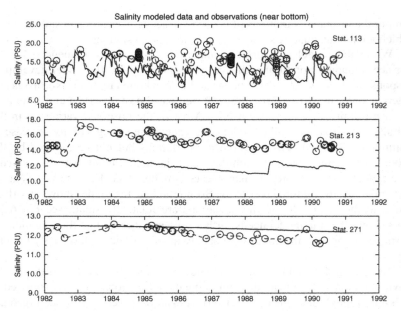

**FIGURE 6.22**   Near-bottom salinity modelled (solid line) and observed (circles) at three stations. The locations of the stations are indicated in Fig. 6.1.

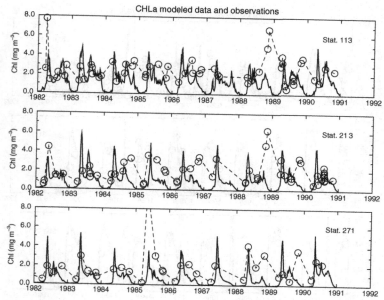

**FIGURE 6.23** Simulated (solid line) and observed (circles) chlorophyll. The locations of the stations are indicated in Fig. 6.1.

observed data are sparse and we have to consider a high degree of patchiness, which might obscure single measurements (see also HELCOM, 1994). Moreover, we used a constant conversion factor (Redfield) to convert model nitrogen to carbon and the ratio C:Chl = 50 to convert carbon to chlorophyll units, which is an oversimplification. Time series of the total model phytoplankton as well as of the functional groups, averaged over the Gotland Sea, are also shown in Figs. 6.29–6.32 in the following section. These figures also display seasonal cycles and interannual variations of the phytoplankton model.

In Fig. 6.24, the simulated and observed concentrations of dissolved inorganic nitrate (DIN), which consists of the sum of nitrate and ammonium, in the model are shown. In this case, the observations are well reproduced by the model. The model DIN near the sea surface follows the annual cycle. The summer minimum and the recovery of DIN during the winter are reproduced within the range of the observations. The near-bottom values of observed and simulated DIN are shown in Fig. 6.25. While in the Arkona Sea and the Bornholm Sea, the simulations reflect the magnitude and variations of observations, in the Gotland Sea, the observed strong variations are not reproduced by the simulations.

Similar results are found for dissolved inorganic phosphorous (DIP). A comparison of modelled and observed near-surface phosphate is shown in Fig. 6.26. The model phosphate roughly displays the observed range and follows the annual cycle.

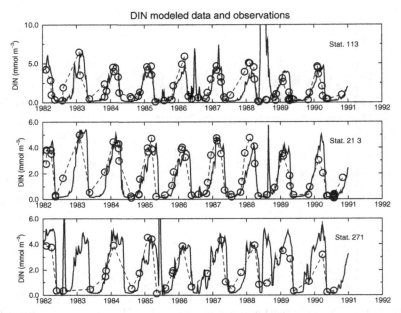

**FIGURE 6.24** Simulated (solid line) and observed (circles) DIN near the sea surface. The stations are indicated in Fig. 6.1.

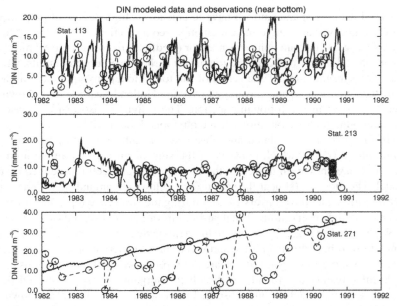

**FIGURE 6.25** Near-bottom values of modelled (solid line) and observed (circles) DIN. The stations are indicated in Fig. 6.1.

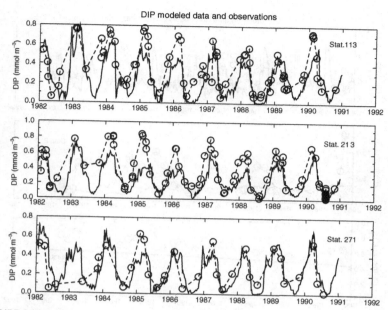

**FIGURE 6.26** Surface phosphate as modelled (solid line) and observed (circles). The stations are indicated in Fig. 6.1.

The model DIP concentrations near the bottom, which are shown in Fig. 6.27, reproduce the annual variations in the Arkona basin but fail to display the near-bottom variations in the Bornholm basin. This can partly be explained by alternating aerobic and anaerobic conditions. While, under aerobic conditions, phosphorus can be bound to a large extent in iron-phosphate complexes in the sediment, huge amounts of phosphate are liberated from the sediment under anaerobic conditions. Low-oxygen concentrations in late summer (compare Fig. 6.28) are correlated with a sudden increase in near-bottom DIP. In the Bornholm sea, the modelled oxygen in the near-bottom water is overestimated, implying DIP concentrations in the model that are too low. In the Gotland Sea, the observed variations of the DIP concentration are smaller. The model does not reproduce the seasonal variability of the DIP and oxygen, but the model results are of the same order of magnitude.

The simulated oxygen near the bottom, shown in Fig. 6.28, compares well with the observed values for the Arkona Sea. However, we find clear discrepancies in the second half of the 1980s in the Bornholm Sea. The slope of the oxygen decrease and the magnitude of the near-bottom oxygen concentration in the Gotland Sea are reasonably well reproduced. However, again the observed variability is not seen in the model.

In summary, for the upper layers of the Baltic, the model simulates the annual cycles of the state variables reasonably well. The simulated surface

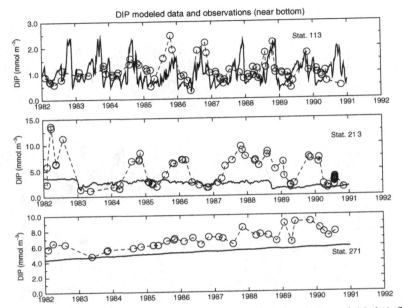

**FIGURE 6.27**    Near-bottom phosphate as modelled (solid line) and observed (circles). The stations are indicated in Fig. 6.1.

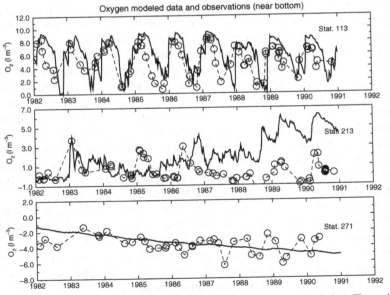

**FIGURE 6.28**    Near-bottom oxygen as simulated (solid line) and observed (circles). The stations are indicated in Fig. 6.1.

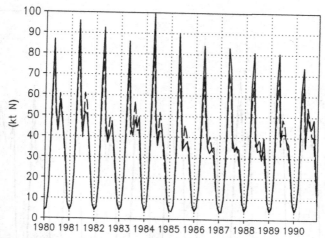

**FIGURE 6.29** Time series of the total phytoplankton model expressed in $10^3$ tons nitrogen in the Baltic Proper for the control run (solid line) and for the simulation with reduced river loads (dashed line).

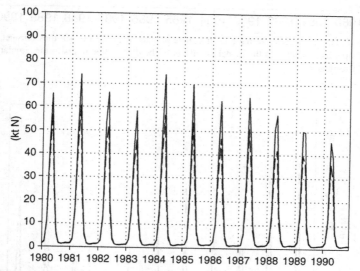

**FIGURE 6.30** Time series of the total diatom biomass expressed in $10^3$ tons nitrogen in the Baltic Proper as simulated in the model (solid line). The dashed line refers to the simulation for reduced river loads.

temperatures are well reproduced by the model and display both the yearly cycles and the interannual variations of the winter minimum, which are important for the timing and spatial pattern of the spring bloom. Annual cycling of dissolved nutrients in the surface waters reflect the observations and the

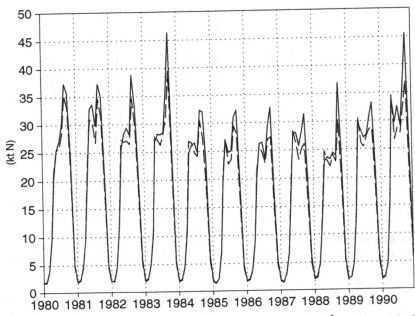

**FIGURE 6.31** Time series of the total flagellate biomass expressed in $10^3$ tons nitrogen in the Baltic Proper as simulated in the model (solid line). The dashed line refers to the simulation for reduced river loads.

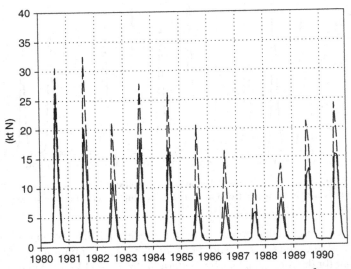

**FIGURE 6.32** Time series of the total cyanobacteria biomass expressed in $10^3$ tons nitrogen in the Baltic Proper as simulated in the model (solid line). The dashed line refers to the simulation for reduced river loads.

expected behaviour. Also the simulated surface values of the dissolved inorganic nutrients are close to the observation.

The depletion of the model oxygen near the bottom is reasonably represented in the Arkona Sea (station 113) and in the central Baltic (station 271) but is clearly too high in the Bornholm Sea (station 213). Thus, the biogeochemical processes in and below the halocline are only partly captured by the model. The simulated near-bottom values of the dissolved nutrients do not fully match the observed data. The too-weak variability in the deep parts of the model indicates that part of the problem can be attributed to the physical component of the model, which obviously underestimates the near-bottom transports. Hence, this discussion highlights the importance of the quality of the circulation model, which is coupled to the biogeochemical model component.

## 6.8  A LOAD REDUCTION EXPERIMENT

We now describe an example of a numerical experiment with the help of the system model. In particular, we wish to study the effect of a reduction of the nutrient loads, nitrate and phosphate, of the rivers by 50% and compare the results with the simulations for realistic river loads, which were discussed in the previous section and serve as control runs. Because the physics is not affected by changed river loads, we focus the discussion on the differences in the chemical and biological state variables. We start with the total inventory of chlorophyll, i.e. the sum of the functional phytoplankton groups expressed in nitrogen. The time series of the basin average of the model chlorophyll of the Gotland sea is shown in Fig. 6.29.

The seasonal cycles are well displayed in both model runs. Starting from low winter values, a strong spring bloom peak develops, followed by moderate summer levels and a less strong, but clearly distinguishable, secondary bloom in autumn. The effects of the load reduction can mainly be detected in the spring and autumn blooms, where the signals of the reduction run are clearly smaller than for the reference run. In summertime, the total phytoplankton biomass is not significantly modified in response to the load reduction.

The seasonal succession of the model phytoplankton is qualitatively well described by the model. This is shown by means of the inventories of model diatoms and flagellates in Figs. 6.30 and 6.31 for the Gotland Sea. The functional group of diatoms exhibits a clear decrease of biomass in response to the load reduction.

This group, which dominates the spring bloom and assumes relatively low levels for the rest of the year, responds to reduced winter nutrient concentrations with significantly smaller spring bloom peaks, see Fig. 6.30. The model flagellates are the most abundant functional group during summer and dominate the secondary bloom in autumn. They depend on a high rate of regeneration of nutrients (microbial loop) and can survive at low nutrient concentrations.

The time series of the total biomass of this group, which is shown in Fig. 6.31, indicates a relatively small effect in response to the smaller river loads.

The functional group of cyanobacteria shows an inverse response to the load reduction, as shown in Fig. 6.32, which displays a time series of the modelled total cyanobacteria biomass in the Baltic Proper. Because of their ability to fix nitrogen, the cyanobacteria biomass increases. A good qualitative check of the model performance is given by the interannual variations of the occurrence of cyanobacteria blooms. Based on satellite data, Kahru (1997) demonstrated that the cyanobacteria accumulation diminished in the Baltic Proper in the mid-1980s, ceased in 1986, 1987 and 1988 and reappeared in 1989 and 1990. The behaviour is well reproduced by the model shown in Fig. 6.32.

In order to highlight the gross effects, we chose the annual means of vertically integrated variables for the year 1990 (after a 10-year simulation) and estimate the percentage of the changes. The results, which are shown in Fig. 6.33, demonstrate that the different state variables respond in different ways to the diminished river loads.

The total chlorophyll model, which consists of the sum of all phytoplankton, diatoms, flagellates and cyanobacteria models, shows a response to the load reduction by a decrease of about 30% in the coastal areas and the western Gotland Sea, while in the central parts of southern and eastern Gotland Sea, no decrease was detected. On the contrary, a slight increase in total chlorophyll occurs.

The most obvious response to the halved nutrient input occurs for diatoms. A remarkable effect of the flagellate biomass in response to the load reductions is found only in coastal areas, whereas a slight increase occurs in the central eastern Gotland Sea. The model cyanobacteria response to the load reduction shows the largest variation. Their biomass is generally increased. In particular, south of the Gotland island, a fourfold increase of the cyanobacteria occurs. Obviously, cyanobacteria benefit from reduced nitrogen inputs due to their ability to fix atmospheric nitrogen. This provides an advantage over the other functional groups of the model phytoplankton. Because the change is expressed in relative units, as percentages, we refer also to Fig. 6.32, which shows the absolute changes of the total cyanobacteria biomass. A substantial increase is found for all model years after 1983; however, the relatively low levels of cyanobacteria biomass in the years 1987 and 1988 occur also in the case of reduced loads.

The dissolved inorganic nutrients within the water column assume lower levels; in particular, a clear decrease of the concentrations is found near the outlets of the great rivers, Fig. 6.33 (bottom). The reduction is stronger for total inorganic model nitrogen, DIN, than for phosphate. This can be attributed to the different pathways of the nutrients. While phosphorus can leave the system only through burial in sediments, nitrogen can also be effectively removed by denitrification.

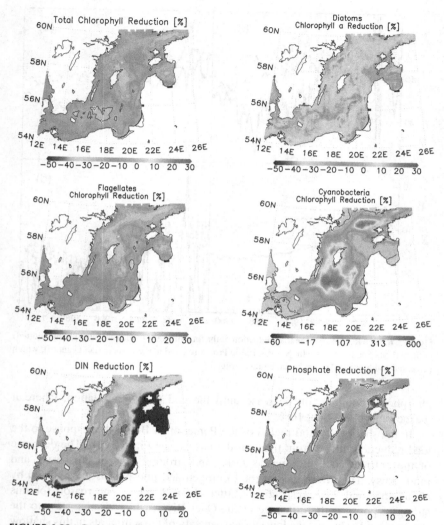

**FIGURE 6.33** Reduction effect on selected state variables after 10 years of integration in terms of annual means for the year 1990 (redrawn from Neumann et al., 2002). (See color plate.)

Next we discuss the effect of the load reduction on the inventory of total nitrogen and phosphorus in the different basins and, as an example of a shallow area, in the Pomeranian Bight, which is located south of the Arkona basin, see Fig. 6.1. The effect of the load reduction in these model regions is shown in Fig. 6.34. In the basins, the nitrogen decreases during a period of several years, see Fig. 6.34(a)–(c). However, this decrease diminishes after about eight years, when the nitrate reaches a plateau at about 20%, while the dissolved inorganic

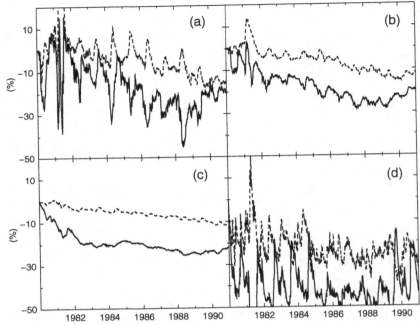

**FIGURE 6.34**    Response to the load reduction of the total nitrogen (solid) and phosphorus (dotted) in the Arkona Sea (a), Bornholm Sea (b), Baltic Proper (c), and in the Pomeranian Bight (d), which is located south of the Arkona basin; compare Fig. 6.1.

phosphorus continues to decrease until the end of the simulation, where it reaches a level of about 10%.

In the shallow coastal region of the Pomeranian Bight, the response to the load reduction differs from those of the basins, see Fig. 6.34(d). The decrease of the nutrient concentrations is faster and stronger than in the basins and more noisy. The inventories of total nitrogen and phosphorus are reduced by about 46% and 73%, respectively, after about 2 years. The reason for this strong response is the influence of the Oder River, which discharges into the Pomeranian Bight, and the low storage capacity of the sediments in this shallow area. It appears that two-thirds of the total nitrogen input from the Oder River leave the Pomeranian Bight and enter the open Baltic Sea, as shown in Fig. 6.35. Note that this applies both to the control and reduction run. As shown in Fig. 6.36, the remaining one-third of the riverine nitrogen load is removed by denitrification. This indicates that the buffer capacity of the shallow Pomeranian Bight can virtually be ignored.

The nitrogen budgets for the Baltic Proper and for the Pomeranian Bight are shown in Fig. 6.36 for both the control run (subplots at the left-hand side) and the reduction experiment (subplots at the right-hand side). In Fig. 6.36, the columns with the label 'river' indicate the averaged yearly input of nitrogen due

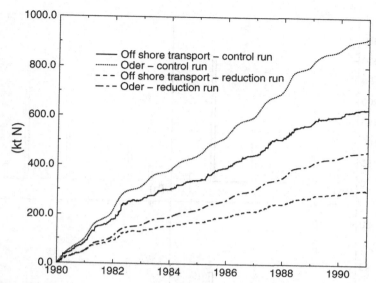

**FIGURE 6.35** Cumulative total nitrogen input from the Oder River (dots: control run, dash-dotted: 50% reduction) and transports from Pomeranian Bight into open sea. The solid line refers to the offshore transport of the control run; the dashed line refers to the reduction experiment.

to the different river loads of both runs. The other columns express mean values of the different contributions to the annual nitrogen budgets in the water column (inventory), in the sediments, changes due to import or export from or to other areas (transports) and fluxes driven by denitrification and nitrogen fixation. The corresponding initial values are subtracted; i.e. zero level corresponds to the initial budget at the start of the simulations.

In the Pomeranian Bight, the reduced river load mainly implies a reduced offshore transport of nitrogen while the other contributions are almost unchanged. The reduced offshore transport assumes the same order of magnitude as the denitrification. In the Baltic Proper, a relatively small decrease of nitrogen is found within the water column, see the column 'inventory' in Fig. 6.36(a) and (b). The nitrogen budget is balanced mainly by a decreased contribution of denitrification in the sediment. The increase due to nitrogen fixation is compensated for by reduced fluxes from sediment and transports. Thus, in shallow areas like the Pomeranian Bight, the effect of the halved river loads is almost completely balanced by a halved offshore transport.

## 6.9 PROJECTION OF FUTURE CHANGES

The projection of expected future changes of marine ecosystems is an important issue and of great interest because the human impact might play a role. Human activities are partly responsible for changes, and projections can help to identify

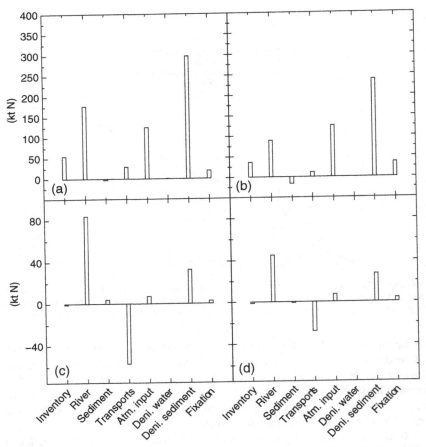

**FIGURE 6.36**    Averaged annual nitrogen budget for the Baltic Proper (a, b) and the Pomeranian Bight (c, d). Panels (a, c) refer to the control run (b, d) to the reduction experiment. The zero level corresponds to the initial budget at the model start. The initial contributions are subtracted from the averages.

specific measures to mitigate unwanted developments. Another demand for the simulation of future trends is the need for sufficient time to allow adaptations to changes. For example, to react on the expectation of rising sea level requires a long lead time to prepare coastal protection facilities.

Simulations with numerical models are the only way to make sound projections for the future. Challenges arise from unknown and uncertain future boundary conditions. On a global scale, the Intergovernmental Panel on Climate Change (IPSS, www.ipcc.ch) regularly assesses past and expected changes in the Earth's climate system.

The bases for the assessment of future development are different representative concentration pathways (RCPs). The RCPs primarily describe greenhouse gas concentrations on a global scale and provide a consistent forcing for global Earth system models. Owing to the coarse spatial resolution of global climate models, regional assessments require a so-called dynamical downscaling with the aid of regional climate models (RCMs). RCMs cover only a part of the globe (e.g. Europe) and are forced at the boundaries by data from global models. The benefit is a much higher spatial resolution; however, RCMs may introduce additional uncertainties. Finally, nutrient loading assumptions are required for the simulation of future regional marine ecosystems scenarios.

Altogether, a number of uncertainties are introduced due to unsureness in underlying assumptions and unknown or unresolved processes in the models. To provide measures of the uncertainties, ensembles of simulations are considered. An ensemble represents a set of simulations with different forcing assumptions, different global and regional models and different model formulations. The spread of the model results represents a measure for the involved uncertainties. The uncertainties usually are different for different state variables in the models. Based on the analysis of uncertainties, for some variables robust trends can be derived, while for other variables no significant trends can be detected.

After the identification of a trend, further simulations may be necessary to elucidate causal relationships and to attribute trends unambiguously to the responsible drivers. Models and forcing can be modified in such a way that only one driving factor changes. Simulations can then help to identify how sensitive the results respond to the driving factors. Examples of ensembles of simulations for the future of the ecosystem of the Baltic Sea are given in, e.g. Neumann (2010) and Meier et al. (2011).

## 6.10 TRACKING OF ELEMENTS

A useful application possibility of ecosystem models is the tracking of elements, e.g. nitrogen or carbon. This implies an increase in the number of state variables and marked and unmarked variables, and the model equations for the considered element have to be doubled (or tripled) because equations are also required for the marked element. The biogeochemistry model only considers the sum of the marked and unmarked part, while the marked tracer is fully available in diagnostics. An element can be marked for several purposes. It can be the entering of an element into the model via a river or just when the element is transferred from one tracer to another. Often, this method is used to determine the source (e.g. river or catchment) of nutrients in a marine ecosystem. This method was first described by Ménesguen et al. (2006). Other applications are given in Neumann (2007) and Radtke et al. (2012).

## 6.11 DISCUSSION

The simulations in the previous sections provided an example of how an advanced ecosystem model can be used to perform numerical experiments for complex marine systems like the Baltic Sea. The reactions of the system to relatively simple changes in the forcing conditions (in particular, diminished riverine nutrient loads) can be complex. The responses cannot be deduced from the inputs in a straight-forward manner. The different state variables change with different time and space scales, and the different parts of the food web respond differently due to the nonlinear nature of the dynamics of the system.

Comparing these model experiments with those performed in Chapters 2 and 3 and based on box models, it is clear that a consistent quantitative description requires three-dimensional coupled models, which resolve the spatial structures associated with mesoscale dynamics and allow adequate description of the characteristic gradients of the system. There are opportunities and also limitations. Many features of the system could be reproduced in a realistic manner, but in some cases further studies are needed to improve the realism of the model. Nevertheless, the capabilities of the models are fascinating and should encourage scientists to cooperate across the disciplines towards further improvements.

# Chapter 7

# Circulation Model, Copepods and Fish

The previous chapter dealt with the coupling of chemical-biological model components, which were developed in Chapters 2 and 3, and full three-dimensional circulation models. The zooplankton was represented by a bulk state variable, which lumped together the biomass of all groups and stages of copepods. A next step in advancing the model system is the inclusion of life cycles of copepods as described in Chapter 4. Such models are important for studies where aspects of the life cycle of copepods, i.e. the propagation through their different stages, have to be considered explicitly.

This type of model can be particularly useful in addressing fish recruitment problems by calculating spatial and temporal changes of the distribution of prey fields for fish larvae. It appears that the general theoretical approach of fisheries models used for stock assessments in marine systems is based on greatly simplified views. Bottom-up developments, ranging from variations in hydrographic conditions to the lower trophic levels of the food chain, are largely ignored. An important element in fisheries modelling is the provision of data from acoustic surveys or catch reports, e.g. biomass data for so-called production models or data of year classes for the analytical models; see Gulland (1974) and Horbowy (1996). Although these data implicitly carry a substantial amount of environmental information, they cannot replace an ecosystem approach. Despite the practical importance of reliable predictions of the development of fish stocks in relation to fisheries mortality, the progress in model development, which includes ecosystem modelling, seems to be relatively slow. There are obvious difficulties in the development of new theories and more complex models, which link ecosystem and fisheries models. One problem is related to the broad range of involved time scales. While the life cycles of fish may span several years, the bottom-up processes evolve at seasonal or annual time scales.

The vision of future-generation models that bridge the gap between higher and lower trophic levels of the marine food web is probably one of the greatest challenges for theoretical research and model development in the coming years.

Introduction to the Modelling of Marine Ecosystems. http://dx.doi.org/10.1016/B978-0-444-63363-7.00007-6

## 7.1   RECRUITMENT (MATCH–MISMATCH)

A first step towards models that link fish development and lower parts of the food web consists of modelling of critical parts of the life cycles of fish. An obviously important example of a critical phase is related to the recruitment success of fish larvae. The problem of fish larvae survival can be attacked by combination of ecosystem models, which simulate bottom-up developments through the food web including stage-resolving zooplankton and particle-tracking models (e.g. Hinrichsen et al., 2001) which calculate the drift of fish eggs and larvae. Such an approach reflects the importance of physical and biological determinants of recruitment variability, which depends on the availability of prey for the larvae. This idea has long been recognized by Hjort (1914) and was extended by Cushing (1975) to formulate the match–mismatch hypothesis, which relates recruitment variability to the timing of the transition from fish eggs to larvae and prey availability. In other words, fish stocks can only reproduce if larvae survive and develop, implying that early stages of zooplankton, i.e. nauplii, are available at the right time and location as food for larvae.

In particular, we here consider the prey part of the problem, i.e. the simulation of the early stage of copepods, the nauplii, which constitute the prey fields for fish larvae. This can be achieved by a combination of the coupled model described in the previous chapter, but with the stage-resolving model described in Chapter 4. Formally, this amounts to the replacement of the bulk zooplankton stage variable by a stage-resolving zooplankton model component. Stage-resolving descriptions of copepods, as outlined in Chapter 4, allow explicit estimations of the evolution and distribution of stages ranging from eggs to adults. This kind of modelling involves growth, development and reproduction of copepods, controlled by food web interactions and by physical transports through advection and mixing by small-scale turbulence.

A first attempt to model such a system with a coupled circulation model and a biogeochemical model component describing the dynamics of nutrients, phytoplankton, stage-resolved zooplankton and detritus was presented in Fennel and Neumann (2003).

## 7.2   COPEPODS IN THE BALTIC SEA MODEL

In order to implant the stage-resolving zooplankton model component into the three-dimensional ecosystem model of the Baltic Sea, we have to be keep in mind that the dynamic signatures of the different stages were designed for an example genera, the *Pseudocalanus*. The characteristic parameters were derived from data of studies of zooplankton in the Baltic Sea (Hernroth, 1985) in conjunction with comprehensive information on *Pseudocalanus* given in Corkett and McLaren (1978). However, in natural systems, and in particular in the Baltic Sea, there are several other calanoid genera, such as *Acartia* and

*Temora*, with slightly varying dynamic signatures and different life histories. While *Pseudocalanus* or *Temora* overwinter mainly as adult stages, the *Acartia* form dormant eggs that rest at the seabed until they are 'switched on' in spring. While *Acartia* is usually abundant in warm waters, the higher stages of *Pseudocalanus* can be found in deeper and more saline parts of the water column.

Downscaling the resolution of groups at the trophic level of zooplankton raises an interesting problem, which was not considered in Chapter 4. When we wish to improve the process resolution in a complex three-dimensional model by the replacement of a bulk variable by stage-resolving variables, we cannot confine the model component to only one group but have to take into account the remaining groups, which belong to the trophic level and were part of the bulk variable.

As a first approach, stages of a 'model copepod' can be considered as aggregated variables that comprise the corresponding stages of several groups of copepods, although their dynamic signatures are somewhat different and their overwintering strategy varies from overwintering adults to dormant eggs. This aggregation is a particularly reasonable approach for systems where the main groups have similar generation times and stage durations, as in the Baltic. This approach is supported by the results of tank experiments with *Pseudo-calanus*, *Acartia* and *Temora*, see Klein-Breteler et al. (1994), which showed an isochronic development with similar stage duration times. To circumvent the problem of different overwintering strategies, we assume that the dormant eggs of *Acartia* start hatching at about the time when *Pseudocalanus* starts to lay eggs. Moreover, our 'model copepod' develops both in colder and warmer areas, where the state variables refers, for example, to *Acartia* in the near-shore regions and in the warmer surface layer but to *Pseudocalanus* in the offshore areas and below the seasonal thermocline. Consequently, the aggregated state variables refer to different genera at different locations.

## 7.3 THREE-DIMENSIONAL SIMULATIONS

In this section, we describe model simulations with stage-resolving zooplankton state variables, where the stages of the model copepod aggregates the corresponding stages of the main groups as outlined in the previous section. The simulations cover a yearly cycle and are performed for the year 1999. The model was driven with meteorological data, which are made available in the frame of the BALTEX[1] project. Data to prescribe river runoff, nutrient loads and atmospheric nutrient deposition were constructed on the basis of climatological data. In particular, data of river loads were not available for the model year. In order to circumvent this problem, we modified data sets from the 1980s under the assumption that a general reduction of loads of 30% applies as

---

1. http://www.gkss.de/baltex/baltex_framell_builder.html

suggested by Lääne et al. (2002). The original data for the 1980s, which were used for the simulations in Chapter 6, are drawn from the Baltic Environmental Database[2] maintained at the University of Stockholm. The initial distribution of temperature and salinity fields were taken from the climatological data set provided by Janssen et al. (1999). The simulation starts on the 1 January 1999 after a spin-up run that was performed to allow currents and stratification to evolve from the initial fields and in response to the forcing. For the initialization of the biogeochemical state variables, data sets of the model run described in the previous chapter were used.

In order to provide initial conditions for the model copepods, we assume an evenly distributed background population of overwintering copepods with the biomass and abundance of $0.3 \, \text{mmol cm}^{-3}$ and $2400 \, \text{ind m}^{-3}$, respectively. This seed population remains in a dormant stage until the phytoplankton spring bloom commences. In response to the food signal, the overwintering model copepods start to grow and lay eggs after reaching the maturing mass. The parameters needed to describe the dynamic signatures of the model copepods are taken from Tables 4.2 and 4.3 in Chapter 4, apart from a modification of the respiration rate. We introduce a background respiration rate, $l_0$, in addition to the respiration rates that were linked to ingestion. Because this background respiration is independent of the ingested food, it follows that the biomass can decrease at depleted food resources while the number of individuals is not affected. The animals will lose weight and starve, as a result, we have to include also a starvation component of the mortality rate that acts on the number of individuals. This will be described in the next section.

### 7.3.1 Time Series of Basin Averages

In order to discuss the results of the model simulations, we start with the consideration of a time series of basin-averaged state variables. We look in particular at the stages of the model copepod and the bulk phytoplankton, which consists of the sum of the three functional groups expressed in chlorophyll units. For the discussion of succession and development of the copepod stages, we confine ourselves to the spatial averages of the state variables for the Bornholm basin, which is an important region for recruitment of cod in the Baltic Sea. The yearly variations of the mean abundance of the model copepod stages and the model chlorophyll for the Bornholm Sea are shown in Fig. 7.1, where the quantities are expressed in relative units, i.e. normalized by the maximum of the corresponding variable. The corresponding development of the mean individual body mass of the model adults, expressed as biomass concentration over abundance, is shown in Fig. 7.2. This figure shows also the mean body mass for the case of a no starvation component of the mortality rate, which will be discussed later.

---

2. http://data.ecology.su.se/Models/bed.htm

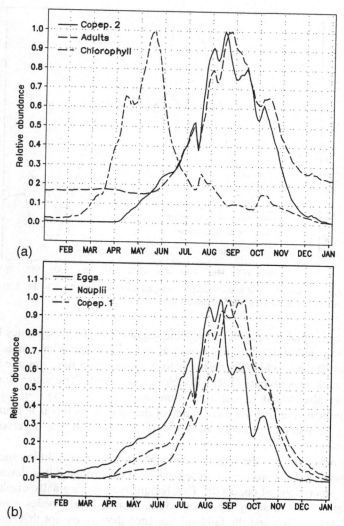

**FIGURE 7.1** Relative chlorophyll concentration and abundance of model copepods averaged over the upper 15 m in the central part of the Bornholm Sea.

The phytoplankton spring bloom starts in April, Fig. 7.1, and the overwintering model adults respond to the food signal with increasing body mass, Fig. 7.2. After a time period of about 1 week, the adults have reached their maturation mass and start to lay eggs. Just after the onset of the spring bloom, the increase in body mass is rather strong, while thereafter a mass level near the maturation mass is established over the whole vegetation period. In the autumn, the individual mass decreases towards the overwintering level.

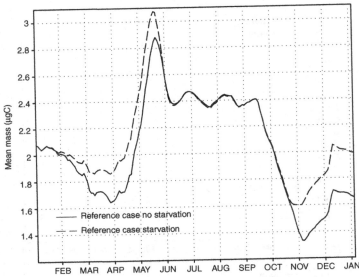

**FIGURE 7.2**   Mean individual mass of adults averaged over the upper 15 m in the central part of the Bornholm Sea, with and without starvation effect.

The egg signals propagate through the stages of nauplii, copepodites 1 and 2, succeeding within 1 week, and arrive at the adult stage after a time period of about 30 days, which corresponds to the generation time observed in the rearing tank experiments (Klein-Breteler et al., 1994). The same generation time was also found in our tank simulation, as shown in Fig. 4.13. The results indicate the internal consistency of the model, but this is not surprising because the parameters in the model were derived from the observed stage durations.

The time series of the egg abundance shows four peaks during the course of the year. The occurrence of these peaks can be attributed to a development of four generations of copepods. This feature is also marginally displayed by the nauplii, but less so by the other stages. Similar yearly cycles were found for the Arkona Sea and the Gotland Sea (not shown), except that the timing is somewhat different due to an earlier spring bloom in the Arkona Sea and a delayed spring bloom in the Gotland Sea.

In order to indicate the importance of the starvation adjustment for the development, the mean individual mass of adults for mortality rates with and without hunger correction is shown in Fig. 7.2.

For depleted food, the basic respiration, $l_0$, which is added to the loss term, $l$, in set (4.36), reduces the biomass while the number of individuals is not affected. This implies that the individuals lose weight and may starve. We take the starvation effect into account by adjusting the mortality, which acts on the abundance, in such a way that the mean individual mass of adults cannot fall below a hunger threshold. As the threshold, we chose the mid-interval mass, $\langle m \rangle_{c_2}$ (Table 4.3), of copepodites 2. To illustrate this adjustment, we consider the model adults in the simple case of no ingestion and, hence, no transfer to the adult stages and no egg laying. This case can be discussed analytically because the dynamical equations (4.36) and (4.41) can be reduced to

$$\frac{d}{dt}Z_a = -(l_0 + \mu_a)Z_a, \text{ and } \frac{d}{dt}N_a = -\mu_a N_a.$$

For a constant mortality rate, the integration from an arbitrary time, $t_0$, yields

$$Z_a(t) = Z_a(t_0)e^{-(l_0+\mu_a)(t-t_0)} \text{ and } N_a(t) = N_a(t_0)e^{-\mu_a(t-t_0)}.$$

As soon as the mean individual mass, $Z_a/N_a$, reaches the starvation threshold, the adjustment of the mortality rate, acting on the abundance, indicates the addition of $l_0$. Without the starvation adjustment, the body mass of overwintering adults becomes unrealistically low.

### 7.3.2 Spatial Distribution

In order to visualize the development of spatial distributions of the model phytoplankton and copepods, we look at a series of snapshots of several state variables, averaged over 2 days and over the upper 10 m. Four snapshots of the total model chlorophyll which refers to the sum of the three functional phytoplankton groups, are shown in Fig. 7.3. The modelled spring bloom starts

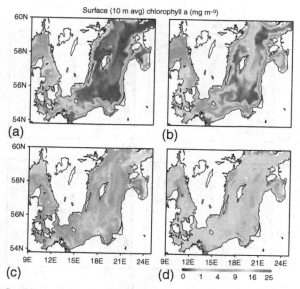

**FIGURE 7.3**  Snapshots of the development of the spatial distribution patterns of the model chlorophyll averaged over a 10-m-thick surface layer for (a) 31 March 1999, (b) 22 April 1999, (c) 22 May 1999 and (d) 10 August 1999. (See color plate.)

in the southwestern Baltic by the end of March and propagates towards the Baltic Proper during April and May, see Fig. 7.3. The strongest signals are found in the coastal areas, in particular, near river outlets. In August, the concentrations are relatively small, with the exception of enhanced signals, which occur locally in several river plumes.

The seed population of adult copepods starts to lay eggs in response to the spring bloom signal of the phytoplankton (not shown). These response patterns emerge, with a certain delay due to the hatching process, also in the model nauplii distribution, which is shown in Fig. 7.4. The model nauplii abundance, i.e. the prey for fish larvae, increases from spring to summer, from about 3000 to about $32,000$ ind m$^{-3}$ in the Arkona and Bornholm sea, while in the central part of Gotland Sea the abundance is smaller. Higher values are found in the summer, with maximum values close to $80,000$ ind m$^{-3}$ in the coastal regions. After the time interval needed to reach the molting mass, the nauplii appear in the copepodites stages, see Fig. 7.5 (copepodites 2 are not shown). The concentration of the model copepodites is smaller than that of nauplii, but the general distribution is similar. The abundance of the adults, which increases 2 months after the spring bloom, shows a similar spatial structure as the other stages, but the abundance is higher than for copepodites because the individuals accumulate in the adult stage, Fig. 7.6.

In summertime, the abundance of all stages approach their maximum level. Spatial patterns of the model copepods for a certain moment in August are

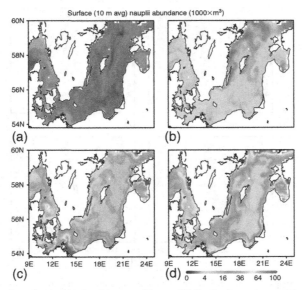

**FIGURE 7.4** Snapshots of the development of the spatial distribution patterns of the model nauplii averaged over a 10-m-thick surface layer for (a) 22 May 1999, (b) 21 June 1999, (c) 21 July 1999 and (d) 22 August 1999. (See color plate.)

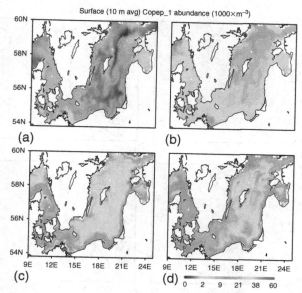

**FIGURE 7.5** Snapshots of the development of the spatial distribution patterns of the model copepodites 1 averaged over a 10-m-thick surface layer for (a) 22 May 1999, (b) 21 June 1999, (c) 21 July 1999 and (d) 22 August 1999. (See color plate.)

shown in Fig. 7.7. It appears that in some areas the spatial distributions of the different stages are similar, e.g. in the eastern Gotland Sea and the Bay of Gdansk. Different distribution patterns of different stages are found, e.g. in the Arkona and Bornholm sea. These similarities and differences indicate that in some cases, the mesoscale circulation is characterized by slowly varying flow patterns, which allow the development of the individuals through the stages within the same defined current patterns. In other areas, the current patterns vary greatly, implying that different stages are transported in different ways during their development. This illustrates that the distribution patterns of the different stages are formed by the mesoscale currents during the temporal development of the stages. If the time scales at which advection changes exceed the duration times of the different stages, we can expect similar spatial distribution patterns. If the time scales of advective changes are smaller than the stage duration times, the different stages occur in different spatial distributions.

Due to the very sparse data sets of zooplankton, it is not an easy task to verify the model results with the help of observations. However, there are data sets that were collected during intense ship campaigns during May and August 1999 in the Bornholm Sea (Möllmann, personal communication). A comparison of the model simulations with these data showed that the observed and simulated abundance are of the same order of magnitude and show similar structures for the total sums of the corresponding stages of the observed groups (*Pseudocalanus, Arcartia* and *Temora*).

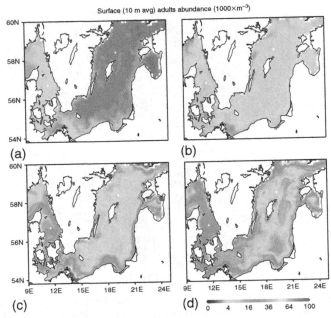

**FIGURE 7.6**    Snapshots of the development of the spatial distribution patterns of the model adults averaged over a 10-m-thick surface layer for (a) 22 May 1999, (b) 21 June 1999, (c) 21 July 1999 and (d) 22 August 1999. (See color plate.)

The results show how three-dimensional models can help us understand complex physical–biological interactions that act on the marine food web. Three-dimensional ecosystem models can provide distributions of prey for fish larvae and contribute to the understanding of the cause–effect chain of the bottom-up control of variations in fish recruitment.

Although the step from bulk zooplankton to stage-resolving zooplankton variables is an important improvement of the model, there are features that are not yet explicitly taken into account. For example, active vertical motion of the model copepods was ignored; i.e. the animals are transported only by currents and turbulent diffusion in a manner similar to a bulk zooplankton description. A background stock of adults is maintained in the model by setting the adult mortality rate to zero at a certain low-abundance threshold, while the other model stages diminish due to the corresponding mortality rates. However, as stated several times, modelling requires simplification, and models capture only a limited set of properties of real systems. For example, in natural systems, some egg laying and development is found also for depleted food resources. The model can be further developed or amended in several ways; e.g. copepods may also graze on detritus and may exercise active ontogenetic or diurnal vertical motion to draw from the food in the surface mixed layer.

Surface (10 m avg) zooplankton (1000×m⁻³)

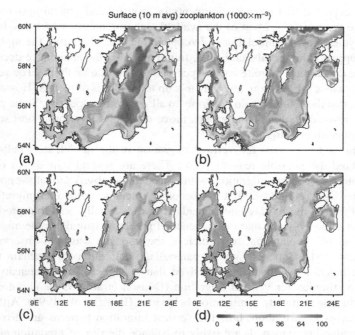

**FIGURE 7.7** Snapshots of the distribution patterns of (a) eggs, (b) nauplii, (c) copepodites 1 and (d) adults for 10 August. (See color plate.)

## 7.4 MODELLING OF BEHAVIOURAL ASPECTS

The term 'zooplankton' refers largely to animals that show individual behaviour, such as active vertical migration. Vertical migration can be stimulated by behavioural patterns, such as foraging, searching for energetic optimum conditions, avoidance of visual orienting predators or turbulence or migration over long distance by using currents.

We can broadly distinguish between diel migration, which refers to vertical motion of copepods with a pronounced daily cycle, and migration patterns on seasonal time scales. Diel migration can be triggered by environmental signals or may refer to ontogenetic behaviour. Stimuli for nondiel migration can be overwintering strategies or the placement in current systems for long-distance migration.

In this section, we focus on diel migration. It is obvious that *in situ* observations of vertical movements are difficult because these involve the general problem of resolving highly variable biological processes in time and space. The common instrument, a net, has a relatively coarse vertical resolution, and the creation of a sufficiently long time series implies a substantial effort. Hence, the number of direct measurements is relatively small. However, it can be expected that new developments in acoustic and optical systems will improve the situation. Most of the available measurements, which resolve the daily cycle, show accumulations of zooplankton animals during the nighttime in the upper, chlorophyll-rich layers. In the daytime, the abundance decreases because the animals migrate into deeper parts of the water column. The vertical travel distance ranges from a few metres up to hundreds of metres. However, it appears that these findings do not apply to all groups of copepods. Some genera show a reverse rhythm, some groups move vertically only for a short season, while other do not migrate at all.

In order to describe the vertical migration in the model, it is helpful to understand the possible reasons for it. There are several hypotheses on the stimulation of vertical migration, with considerable observational support. As stated by Dagg et al. (1998), it has become widely accepted that migration to dimly lit areas during daylight is made to avoid visually orienting predators. There is some evidence that the presence of predators stimulates the migration behaviour of zooplankton. Nevertheless, the opposite was also observed for small copepods. Copepods with a camouflage coloration may remain in the surface layer. It has been hypothesized that changes in light intensity may act as a stimulus for vertical migration (Dodson et al., 1997). The need for food can also stimulate a migration behaviour (Dagg et al., 1998). Although not much is known about how the vertical migration patterns are driven by food availability, copepods are likely to balance the risk of predation and the necessity of gaining energy from food. In this sense, the avoidance of predation and the need for food may control patterns of diel migration. Other hypotheses claim that avoidance of turbulence or Ultraviolet light are important factors. Due to the variety of migration behaviours, it is clear that migration patterns have group-specific or even stage-specific aspects, and there is no universal description that applies to all copepods. However, a combination of the two aspects, avoidance of predation and the necessity of gaining energy from food (Batchelder et al., 2002; Dagg et al., 1998), may represent the driving forces for a wide range of behavioural patterns of copepods.

## 7.4.1 Vertical Motion

We consider a model copepod that performs diurnal vertical motion and discuss ways how this property can be described theoretically. We follow the most straightforward approach, where the responses to environmental signals force the behaviour. As environmental parameters, we choose light, which serves as a trigger for avoidance of visually orienting predators, e.g. Dagg et al. (1998), and food availability. Let our model copepod prefer a certain relatively low light intensity, $I_0$, say, $0.1 \, \text{W m}^{-2}$. The depth where this level is assumed will vary over the course of the day. During daytime, the light level, $I_0$, is found in deeper parts of the water column, while it will move upwards towards the evening when the light grows dim. Assuming that the copepods tend to follow this light level, we can relate the vertical speed of copepods to the irradiation as

$$w_{light} = w_{light}[I_0 - I(z)], \tag{7.1}$$

where $I_0$ is the preferred light level and $I(z)$ is the light at depth $z$, which is the current position of the animal. The functional relationship can be adjusted to observations or plausible arguments. The vertical speed will be negative (downward) if the light level, $I(z)$, exceeds $I_0$ and positive (upward) if $I(z)$ is smaller than $I_0$. An approach could be, for example,

$$w_{light} = w_{max} \text{sign}[I_0 - I(z)] \frac{[I_0 - I(z)]^2}{I_h + [I_0 - I(z)]^2}, \tag{7.2}$$

where $I_h$ is a 'half-saturation' constant and $\text{sign}(x) = \pm 1$ for positive or negative arguments, respectively.

As the second migration-forcing mechanism, we consider the availability of food, e.g. phytoplankton, $P$, expressed as biomass concentration in carbon or nitrate units. As a plausible approach, we let the copepods sense the food gradient and define the corresponding vertical velocity as a function of the derivative with respect to $z$,

$$w_{food} = w_{food}[P_z(z)]. \tag{7.3}$$

Here $P_z(z)$ is the food concentration gradient at depth $z$, i.e. the position of the copepod. An approach similar to Equation (7.2) is

$$w_{food} = w_{max} \text{sign}(P_z) \frac{P_z^2}{k_{food}^2 + P_z^2}, \tag{7.4}$$

with a half-saturation constant, $k_{food}$, that controls the ability of the zooplankton to sense food gradients. This definition ensures that the speed, $w_{food}$, is always towards a local maximum of the food concentration. Apart from phytoplankton, the food concentration can include other components, such as detritus or micro-zooplankton, weighted by a preference or food-quality parameter.

Combining both contributions with active vertical motion shows that a movement towards a preferred light level, as well as towards the food maximum, involves conflicts. During daytime, the preferred light level is found in deep water, while high food concentrations usually stay near the surface or in the thermocline. This example shows that the animals have to decide between their conflicting preferences. The choice amounts to risk of death by predation or starvation. To solve this problem, we can take the state of nutrition of the copepods into account. This can be done in a straightforward manner for individual-based models, where the life cycle of each animal is followed. For our model copepods this can be related to the mean individual mass. If the animals are hungry, the urge for food becomes more important than the risk of potential predation. In the model, these ideas can be included by introducing a dynamically developing index of nutritional state, $f$, which is related to the mean individual mass of the copepods. We can define the index such that it varies between zero and unity, where values near zero mean the copepod is close to starvation and values near to unity refer to a good nutrition status. With the notations for the mean individual mass and the molting mass introduced in Chapter 4, we can define a nutritional state index for the stage, $i$, as

$$f_i = \theta \left( \overline{m}_i - \frac{X_{i-1}}{2} \right) \frac{\overline{m}_i - \frac{X_{i-1}}{2}}{X_i - \frac{X_{i-1}}{2}}; \qquad \frac{X_{i-1}}{2} \leq \overline{m}_i \leq X_i, \qquad (7.5)$$

where the step function was introduced to avoid negative values. This choice means that the search for food dominates the behaviour if the mean individual mass is equal to or less than half the molting mass of the previous stage.

Then both migration velocities can be combined with weight factors, $f$ and $1 - f$, as

$$w = f w_{\text{light}} + (1 - f) w_{\text{food}}. \qquad (7.6)$$

Clearly, in order to construct expressions that mimic behavioural patterns, we can take into account other environmental parameters that are computed in the three-dimensional model. Obvious examples are temperature, salinity and oxygen. The relative importance of the different parameters depends on how the considered genera respond to environmental signals and how important these responses are in the context of a given problem. The formulation of the processes in the model requires information on the properties of the considered copepods. There are still large gaps in the recent knowledge on behaviour of specific genera, and close cooperation among observations, experiments and theoretic descriptions is essential to developing realistic models.

An example of the proposed approach is given in Neumann and Fennel (2006) for dominating copepod species in the Baltic Sea. The focus in this study is on ontogenetic (seasonal) vertical migration rather than on diel vertical migration. In addition to the nutritional state temperature, salinity and oxygen are considered to estimate vertical migration. Different behavioural rules for copepod species and stages generate different vertical distribution patterns of copepodes.

### 7.4.2 Visibility and Predation

Most ecosystem models truncate the food web at the trophic level of zooplankton; therefore, the parameterization of mortality is a crucial point. There are a variety of determinants causing death, ranging from environmental conditions to availability of food to predation. In order to simulate the zooplankton mortality, both natural mortality and death due to predation have to be included. Moreover, different stages can cope with unfavourable conditions in different ways. Adult copepods, e.g. can escape bad conditions, whereas nauplii are less mobile and more restricted to their local environment.

Given the complexity of zooplankton mortality, the use of constant mortality rates is not satisfactory. In order to implicitly consider features of the truncated higher trophic levels, we may assume that a certain fraction of the death rate is due to visual-orienting predators. To this end, we split the mortality rate into a constant part and a light-dependent contribution. It was suggested by Fiksen et al. (1998) that a light intensity of about $1\,W\,m^{-2}$ is the lowest level where predators, i.e. fish including larvae, can sense prey visually. This knowledge can be used to construct the variable part of mortality in such a way that it vanishes for light below this critical level. As a functional shape, we may choose a limiting function of the Michaelis–Menten type. Then the mortality rate can be written as

$$\mu_i = \mu_0^{(i)} + \mu_v^{(i)}(z), \tag{7.7}$$

where

$$\mu_v^{(i)}(z) = C_i \frac{I(z)}{I_h + I(z)}, \tag{7.8}$$

with $\mu_0^{(i)}$ being the constant part of the mortality, $I(z)$ the light level at depth $z$, and $I_h$ the critical light level for visibility of prey. An example for the dependence of the mortality rates on water depth for different light intensities at the sea surface is sketched in Fig. 7.8. This approach mimics a daily cycle of predation and implies that during daytime, the mortality rates in the upper part of the water column are higher than in the dark, deeper water. This approach is supported by studies of the stomach contents of predators, which showed a clear feeding preference during daytime. Moreover, a light-dependent part of the mortality rates implies an annual cycle of the mortality rates in the temperate latitudes. From the viewpoint of model development, approach (7.8) has attractive features. However, whether or not this model formulation applies to a problem at hand also depends on the local condition of the considered systems and of the behavioural patterns of the relevant predators.

## 7.4.3 Individual-Based Versus Population Models

The mathematical formulation of behavioural patterns, as discussed in the previous sections, is obviously easier in the frame of individual-based models, where the status of individuals is tracked during their life cycle. As a rule IBMs are of increasing importance for increasing trophic levels (Batchelder et al.,

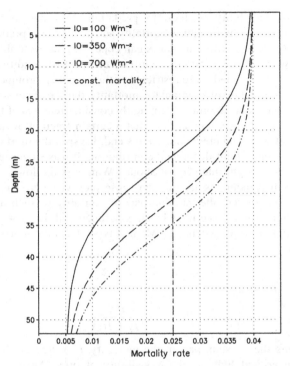

**FIGURE 7.8**  Dependence of zooplankton mortality on depth for different irradiance $I_0$ at sea surface.

2002; DeAngelis and Gross, 1992). There are certainly many problems that require an individual approach, in particular if we deal with systems of low abundance or if the genetic variation among a set of individuals is studied to understand adaptation processes. For systems with high abundance, however, it is vital to consider groups of copepods or other genera and characterize their properties by the behaviour of mean individuals. This implies statistical consideration: We do not require that all individuals behave in exactly the same way; we rather say that on average the individual behave like a mean one. In this chapter, it is shown that stage-resolving zooplankton models have the potential to take into account behavioural patterns in a statistical sense. Although there are always individuals that behave differently from the mean individual, the statistical averages over all individuals, i.e. the population, follows objective rules, provided the numbers of the individuals are sufficiently high.

## 7.4.4  Water-Column Models

In order to illustrate some of the outlined behavioural aspects, we need a model with high vertical resolution because these behavioural features cannot be described in box models discussed in Chapter 4. We choose a one-dimensional water-column model that horizontally covers only a few grid boxes of our full three-dimensional model. Such a model is easy to run and provides a workbench model to test cases that require a high vertical resolution. We extend the stage-resolving zooplankton model described in Chapter 4 by the processes outlined in the two previous sections.

The zooplankton stages, except the eggs, are able to move actively up and down in the water column according to Equations (7.2) and (7.4). The maximum swimming speed, $w_{max}$, is adapted to the body size, which in turn corresponds to the mass intervals of the stages. It ranges from $80 \, m \, day^{-1}$ for the adults to $8 \, m \, day^{-1}$ for the nauplii stage. For simplicity, we assume that the same behavioural strategy applies to all stages. The vertical migration is forced by a combination of two aspects: the need for food and escaping from the surface waters to avoid predation by visual-orienting predators. Sometimes both mechanisms can be in conflict, so we use the nutritional state index, defined by Equation (7.5), to estimate the relative weight of the two behavioural rules. The index is determined by the mean individual mass, i.e. the ratio of the state variables biomass and abundance. For a low level of the mean individual mass, the need for food dominates the fear of predation.

First we run the model without vertical migration. Some results of this reference case are shown in Fig. 7.9. The annual cycle of the vertically averaged total zooplankton biomass is shown in upper left panel, while the upper right panel shows how the nutritional state index for the adult mean individual mass varies. The lower panel shows the vertical distribution of the adult biomass concentration during a full model year and during the month of June. The isolines in the lower panels indicate the temporal variation of the vertical phytoplankton concentration expressed as chlorophyll.

In the first experiment, case 1, the copepods migrate vertically in response to the light, according to Equation (7.2). The results are shown in Fig. 7.10. As an obvious effect of escaping from the surface during daytime, the total zooplankton biomass increases. Note that the activity of predators is implicitly controlled through the light-dependent mortality, as illustrated in Fig. 7.8. The nutritional state index increases as well. The index is not used yet to control the vertical motion. Compared with the reference run, the vertical distribution of the adult biomass concentration shows a higher biomass below the upper mixed layer. The higher temporal resolution in the lower right panel shows a well-pronounced diel migration pattern.

In the next experiment, case 2, we now consider both the light and hunger aspects to force vertical migration as defined in Equation (7.6), while in the case 1 experiment the copepods were forced to go upward, regardless of their nutrition status. The results are shown in Fig. 7.11. If the copepods are in good shape, they remain in deeper waters. Consequently, the main difference from the results of case 1 is the more pronounced accumulation of the adults below the surface layer in a depth of about 40 m. The daily cycle is clearly displayed,

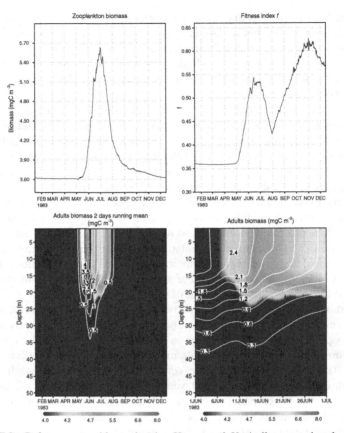

**FIGURE 7.9**    Reference case without migration. Upper panel: Vertically averaged total zooplankton biomass concentration (left). Nutritional state (fitness) index for the mean individual adult (right). Lower panel: Variation of the vertical distribution of the adult biomass concentration: (left) annual; (right) during June. The variation of chlorophyll concentration is also indicated by isolines. (See color plate.)

but during the night the concentration in the surface layer is smaller than in the other cases.

In this model configuration, vertically migrating zooplankton gain a benefit which results in a higher biomass. Slight changes of rules to force the migration speed can change the vertical distribution patterns of the model zooplankton. In case 2, the probability of finding zooplankton in deeper layers is higher than in case 1. Thus, small differences in the formulation of vertical migration can be assigned to different genera to find their potential ecological niches. However, for a sound validation of such models, more detailed knowledge

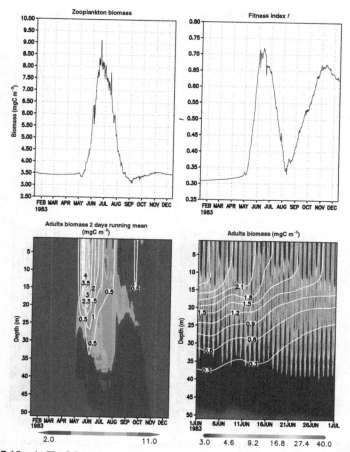

**FIGURE 7.10**    As Fig. 7.9, but for experiment case 1 with vertical migration driven only by light-dependent avoidance behaviour. (See color plate.)

of the behaviour is required. Nevertheless, these model experiments help us to understand quantitatively how behaviour can create distribution patterns through reaction and adaptation to environmental conditions. This can support the development of new hypotheses and stimulate new experimental investigations.

We also have to be aware that this model is still truncated at the trophic level of zooplankton, and predators are taken into account in terms of parameterization of mortality rates. The main predators of zooplankton, i.e. fish larvae and fish, can display complex behaviour and compensate for zooplankton defense strategies. In the next sections, we present an approach to how the model can be extended to include explicitly higher trophic levels. Among other aspects, this

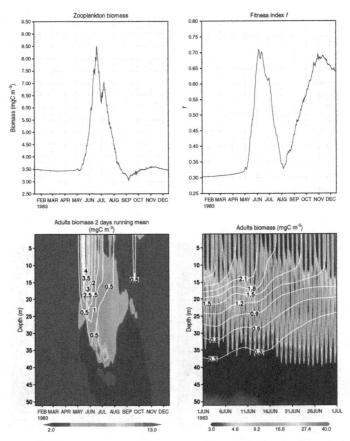

**FIGURE 7.11**   As Fig. 7.9, but for experiment case 2 with vertical migration driven by both light-dependent avoidance behaviour and hunger. (See color plate.)

extension permits the replacement of crude zooplankton mortality approaches by a more mechanistic approach.

## 7.5   FISH IN A THREE-DIMENSIONAL MODEL

The next step in expanding the model towards the higher tropic level is the inclusion of the fish model considered in Section 4.4 into the three-dimensional coupled physical–biological model. Such a three-dimensional model allows simulations of marine food webs to explore effects of varying fishing pressure and environmental and/or climate changes on marine ecosystems. Spatially explicit three-dimensional models can help identify biogeochemical feedback of fish and analyse connectivities of different habitats through material transports by fish. In this section, we discuss how the mass-class structured fish model, developed as a box model in Section 4.4, can be embedded into our three-dimensional physical–biogeochemical model system. There are several different approaches to end-to-end models; an overview can be found in Travers et al. (2007). The long life cycles of fish, which can span from a couple of years to decades, implies the need of longer simulations, ideally over decades as done for the box model in Section 4.4.6. Spatially explicit models with moving fish simulations over many years would require substantial efforts and are not yet available. The discussion in the following sections is largely based on a four-year run described in Radtke et al. (2013).

## 7.5.1 Coupling Upper and Lower Food Webs

Our aim is to embed the fish model component into a three-dimensional nutrient–phytoplankton–zooplankton–detritus (NPZD) model, in particular the model system Ecological ReGional Ocean Model (ERGOM), as described in Section 6.5.2, to generate a nutrient–phytoplankton–zooplankton–fish–detritus (NPZFD) model. In the biogeochemical model ERGOM, the zooplankton mortality also plays the role of truncation parameter, which implicitly includes the consumption of zooplankton by fish. Now we reduce the zooplankton mortality rate by two-thirds but let the zooplanktivorous fish act on the model zooplankton; i.e. instead of a more or less constant mortality rate in the truncated model, the consumption of zooplankton by fish now dominates the mortality of zooplankton. This provides the link of the lower food web to fish, while the feedback of fish to the lower food web is represented by respirational products of fish, which flow into the ammonium pool, as well as egestion, excretion and dead fish, which enter the detritus pool. This amounts to three pathways of mass fluxes between lower and higher parts of the food web.

Any description of fish in a spatially explicit three dimensional model implies that some aspects of fish behaviour, such as swimming, foraging, avoidance and migration to spawning grounds, have to be considered. We introduce a set of simplifications to make the problem tractable. Because fish can move up and down in a certain part of the water column, we chose the upper 60 m to find food, but we are not interested in an explicit resolution of small-scale motion or in the details of vertical movement. Instead, we work with a vertically integrated representation of fish. Within a certain time step, say, 1 day, fish can virtually scan a certain part of the water column that is of interest with respect to food. This approach amounts practically to a two-dimensional, horizontal fish model, where the growth rates of the zooplanktivores, i.e. herring, sprat and small cod, are controlled by the vertically averaged zooplankton concentration in the upper 60 m of the water column. While the consumption of zooplankton by fish changes the total zooplankton biomass, the shape of the vertical profile of the zooplankton concentration is maintained. In other words, the fish per area interacts with the vertically integrated zooplankton, while the zooplankton distribution remains three-dimensional.

## 7.5.2 Horizontally Moving Fish

Following Radtke et al. (2013), the horizontal motion of fish is governed by two *ad hoc* rules:

- Fish follow the food.
- Fish migrate to the spawning areas during the spawning periods.

Sprat, herring, and cod swim to locations of the best food conditions. The control parameters of the swimming speeds for the different mass classes (size classes) are listed in Table 7.1, along with the mass ratios of consumers and food. Moreover, the fish is instructed to migrate to the spawning areas that are

**TABLE 7.1** Swimming Speeds

| Species | Min Mass (g) | Length (cm) | Max Speed (length s$^{-1}$) | Max Speed, $v_{x_i}^{swim}$ (km day$^{-1}$) |
|---------|--------------|-------------|------------------------------|---------------------------------------------|
| Cod     | 5            | 10.8        | 0.8                          | 7.5                                         |
|         | 30           | 17.9        |                              | 12.3                                        |
|         | 100          | 24.5        |                              | 16.9                                        |
|         | 200          | 34.6        |                              | 23.9                                        |
|         | 800          | 48.5        |                              | 33.5                                        |
|         | 1500         | 73.8        |                              | 51.0                                        |
|         | 10000        |             |                              |                                             |
| Herring | 5            | 9.6         | 1.2                          | 10.0                                        |
|         | 10           | 12.9        |                              | 13.4                                        |
|         | 30           | 17.4        |                              | 18.1                                        |
|         | 60           | 22.8        |                              | 23.6                                        |
|         | 150          | 28.7        |                              | 29.8                                        |
|         | 240          |             |                              |                                             |
| Sprat   | 5            | 10.8        | 1.0                          | 9.3                                         |
|         | 10           | 12.9        |                              | 11.2                                        |
|         | 15           | 14.5        |                              | 12.5                                        |
|         | 20           | 16.7        |                              | 14.4                                        |
|         | 35           |             |                              |                                             |

The choice of maximum swimming speeds of cod, herring and sprat related to body length. The maximum speed expressed in body length per second is constant for all mass classes of each fish species. The maximum swimming speeds, $v_{x_i}^{swim}$, in the unit km day$^{-1}$ is used in the model. Note that the smallest mass class is not listed because recruits do not swim actively.

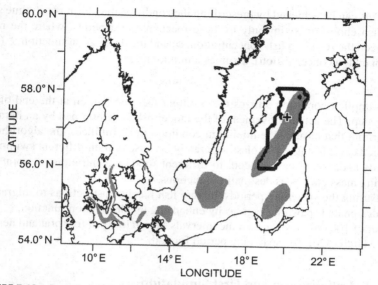

**FIGURE 7.12**   Spawning areas (grey) of cod and sprat in the Baltic Sea, after Bagge and Thurow (1993). The cross indicates the station BMPJ1 for later reference. The thick solid line marks the 80m isobath.

the deep basins for sprat and cod, Fig. 7.12, and to shallow coastal areas for herring.

To estimate the swimming speed of the different mass classes, we use the geometric average for the mass interval $m^{geo} = \sqrt{m_{min}m_{max}}$ and relate this to a length. All fish of the same species are assumed to swim at a constant relative speed (body lengths per second). In order to quantify the speed with the help of data from the literature, it is important to distinguish between an average speed that fish are able to maintain during the day and the maximum speed measured, e.g. in escape situations. Extreme high swimming speed of up to 18.6 body lengths per second are reported for herring (Misund and Aglen, 1992). In Radtke et al. (2013), the swimming speeds of $0.8 \, \mathrm{bl\,s^{-1}}$ and $1.0 \, \mathrm{bl\,s^{-1}}$ are used for cod and herring, respectively, as suggested by Fernö et al. (2011) and Huse and Ona (1996). For sprat, the critical swimming speeds are reported to be slightly higher than for herring (Turnpenny, 1983). Hence, we also chose a slightly higher average speed of $1.2 \, \mathrm{bl\,s^{-1}}$.

The implementation of fish movement in the model is based on a nonlocal algorithm (Radtke et al., 2013). The method provides a mechanism that fish swim with speeds up to $v_{x_i}^{swim}$ to match destination functions. During one time step, $\Delta t = 1$ day, the fish can reach all grid points within a circle of the radius $\rho = \Delta t \, v_{x_i}^{swim}$. The destination function for the zooplanktivores is given by the zooplankton distribution, while the distributions of sprat and herring serve as destination function, for the predator fish. Thus, the model arranges the fish proportional to the distribution of food.

This can be achieved by prescribing the condition that during each time step the fish choose to swim only to those reachable grid points, where the ratio between the resulting fish concentration, $\phi$, and the destination function, $\zeta$ (e.g. zooplankton concentration), assumes a minimum, i.e.

$$\phi/\zeta = \text{min.} \tag{7.9}$$

This implicit condition utilizes the resulting fish distribution at the end of the time step. The actual movement of the model fish is determined by an iterative algorithm that ensures the minimum condition to be fulfilled. The algorithm is applied to each single size class separately because of their different swimming speeds and, in the case of cod, to different food concentrations of sprat and herring mass classes as destination functions.

During the spawning periods, mature fish receive instructions to migrate to the designated spawning regions by enhancing the destination function, $\zeta$, by a factor of 100. We have chosen the intervals March to April for sprat and herring and March to May as spawning period for cod.

### 7.5.3 Initialization and First Simulation

The initial fish distribution is chosen to match the average zooplankton distribution, which was calculated in pre-runs with the biogeochemical model ERGOM with a truncated food web, as described in Section 6.6. The underlying assumption is that fish follows the food, and hence, the general pattern of the zooplankton matches the general prey fish distribution. Similarly, the predators follow the prey and their spatial distributions will match them as well. In the first experimental simulations, the model system was integrated over a time period of 4 years, 1980–1983 (Radtke et al., 2013). The spreading of the horizontal distributions of the vertical averaged zooplankton biomass is shown in Fig. 7.14. The zooplankton development starts in the southern part of the Baltic and propagates northward. Comparing these patterns with the annual cycle of phytoplankton shown in Figs. 6.10 and 6.11, reveals that the zooplankton distribution follows the propagation of the phytoplankton bloom. It is clear that this propagation mainly represents a phase speed, which reflects the different timings of the spring blooms in different parts of the Baltic but does not reflect a physical movement of zooplankton, apart from a certain drift of zooplankton by currents.

The dynamics of the total model zooplankton biomass of the whole Baltic Sea is shown in Fig. 7.13. Although these plots include the signals from both the southern and northern parts of the Baltic, i.e. from part with earlier and later onsets of zooplankton development, it appears that the total model zooplankton biomass increases already in April, which is a little too early. This is typical for model configurations with a bulk zooplankton state variable. As discussed in Chapter 4, in a stage-resolving zooplankton model, the animals propagate through the different stages, and hence, the biomass development is slower.

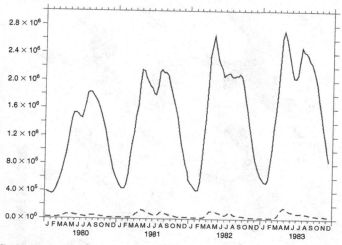

**FIGURE 7.13**    Time series of the modelled biomass of bulk zooplankton in tons. The dashed line refers to the zooplankton mass in the deep area of the Eastern Gotland Basin, see Fig. 7.12, after Radtke et al. (2013).

The total biomass peaks in August and declines in October. Observational estimates of the total zooplankton biomass are difficult to obtain. One way is to use fish data and assess the zooplankton biomass consumed by zooplanktivorous fish. For the Baltic Sea, Arrhenius and Hansen (1993) applied this method, and their findings indicate a later start of the zooplankton development and higher peak levels, of up to $10^7$ tons, which exceeds those shown in Fig. 7.13.

### 7.5.4  Results and Discussion

The model was run for 4 years to show the development of spatial distributions of cod, sprat and herring in the three-dimensional model. A selection of 10-day means of fish biomass is shown for herring, sprat and cod in Figs. 7.15–7.17, respectively. During their spawning period, March to April, herring stays in the shallower near-coastal areas but spreads over wide areas of the Baltic Sea during the summer. The horizontal migration of herring aims for regions where zooplankton had developed in response to the phytoplankton blooms. In September, see Fig. 7.15, the zooplankton biomass decreased and substantial amounts can mainly be found in coastal areas where herring accumulates. A part of the model herring migrates northward into the Bothnian Bay, where zooplankton development has progressed during the summer. During March and April, the model sprat concentrates in the spawning areas located in the central parts of the deep basins of the Baltic Proper. After the spawning period, sprat spreads over larger areas to follow the food and also migrates into the northern parts of the Baltic. Nevertheless, a significant part of the stock stays close to the spawning areas in the central Baltic. Similar to as sprat during the

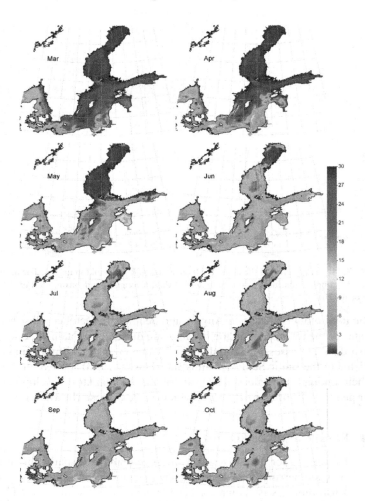

**FIGURE 7.14**   Distribution of the vertically integrated zooplankton model, averaged over the first 10 days of each month, from March to October in 1980 ($t\,km^{-2}$), after Radtke et al. (2013). (See color plate.)

time span of March and April, the model cod aggregates in the spawning areas from March to May. The search for food starts in June when the instruction to stay in the spawning areas no longer applies and the fish begin to aim for food with as the highest priority.

Contrary to the annual development of the zooplankton distribution, Fig. 7.14, the distribution patterns of fish, Figs. 7.15–7.17, are basically controlled by the horizontal migration of fish. As listed in Table 7.1, the swimming speeds are rather high so that the larger mass classes can cross whole basins in a span of a week or two. Thus, fish can relatively quickly reach locations where food is available, and this justifies, to some degree, the use of

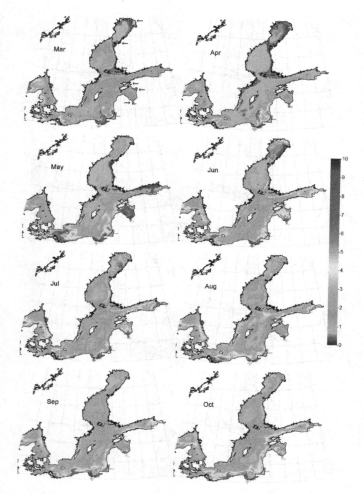

**FIGURE 7.15** Distribution of the total herring biomass averaged over the first 10 days of each month, from March to October in 1980 $(t\,km^{-2})$, after Radtke et al. (2013). (See color plate.)

box models for qualitative discussions of the temporal developments of fish stocks. However, in box models, the fish can use food items immediately, while in a spatially resolved model, there is a delay caused by the time that fish need to find the food.

## 7.5.5 Fish and Biogeochemistry

The question of whether fish may have any effect on the biogeochemistry of the sea may need to be addressed. We expected that this would not be the

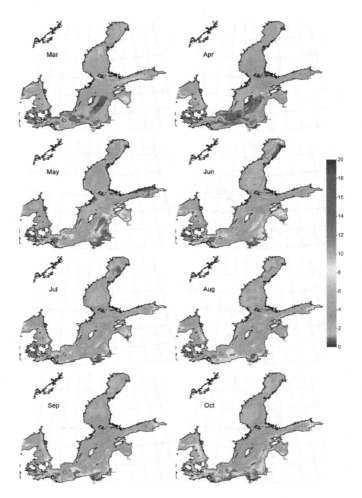

**FIGURE 7.16** Distribution of the total sprat biomass averaged over the first 10 days of each month, from March to October in 1980 (t km$^{-2}$), after Radtke et al. (2013). (See color plate.)

case because otherwise the truncated biogeochemical models, which end at the zooplankton mortality, are not well justified. However, there are two issues that may potentially affect biogeochemical fluxes:

- In a full nutrient-to-fish model, the mortality of zooplankton varies spatially because of fish migration. If fish went to areas with abundant food, the zooplankton mortality becomes small in locations where the zooplankton concentration is relative low. This may have effects in comparison to the truncated models with a constant zooplankton mortality.

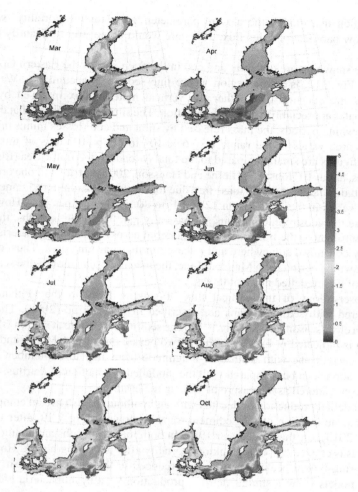

**FIGURE 7.17** Distribution of the total cod biomass averaged over the first 10 days of each month, from March to October in 1980 (t km$^{-2}$), after Radtke et al. (2013). (See color plate.)

- When fish aggregate in spawning areas, it can contribute to the pools of dissolved inorganic nutrients (DIN) and particular organic nutrients (PON) by respiration and excretion.

Because the food levels are low during the spawning time of sprat and cod in the spawning area the excretion products of the fish mainly consist of material that was eaten earlier, outside the spawning region. A measure to show the net effect on material fluxes caused by fish is the mass loss of fish minus the uptake through consumption of zooplankton in the spawning areas. Without an explicit fish model, the closure parameter 'zooplankton mortality' would somehow parameterize this flux, and there would be no way to quantify these effects.

The spawning area of sprat and cod in the deep part of the Eastern Gotland Basin, Fig. 7.12, is also a region where fine sediments accumulate. We may address the question of whether the vertical material flux imposed by fish contributes appreciably to the sedimentation. The simulation showed that during the spawning periods, the mass released by sprat and cod (losses minus uptake) in this area varies in the range of $0.6 \times 10^3$ to $1.5 \times 10^3$ tons of nitrogen. These figures are small compared to the total $N$ loads to the Baltic Sea (from all sources, about $10^6$ tons, e.g. Hjerne and Hansson, 2002) but are still one order of magnitude smaller than the mass flux due to nitrogen fixation in the same area, which varies in the model from $12 \times 10^3$ to $20 \times 10^3$ tons per year. However, because the geostrophic currents are basically parallel to the isobaths, the fish can contribute to the transport of fine material across isobath, i.e. material that was by consumed somewhere and released in the spawning area. Thus, during the spawning period from March to June, the fish model can modify the seasonal cycle of the modelled nutrient fluxes.

Observations of the vertical flux of sinking PON in the Gotland Sea measured with sediment traps are described in Leipe et al. (2008). The data show a clear seasonality, see Fig. 7.18. The sedimentation rate at station BMPJ1, which is indicated in Fig. 7.12, shows two peaks, one during March and April and a second one with a stronger magnitude from July to September. These observations can be compared with the modelled production of detritus, which can be considered as a measure of sinking material.

Modelled production of detritus with and without the fish model component at the same location of the sediment traps is shown in Fig. 7.19, after Radtke et al. (2013), for the model year 1981. In both models, a substantial amount of detritus is created by phytoplankton mortality after the spring bloom. However, the spring-bloom peak is significantly reduced in the model with fish. While both models show a similar detritus production by a cyanobacteria bloom in autumn, the flux of zooplankton to detritus has been changed significantly. In spring, the losses of zooplankton to detritus are strongly diminished if zooplankton is eaten by fish. The fish mortality balances this decrease in detritus generation only partly. When the larval fish left their spawning region in summer, the zooplankton mortality is small, and an increase of zooplankton during fall is well pronounced in the new model. Note that zooplankton mortality was reduced to one-third of the value of the truncated model, while fish impose a strong predation mortality on zooplankton. The better agreement of the detritus flux in the full model with the seasonal pattern of the measured PON fluxes in Fig. 7.18 is obviously due to the role of fish.

It is likely that the sedimentary fluxes caused by fish are even somewhat underestimated by the model in Radtke et al. (2013) because only a certain amount of the detritus arrives at the bottom. The duration of the downward

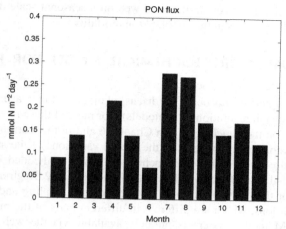

**FIGURE 7.18** Vertical flux measured in a sediment trap experiment at station BMPJ1 in the Central Gotland Sea (57.3°N, 20.0°E, indicated in Fig. 7.12) in 180-m depth (data courtesy of Falk Pollehne). Shown are the monthly means from 1996 to 2011.

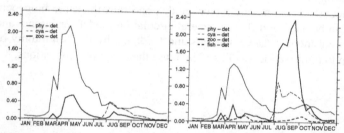

**FIGURE 7.19** Modelled production of detritus by nitrate-limited phytoplankton (thin solid line), cyanobacteria (thin dashed line), zooplankton (thick solid line) and fish (thick dashed line): (a) for the model without fish and (b) for the model including fish (mmol N m$^{-2}$ day$^{-1}$). The numerical values correspond to the grid cell at the location of the station BMPJ1 in the Central Gotland Sea (57.3°N, 20.0°E, indicated in Fig. 7.12) for the model year 1981, after Radtke et al. (2013).

motion is controlled by the sinking velocity, which usually varies with the size classes of particles, while in the model all detritus sinks at the same speed. Some part of the sinking material will already be mineralized within the water column. In particular, slowly sinking particles are more likely to be remineralized before they reach the bottom. In reality, fish fecal pellets will sink quite fast and contribute more to the sedimentation of PON as other items of detritus.

The analysis of the simulation indicates that the transport of mass through moving fish is a mechanism of connecting the depositional areas in the central Baltic with the coastal and shallower areas of the sea. The fish stock dynamics also modify nutrient fluxes in the food web on a seasonal scale due to spatial and temporal variations in the zooplankton mortality.

## 7.6 ERGOM: A BIOGEOCHEMICAL MODEL FOR REGIONAL SEAS

In the first chapters of this book, we discussed models of increasing complexity but without spatial resolution (box models). For most of these model, the codes are provided on the attached CD. In Chapter 6 and in this current chapter, the models are more complex because the spatial variations of the state variables are taken into account. The biogeochemical model embedded in circulation model is called ERGOM and is described in Section 6.5.2. Regional aspects are important for the treatment of, e.g. bentho–pelagic coupling and suboxic and anoxic conditions, which are different in different parts of the marine system. The ERGOM model system is publicly available via the web portal www. ergom.net and can freely be used for research and education. ERGOM was basically designed for the coupling of a biogeochemical model with a three-dimensional circulation model. However, for learning, testing and teaching it can also run in a zero- or one-dimensional environment. ERGOM is formulated independently from the circulation model, and thus the same biogeochemical model can be used in different general circulation models or in one- and zero-dimensional models. Developments and feedback by the community of users can help to improve the model and increase the range of applications.

# Chapter 8

# A Brief Introduction to MATLAB

## 8.1 FUNDAMENTALS

MATLAB is a widely used interactive software package for scientific and engineering numeric computations. It integrates numerical mathematics, signal processing and graphics in a user-friendly environment. The name 'MATLAB' stands for 'matrix laboratory'. MATLAB is a registered trade name of MathWorks, Inc., which provides a wealth of information on its website[1]. The software package comes with many toolboxes that provide powerful tools for data manipulation, such as interpolation or spectra analysis. Here we give only a brief discussion of those parts that are needed to solve the equations of the simple ecosystem models used in Chapters 2, 3, and 4 and to visualize the results.

To communicate with MATLAB, we start at the MATLAB prompt in the command window. MATLAB has a help function that explains all MATLAB commands by typing **help** and the command. The essential objects are matrices. A matrix is an array of numbers. Scalars are 1-by-1 matrices and vectors are matrices with only one row or column. The easiest way to enter matrices is to write an explicit list at the MATLAB prompt. The list of elements is separated by blanks and surrounded by '[' and ']'. A vector V for example, can be entered by the statement

$$V = [1\,2\,3]$$

which results in the output

$$V =$$

$$1\ 2\ 3$$

The vector V, which has the dimension (1,3), is saved in the workspace for later use. A matrix can be typed by several lists of element separated by semicolons. A matrix M of the dimension (3,3) is, e.g.

$$M = [1\ 2\ 3;\ 4\ 5\ 6;\ 7\ 8\ 9]$$

---

1. http://www.mathworks.com

Introduction to the Modelling of Marine Ecosystems. http://dx.doi.org/10.1016/B978-0-444-63363-7.00008-8

The output passed to the workspace is

$$M =$$
$$\begin{matrix} 1 & 2 & 3 \\ 4 & 5 & 6 \\ 7 & 8 & 9 \end{matrix}$$

Matrix elements can individually be referenced with indices in side parentheses. For example M(1,1) returns

$$ans =$$
$$= 1$$

and M(2,3) gives

$$ans =$$
$$= 6$$

The command $M(2,3) = 10$ replaces the element with the indices (2,3) by the number 10 and typing M gives

$$M =$$
$$\begin{matrix} 01 & 2 & 3 \\ 4 & 5 & 10 \\ 7 & 8 & 9 \end{matrix}$$

Larger matrices can be constructed using little matrices. For example, we can attach a row to the matrix M by adding a vector A,

$$A = [20\ 30\ 40]$$

as

$$M = [M; A]$$

The result is

$$M =$$
$$\begin{matrix} 1 & 2 & 3 \\ 4 & 5 & 10 \\ 7 & 8 & 9 \\ 20 & 30 & 40 \end{matrix}$$

A useful command to create vectors with defined increment is

$$x = 1{:}3$$
$$1\ 2\ 3$$

The default step is one. An arbitrary step, e.g. 0.1 can be prescribed by x=1:.1:3. The MATLAB functions **ones**(m,n) and **zeros**(m,n) create a $m$-by-$n$ matrix of ones and zeros, respectively. The command **size**(M) returns the number of rows and columns of a matrix M as a two-row vector.

## 8.1.1 Matrix and Array Operations

The transpose of a matrix can be obtained by adding a prime, ', to the name of the matrix. Let W=V', and V be the vector [1 2 3], then

$$V =$$

$$1\ 2\ 3$$

and

$$W =$$

$$1$$

$$2$$

$$3$$

Addition and substraction of matrices are denoted by + and −. These operations are array operations and are only defined when the matrices have the same dimension. An exception is the addition or substraction of a scalar. For example, M+4 adds 4 to each element of M.

Multiplication of matrices is denoted by *. The operation is only defined when the inner dimension of the two matrices are the same; i.e. M*N is permitted when the second dimension (number of columns) of M is the same as the first dimension (number of rows) of N. An exception is the multiplication by a scalar. For example, M*4 multiplies each element of M by 4.

Try the multiplication of our two vectors, V and W,

$$V * W = 14$$

while the multiplication of W and V gives

$$W * V =$$

| 1 | 2 | 3 |
|---|---|---|
| 2 | 4 | 6 |
| 3 | 6 | 9 |
| 20 | 30 | 40 |

An important operation is the elementwise multiplication, which is indicated by a dot before the star, .*. For example,

$$V. * V =$$

$$1\ 4\ 9$$

or

$$W. * W =$$

$$1$$

$$4$$

$$9$$

Similarly, the elementwise division is enabled by '. /'. If we wish to calculate the inverse elements of a matrix, M, the command 1./M will give an error message, which says that the matrix dimensions must agree. The scalar, 1, in the nominator is treated like a 1-by-1 matrix. To get the correct result, we have to define a matrix with unit elements of the same dimension as M. This can be achieved with the commands **ones** and **size** as **ones**(**size**(M)).

The matrix elements can also be defined by functions, e.g.

$$X = [\exp(-2), \mathbf{sqrt}(3), \mathbf{sin}(\mathrm{pi}/4)]$$

Let a vector $x$ be defined by x = 1:10, then the $\sqrt{x}$-function, **sqrt**(x), creates a vector of the same size as x, where the elements are the values of the function for the elements of x. The step function $\theta(x)$ can be realized by a relational operator in parenthesis, (x > 0), which return one for positive arguments x and zero for negative values.

Apart from referencing individual elements of a matrix by enclosing their subscripts in parentheses, we can also use vector subscripts, which allows us to access submatrices. For example, let M be a 10-by-10 matrix. Then

$$M(1:5, 10)$$

returns a 5-by-1 submatrix, i.e. a column vector, that consists of the first five elements of the tenth column, while

$$M(1:4, 1:4)$$

specifies the 4-by-4 submatrix of the elements of the first four columns and rows. Using the colon by itself as a subscript refers to all elements of the corresponding column or row. For example, the statments

$$M(:, 1) \text{ and } M(:, 5)$$

return the first and the fifth column.

## 8.1.2  Figures

MATLAB has a sophisticated graphic toolbox that is very helpful in visualizing formulas and data. The easiest way to become familiar with the syntax of the MATLAB commands is to go through an example. To this end we reproduce Fig. 2.3, which shows the functions

$$P = P_0 \frac{P_0 + N_0}{P_0 + N_0 \exp[-k'(N_0 + P_0)t]},$$

and

$$N = N_0 \frac{(N_0 + P_0)\exp[-k'(N_0 + P_0)t]}{P_0 + N_0 \exp[-k'(N_0 + P_0)t]},$$

versus time. Note that $N = P_0 + N_0 - P$.

First we specify the time interval from zero to the final time, tfin, with a step of 0.01; i.e. type in

$$\text{tfin} = 8; t = 0 : .01 : \text{tfin};$$

The output on the screen can be suppressed by the semicolon at the end of the line. We specify the constants by typing the list

$$\text{N0} = 5; \text{P0} = .5; r = 1; k=r/\text{N0};$$

The next step is to type the two formulas,

```
PP=P0+N0*exp(−k*(P0+N0)*t);
P=P0*(P0+N0)*ones(size(t))./PP;
N=P0+N0-P;
```

Note in the second line we generated a matrix with unit elements of the size of t. Now we are prepared to start with the graphical representation. First, we type at the MATLAB prompt

$$f = \textbf{figure}$$

which produced a figure window, where 'f' is an optional figure handle, which can be useful for further manipulations. The graphical representation will be invoked with the **plot** command. Typing

$$\textbf{plot}(t,N,t,P)$$

produces the graph shown in Fig. 8.1. If we wish to draw the graphs in different styles, we can specify the line style by a string command after the corresponding variable. For example, **plot**(t,N,t,P, '-.') plots the second curve in dash-dotted mode. Other options are ':' and '- -' for dotted and dashed lines. Solid lines, which are indicated by the symbol '-', are the default style. Further information can easily be obtained by typing **help print**. Thus, we type

$$\textbf{plot}(t,N,t,P, \text{'-.'})$$

The axis can be manipulated with the **axis** command. In particular,

$$\textbf{axis}([X_{\min} X_{\max} Y_{\min} Y_{\max}])$$

specifies the range of the $x$ and $y$-axes. We may type

$$\textbf{axis}([0 \text{ tfin } 0 \text{ } 6])$$

In order to give labels to the axis, we use the commands **xlabel**('string') and **ylabel**('string') and type

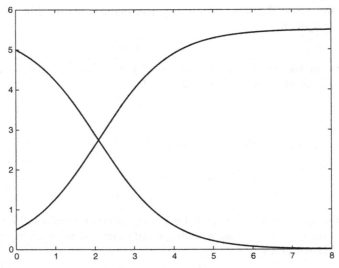

**FIGURE 8.1**   First step to create Fig. 2.3.

xlabel('t/d')
ylabel('N & P').

In Fig. 2.3, we draw lines to indicate the initial values of $P_0$, $N_0$ and the sum $P_0 + N_0$. This can be done with the **line**(X,Y) command, which draws a line from the start and end points specified in the vectors X and Y.

h1=**line**([0 tfin],[P0 P0])
h2=**line**([0 tfin],[N0 N0])
h3=**line**([0 tfin],[N0+P0 N0+P0]).

Here, the handles h1, h2 and h3 were assigned to each **line** command. Then our graph changes as shown in Fig. 8.2.

The style of the lines can be modified by using the **set** command. This command allows us to call each line with the handle and change the line style. The syntax is

set(h1,'LineStyle','- -')
set(h2,'LineStyle','- -')
set(h3,'LineStyle','- -')

Moreover, we may wish to add a legend to the plot. The command **legend**('N','P') inserts a box for the strings that refer to the variables P and N at a certain location. The location of the box can be manipulated by an optional number. For example, adding a zero selects an automatic 'best' placement with least conflict with the curves. For other options and additional information, type

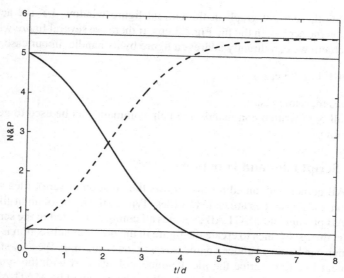

**FIGURE 8.2**   Development of Fig. 2.3.

**help legend**. To proceed, we may add the command

   **legend**('N','P',0).

Text can be inserted with the **text**(X,Y,'string') command, which specifies the location, X and Y, in the coordinates of the figure and inserts the string. We can type

   **text**(5,P0+.2,'P_0')
   **text**(2,N0+.2,'N_0')
   **text**(2, P0+N0+.2, 'P_0+N_0')

Now the graph shown in Fig. 2.3 is produced. Usually, we wish to send the figure to a printer or save it into a file. This can be done with the **print** command. The line

   **print**

sends the current figure to the local printer.

Print into a file has the syntax print -d[device] [file-name]. For an encapsulated Postscript we type

   **print** -deps Fig.2.3.eps

This prints the current graph in the current figure window, i.e. the activated window on the screen, in the file Fig.2.3.eps. If there are several figure windows on the screen, we can directly address a figure by its handle, in our case,

**print**(f,'Fig.2.3.eps')

This gives the same result.

For all bold written commands, the help command can be used to get more information.

## 8.1.3 Script Files and Functions

MATLAB comes with an editor that allows the creation of script files with the extension .m. Those files are called m-files. Type in the name of an m-file at the MATLAB prompt and MATLAB executes all commands listed in the script. For example, the list of lines to create the figure in the previous subsection is a script file, which can be saved under a name, say, Fig_generation.m. We may start with a comment line to describe the file. Comments lines start with the symbol %, which has the effect that text behind it will not be executed by MATLAB.

```
% Creation of Fig. 2.3
% parameters
N0=5; P0=.5;  % mmol N/m3
r=1;       %1/day
tfin=8;t=0:.01:tfin; % time interval
k=r/N0;    % 1/day/mmol N/m3
PP=P0+N0*exp(-k*(P0+N0)*t);
P=P0*(P0+N0)*ones(size(t))./PP;
N=P0+N0-P;
f=figure;
plot(t,N,t,P, '-.');
axis([0 tfin 0 6]);
xlabel('t/d'), ylabel('N and P');
h1=line([0 tfin],[P0 P0]);
h2=line([0 tfin],[N0 N0]);
h3=line([0 tfin],[N0+P0 N0+P0]);
set(h1,'LineStyle','- -');
set(h2,'LineStyle','- -');
set(h3,'LineStyle','- -');
legend('N','P',0);
text(5,P0+.2,'P_0');
text(2,N0+.2,'N_0');
text(2, P0+N0+.2, 'P_0+N_0');
print(f,'Fig.2.3.eps')
```

Note that the semicolon, ';', at the end of each line suppresses the display of the results of the executed line on the screen. Typing the command **help** followed by the name of the m-file displays all comment lines at the beginning of the m-file until the first command line is reached. For example,

**help** Fig_generation

displays the first two lines of our script,

Creation of Fig. 2.3
parameters

while typing

Fig_generation

runs the code and creates the Fig. 2.3.

Functions are special m-files that differ from script files in that arguments can be passed. The first line of a function file starts with the word 'function'. For example, we can write a limiting function for nutrient uptake (2.10), i.e.

$$f(N) = \frac{N}{k_N + N}$$

as a function file.

```
function f=MiMe(N)
% Michealis Menten formula
k_N=.3;
f=N./(k_N+N);
```

Note that the elementwise division, './', is only required if the argument, $N$, is a vector. After saving the file as MiMe.m, we can type at the MATLAB prompt the name of the file with an argument. Typing MiMe(.3) MATLAB returns ans=0.500, or MiMe(1) gives ans= 0.7692. If we wish to plot the function over the range from 0 to 2, we may type, e.g. the line x=0:.1:2;plot(x,MiMe(x)). Typing

**help** MiMe

displays all comments starting with second line, in our case,

Michealis Menten formula

As a second example, we write a code to estimate the normalized length of the day using the formula (5.2). We choose the latitude of 54°N.

```
function Delta_d=daylength54(t)
% Calculation of the length of the day
```

```
% based on astronomical formulas
% ————————————————————————
latitude=54;
lat=latitude*2*pi/360;
% zenith angle
% Fourier series (Spencer)
A=[.006918 -.399912 -.006758 -.002697];
B=[.070257 .000907 .001480];
G=2*pi/365*t;
% inclination
inc=A(1)+A(2)*cos(G)+A(3)*cos(2*G)+A(4)*cos(3*G);
inc=inc+B(1)*sin(G)+B(2)*sin(2*G)+B(3)*sin(3*G);
% length of the day
arg=-sin(lat)*sin(inc)/cos(lat)./cos(inc);
Delta=24/pi*acos(arg);    %in hours
Delta_d=Delta/24;         % normalized
```

For example, for the day 75 we type daylength54(75). The code returns

$$\text{ans} =$$

$$= 0.4874$$

## 8.2 ORDINARY DIFFERENTIAL EQUATIONS

In various parts of the book, we dealt with systems of differential equations of the type,

$$\dot{x} = \frac{\mathrm{d}}{\mathrm{d}t}x = f(t, x).$$

Here, $x = x(t)$ is a function of time, $t$. The differentiation with respect to $t$ is often noted by a dot over the function. A unique solution requires the prescription of initial conditions,

$$x(t_0) = x_{in}.$$

Because the differential quotient is defined by the limit of

$$\frac{x(t + \Delta t) - x(t)}{\Delta t} \approx \frac{\mathrm{d}}{\mathrm{d}t}x = f(t, x),$$

for small time intervals $\Delta t \to 0$, we can find the solution by

$$x(t + \Delta t) \approx x(t) + \Delta t \cdot f[t, x(t)].$$

This so-called Euler's method is the simplest approach, which may work if the time steps $\Delta t$ are sufficiently small, i.e. small compared to the typical rate in the problem at hand. In some cases, it might be useful to apply this simple method. However, MATLAB provides an advanced solver that allows us to choose the required accuracy of the solution. We choose the example of 'ode23', where 'ode' stand for ordinary differential equation. The structure of the solver, which integrates such an initial value problem from the initial time, t0, to the final time, tfinal, is

$$[t, x] = \text{ode23}('\text{ode} - \text{name}', \text{tspan}, x_0).$$

Here, ode23 is the name of the solver, tspan is a vector specifying the interval of integration, x0 is the vector of initial condition for the problem and 'ode-name' is the name of a function file that describes the system of ordinary differential equations. The output argument t is a column vector. The elements are the time points at which the solution was calculated. The second output argument $x$ is the solution matrix, where the rows correspond to the solution at the corresponding time points, i.e. the elements of $t$.

There are several choices for *tspan*. If we set tspan = [t0 tfinal], the solver returns the solution evaluated at every integration step. For tspan vectors with more than two elements, the solver returns solutions at the time points give in tspan. For example, if we wish to see the solution on integer values of the time interval, we set tspan = [t0 : tfinal]. The choice of the time interval does not affect the accuracy of the solution because the solver uses internal time steps that traverse the interval from the first to the last element of the vector tspan. We note that MATLAB provides several solvers, e.g. 'ode45' or 'ode113', that might be useful for different cases. We consider only one solver here and refer to the MATLAB manuals for more information. A quick overview of the set of solver can be obtained by typing
**help** ode23.

In summary, to solve the model equations, a script m-file and a function file are needed. The script file prescribes the initial conditions, integration interval and the used solver and creates figures to visualize the results. The function file define the differential equations that constitute the model. In order to illustrate the procedure, we use the MATLAB scripts to solve the simple problem discussed in Chapter 2. The model is based on Equations (2.25)–(2.28). The model run produce the results shown in Fig. 2.11.

```
% simple box model of a marine ecosystem
%===============================
% t0 initial time of simulation
% tf final time
% state vector [N P Z D]
% initial vector x_in
%
t0=1; tf=365;
tspan=[t0 tf];
x_in=[0.99 .01 .01 .99]';
% call ordinary differential equation solver
[t,x]=ode23('NPZDfunc',tspan,x_in);
```

```
% visualize the output
f1=figure
subplot(2,1,1)
plot(t,x(:,1),'-',t,x(:,2),'–',t,x(:,3),'-.');
axis([t0 tf 0 1.1]); xlabel('t/d');
ylabel('N P Z');
title('nutrient-phytoplankton-zooplankton cycle, upper box');
legend('N','P','Z',0);
subplot(2,1,2)
plot(t,x(:,4));
axis([t0 tf 0 2]); xlabel('t/d');
ylabel('D');
title('detritus cycle in lower box');
```

We may save this file under the name 'NPZDdyn.m'. The result of integration consists of a column vector t, which is specified by the length of time interval and by the number of points needed for the integration. The solution x is a matrix with four columns, one for each state variable, while the number of rows is the same as the number of elements of t. In order to plot the different state variables, we use the colon and the number of the column as subscripts to cut the state variable out of the matrix x. For example, in the plot command the, expression x(:,1) refers to the first state variable, the nutrients, N.

Before the programme can be run, we have to define the function file that describes the model and that was called as the string 'NPZDfunc' in the ode23-solver routine. This function file, which contains the code for Equations (2.25)–(2.28), can be written as,

```
function xdot=NPZDfunc(t,x)
% function corresponding to NPZDdyn
%*******************************************
%      PARAMETER LIST
%---------------------------------------------
% phytoplankton
k_N=.1;        % half saturation constant
r_max=1;       % maximum uptake rate of phytoplankton
LPN=.01;       % respiration / extracelluar release
LPD=.02;       % loss through sedimentation or mortality
% zooplankton
g_max=.5;      % maximum grazing rate;
I=1.2;         % Ivlev Konstante
LZN=.01;       % in upper layer recycled material
LZD=.02;       % loss rate of zooplankton
x0=.01;        % background plankton level
%-------------- process control ---------------
d0=daylength54(75);   % threshold, length of the day
```

```
dl=daylength(t);
theta=(dl > d0);        % switch on "biology"
A_mix=.5;               % winter mixing
r_max=r_max*theta;
uptake=r_max*x(1)*x(1)/(k_N+x(1)*x(1));
g_max=g_max*theta;
grazing=g_max*(1-exp(-I*x(2)*x(2)));
theti=1-theta;          % switch off biology
A_mix=theti*A_mix;
%**********************************************
% M O D E L   E Q U A T I O N S
%─────────────────────────────────────────────
xdot(1)=-x(2)*uptake+LZN*(x(3)-x0)+LPN*(x(2)-x0)...
+A_mix*(x(4)-x(1));
xdot(2)= x(2)*uptake-(LPN+LPD)*(x(2)-x0)-x(3)*grazing;
xdot(3)=x(3)*grazing-(LZD+LZN)*(x(3)-x0);
xdot(4)=LZD*(x(3)-x0)+LPD*(x(2)-x0)...
-A_mix*(x(4)-x(1));
xdot=xdot';
```

Note that for the first and fourth element of xdot, two lines were used for the corresponding equations. If more than one line is needed, we can continue the statement by writing the continuation command, three dots ('...'), at the end of the line and continue on the next line. Then save this function file under the name 'NPZDfunc.m' in the same directory as 'NPZDdyn.m'. Each step time the function file 'NPZDfunc.m' is called in the solver ode23 in the file 'NPZDdyn.m', it returns the vector 'xdot' at the corresponding time $t$. Now we can type 'NPZDdyn' and get as output the matrix of the state variables for all time steps and the figures of the state variables versus time that correspond to Fig. 2.11.

Because the solvers are rather complex, we also show how the solution can be obtained with the Euler method. Here, we allocate the vector for the time variable and the matrix for the state variables with the aid of the command **zeros**. Note that **zeros**(n,4) defines a matrix(n,4) with zero elements and **zeros**(n,1) creates a vector of length $n$ with zero elements.

```
% NPZDdyn_Euler.m uses the Euler method
% simple box model of a marine ecosystem
% generates Fig. 2.11
%==============================
% t0 initial time of simulation
% tf final time
% state vector [N P Z D]
% initial vector x_in
```

```
%
t0=1; tf=365;
tspan=[t0 tf];
x_in=[0.99;.01;.01;.99];
% Euler method % define time steps
timestep=0.05;
n_timesteps=365/timestep;
x1=x_in;
x=zeros(n_timesteps,size(x_in,1));
t=zeros(n_timesteps,1);
for i=1:n_timesteps;
t1=(i-1)*timestep+t0;
xdot=NPZDfunc(t1,x1);
x1=x1+timestep*xdot;
x(i,:)=x1;
t(i)=t1;
end
% visualize the output
f1=figure
subplot(2,1,1)
plot(t,x(:,1),'-',t,x(:,2),'--',t,x(:,3),'-.');
axis([t0 tf 0 1.1]); xlabel('t/d');
ylabel('N P Z');
title('nutrient-phytoplankton-zooplankton cycle, upper box');
legend('N','P','Z',0);
subplot(2,1,2)
plot(t,x(:,4));
axis([t0 tf 0 2]);
xlabel('t/d');
ylabel('D');
title('detritus cycle in lower box');
```

Running this m-file again reproduces Fig. 2.11. It is interesting that larger time steps, e.g. timestep=0.1 or even timestep=0.5 still gives stable results and reproduces Fig. 2.11. However, if the time step is further increased, the method becomes unstable.

## 8.3  MISCELLANEOUS

We describe some further statements that are useful for the MATLAB programming in the context of this book. Ordinarily, each MATLAB function, defined by an m-file, has its own local variables, which are separate from those of other functions and from those of the base workspace. However, if a variable needs to be defined in several functions, this variable must be declared as global in each corresponding m-file. The line

global X Y Z

defines X, Y, and Z as global in scope. All m-files that declare the global variables share a single copy of that variable, and any assignment to that variable, in any function, is available to all the other functions. Stylistically, global variables should have long or not too simple names to avoid confusion with local variables, although this is not required. Another nice feature is the use of cells, which are defined as sections of the code embedded between two lines and that start with the symbol '%%'. Cells can be run separately from the MATLAB editor with the button 'run section'. This is particularly useful for larger m-files.

In order to display a text on the screen to indicate some properties of the running m-files, the text must be defined as a string, e.g. X='example', then the statement **disp(X)** displays

example

on the screen.

A useful statement for labelling plots with the commands **title**, **xlabel**, **ylabel** is the string conversion command **num2str**, which can be read as 'number to string'. For example, T = num2str(X) converts the matrix X into a string representation of T with about four digits and an exponent if required.

Sometimes we have data specified at several points and wish to interpolate these data at a set of other points. For data interpolation, we can use the m-file **spline**, which is based on a cubic spline interpolation. For example, Yint = spline(X,Y,Xint) interpolates the underlying function Y to find the values of Yint at the points specified in the vector Xint. The vector X specifies the points at which the data Y is given.

# Appendix

# Content of the Booksite

The book has a web site, http://booksite.elsevier.com/978044633637/, on which two types of complementary material are provided:

- Animated pictures
- MATLAB files

A selected number of movie files shows the temporal and spatial variations of a subset of state variables and illustrate the performance of the three-dimensional model. The MATLAB programs comprise most of the relatively simple models discussed in Chapters 2, 3 and 4 of this book. The booksite is structured as follows: In the root are two folders: movies and MATLAB. In the movie folder, there are subdirectories named 'chlorophyll', 'DIN' and 'PO4'. Each of these folders contains movie files for the corresponding state variables. For each simulation year, a separate folder named by the year, i.e. '1980'–'1990', is provided. All movies show the simulated surface concentrations of the corresponding state variables. The shown chlorophyll animations refer to the sum of the functional groups of primary producers, which are expressed in chlorophyll units. The variable DIN (dissolved inorganic nitrogen) refers to the sum of model nitrate and ammonium, while PO4 corresponds to the phosphate model state variable. To start the movies you have to choose a state variable and then click on the year.

The MATLAB directory on the booksite is structured in four subdirectories: one each for Chapters 2 and 3 and two for Chapter 4, called 'chap4_zoo' and 'chap4_fish'. The MATLAB files listed in the subdirectories refer to most of the models discussed in the first four chapters and produce, in particular, many of the corresponding figures shown the chapters. All files are listed in Tables A.1–A.3, where the script files, i.e. m-files, the corresponding function files, options and output, in form of the generated figures, are indicated. For example, the MATLAB file NPZDdyn.m calls the function NPZDfunc.m, which in turn involves the function daylenght54.m and generates the output shown in Fig. 2.11. The same applies for the file NPZDdyn_Euler.m, where the Euler's method is used instead of a MATLAB ode-solver.

**TABLE A.1** List of M-files Used in Chapter 2

| M-file | Functions involved | Comments |
|---|---|---|
| NPZDdyn.m | NPZDfunc.m | Fig. 2.11 |
| NPZDdyn_Euler.m | daylength54.m | |
| NPZDdyn_1.m | NPZDfunc_1.m daylength.m | Fig. 2.14 |
| NPZD_eutro.m | NPZDfunc_eutro.m daylength.m | OP_ti_on=1 Fig. 2.17 |
| | | OP_ti_on=2 Fig. 2.18 |

Open the subdirectory 'chap2' in the MATLAB directory. To run the models, type in the name of the m-files listed in the left column at the MATLAB prompt, or press the run button from the MATLAB editor.

For the application of the m-file NPZD_eutro.m, it is necessary to open the file with the MATLAB editor and toggle between the options:

'N supply at the rate $Q^{ext} = 2 \times 10^{-3}$ day$^{-1}$ for 3 years', or
'N supply at the rate $Q^{ext} = 2 \times 10^{-3}$ day$^{-1}$ for 3 years and removal at the same rate for the next 3 years'.

To choose between the options, we have to enable or disable one of the statements 'OP_ti_on=1' or 'OP_ti_on=2' with the aid of the comment sign %, see the lines 20–22 of the file NPZD_eutro.m.

The m-files in the subdirectory 'chap3' involve more complex choices of different cases, which are invoked by several options. The examples comprise the cases with or without external nitrogen or phosphorous loads and the cases are split between subcases with and without nitrogen fixation by cyanobacteria. The variable 'OPTION' is assigned to one of the five cases at line 22 in the m-file SN_FGZD.m and at line 28 in SN_FGZDOx.m, while a second variable 'Op_Tio_n' has to be set to the numbers 1 or 2 to invoke nitrogen fixation ('Op_Tio_n=2') or not ('Op_Tio_n=1'). The output in terms of figures related to the options is listed in Table A.2.

The m-files used in Chapter 4 are more complex. But with the experiences of the codes used in the preceding chapters, it is a straightforward procedure to handle the codes. An overview about the codes and the options and scenarios is given in Table A.3. The direct way to start with the model is to select a file in the left column of Table A.3 and open the file with the MATLAB editor. Then the options and scenarios can be chosen and the model can be started by the 'run' button.

**TABLE A.2  List of M-files Used in Chapter 3**

| M-file | Functions involved | Comments |
|---|---|---|
| SN_FGZD.m | SN_FGZDfun.m daylength.m | OPTION=1: Fig. 3.6, Fig. 3.7, Fig. 3.9 |
| | | OPTION=2: Fig. 3.10, Fig. 3.11, Fig. 3.12 |
| | | OPTION=3: Fig. 3.13, Fig. 3.14, Fig. 3.15 |
| | | OPTION=4: Fig. 3.16, Fig. 3.17 |
| | | OPTION=5: Fig. 3.18, Fig. 3.19 Fig. 3.20, Fig. 3.21 |
| SN_FGZDOx.m | SN_FGZDOxfun.m daylength.m | OPTION=1: Fig. 3.23, Fig. 3.24 Fig. 3.25, Fig. 3.26 Fig. 3.27 |
| | | OPTION=2: Fig. 3.28, Fig. 3.29 Fig. 3.30 |
| | | OPTION=3: Fig. 3.31, Fig. 3.32 Fig. 3.33 |
| | | OPTION=4: Fig. 3.34, Fig. 3.35 Fig. 3.36 |
| | | OPTION=5: Fig. 3.37, Fig. 3.38 Fig. 3.39, Fig. 3.40 Fig. 3.41 |

Open the subdirectory 'chap3' in the MATLAB directory. To run the models, open the m-files listed in the left column with the MATLAB editor, toggle an option and press the run button.

**TABLE A.3** List of M-files Used in Chapter 4

| M-file | Functions involved | Comments |
|---|---|---|
| Tank.m | Tank_fun.m<br>PCpara.m, zfermi.m<br>fngraz.m, fc1graz.m<br>fc2graz.m, mtrans.m | Options: F_Egg_S=0,<br>T_water=8<br>Fig. 4.9, Fig. 4.9(b), (a) |
| | | Option: T_water=15<br>Fig. 4.11 |
| | | Options: F_Egg_S=1<br>T_water=15<br>Fig. 4.12, Fig. 4.13<br>Fig. 4.14 |
| ExpBaltic.m | ExpBaltic_fun.m<br>PCpara.m, zfermi.m<br>fngraz.m, fc1graz.m | Option: Mor_TALi_ty=1<br>Fig. 4.15, Fig. 4.16<br>Fig. 4.17, Fig. 4.18 |
| | fc2graz.m, mtrans.m<br>daylength.m, MiMe.m | Option: Mor_TALi_ty=2<br>Fig. 4.19 |
| | LossD.m | Option: Mor_TALi_ty=3<br>Fig. 4.20 |
| **M-file** | **Functions involved** | **Comments** |
| N2F_scenarios.m | N2F_fun.m<br>Mini.dat | Scenario 1<br>Fig. 4.23, Fig. 4.24 |
| | Temperature.dat | Scenario 2<br>Fig. 4.25 |
| | | Scenario 3<br>Fig. 4.26, Fig. 4.31(a) |
| | | Scenario 4<br>Fig. 4.27, Fig. 4.31(b) |
| | | Scenario 5<br>Fig. 4.28 |
| | | Scenario 6<br>Fig. 4.29, Fig. 4.30 |

Open the subdirectories 'chap4_zoo' or 'chap4_fish' in the MATLAB directory. To run the models, open the corresponding m-files listed in the left column wit the MATLAB editor, toggle option or scenarios and press the run button.

# Bibliography

Aksnes, D.L., Lie, U., 1990. A coupled physical–biological pelagic model of a shallow sill fjord. Estuar. Coast. Shelf Sci. 31, 459–486.

Aksnes, D.L., Ulvestad, K.B., Balino, B.M., Berntsen, J., Egge, J.K., Svendsen, E., 1995. Ecological modeling in coastal waters: towards predictive physical–chemical–biological simulation models. Ophellia 41, 5–36.

Arrhenius, F., 1998a. Growth and seasonal changes in energy content of young Baltic Sea herring (*Clupea harengus* L.). ICES J. Mar. Sci. 53, 792–801.

Arrhenius, F., 1998b. Variable length of daily feeding period in bioenergetics modelling: a test with 0-group baltic herring. J. Fish Biol. 52, 855–860.

Arrhenius, F., Hansen, S., 1993. Food consumption of larval, young and adult herring and sprat in the Baltic Sea. Mar. Ecol. Prog. Ser. 17, 125–137.

Arrhenius, F., Hansen, S., 1996. Food intake and seasonal changes in enrgy content of young Baltic Sea sprat (*Sprattus sprattus* L.). ICES J. Mar. Sci. 55, 319–324.

Backhaus, J.O., 1985. A three-dimensional model for simulation of shelf sea dynamics. Dtsch. Hydrograph. Z. 38, 165–187.

Bagge, O., Thurow, F., 1993. The Baltic cod stock, fluctuations and the possible causes. In: ICES C.M.1993/CCC Symposium/No. 14.

Bakun, A., Nelson, C.S., 1991. The seasonal cycle of wind-stress curl in the sub-tropical eastern boundary current regions. J. Phys. Oceanogr. 21, 1815–1834.

Baretta-Bekker, J.G., Baretta, J.W., Ebenhöh, W., 1997. Microbial dynamics in the marine ecosystem model ERSEM II with decoupled carbon assimilation and nutrient uptake. J. Sea Res. 38, 195–211.

Batchelder, H.P., Edwards, C.A., Powell, T.M., 2002. Individual-based models of copepod population in coastal upwelling regions: implications of physiologically influenced diel vertical migration on demographic success and nearshore retention. Prog. Oceanogr. 53, 307–333.

Batchelder, H.P., Miller, C.B., 1989. Life history and population dynamics of *Metridia pacifica*: results from simulation modelling. Ecol. Modell. 48, 113–136.

Batchelder, H.P., Williams, R., 1995. Individual-based modelling of the population dynamics of *Metridia lucens* in the North Atlantic. ICES J. Mar. Sci. 52, 469–482.

Black, K.P., Vincent, C.E., 2001. High-resolution field measurements and numerical modelling of intra-wave sediment suspension on plane beds under shoaling waves. Coast. Eng. 42(2), 173–197.

Bleck, R., Rooth, C., Hu, D., Smith, L., 1991. Salinity driven thermocline transients in a wind- and thermohaline forced isopycnic coordinate model of the North Atlantic. J. Phys. Oceanogr. 22, 1486–1505.

Blumberg, A.F., Mellor, G.L., 1987. A discription of a three-dimensional coastal ocean circulation model. In: Heaps, N. (Ed.), Three-Dimensional Coastal Ocean Models. American Geophysical Union, Washington, DC, pp. 1–16.

Bodungen, B., Bröckel, K., Smetacek, V., Zeitschel, B., 1981. Growth and sedimentation of the phytoplankton spring bloom in the Bornholm Sea. Kieler Meeresforsch. Sonderh. 5, 49–60.

Bryan, K., 1969. A numerical method for the study of the circulation of the world ocean. J. Comput. Phys. 4, 347–376.

Carlotti, F., Giske, J., Werner, F., 2000. Modeling zooplankton dynamics. In: Harris, R., Wiebe, P., Lenz, J., Skjoldal, H.R., M. Huntley (Eds.), ICES Zooplankton Methodology Manual. Academic Press, San Diego, pp. 571–667.

Carlotti, F., Nival, P., 1992. Model of copepod growth and development: moulting and mortality in relation to physiological processes during individual moult cycle. Mar. Ecol. Prog. Ser. 83, 219–233.

Carlotti, F., Sciandra, A., 1989. Population dynamics model of Euterpina acutifrons (Copepoda: Harpacticoida) coupling individual growth and larval development. Mar. Ecol. Prog. Ser. 56, 225–242.

Caswell, H., John, A.M., 1992. From the individual to the population in demographic models. In: DeAngelis, D.L., Gross, L.J. (Eds.), Individual-Based Models and Approaches in Ecology. Chapman and Hill, New York, pp. 36–66.

Chelton, D.B., deSzoeke, R.A., Schlax, M.G., Naggar, K.E., Siwertz, N., 1998. Geographical variability of the first baroclinic Rossby radius of deformation. J. Phys. Oceanogr. 28, 433–460.

Corkett, C.J., McLaren, I.A., 1978. The biology of Pseudocalanus. In: Russell, F.S., Yonge, M. (Eds.), Advances in Marine Biology. Academic Press, London, New York, San Francisco, Boston, Sydney, Tokyo, San Diego, pp. 2–211.

Cushing, D.H., 1959. The seasonal variations in oceanic production as a problem in population dynamics. J. Conseil. Exp. Mer. 24, 455–464.

Cushing, D.H., 1975. Marine Ecology and Fisheries. Cambridge University Press, Cambridge.

Dagg, M., Frost, B., Newton, J., 1998. Diel vertical migration and feeding in adult female Calanus pacificus, Metrida lucens and Pseudocalanus newmani during a spring bloom in Dabob Bay, a fjord in Washington USA. J. Mar. Syst. 15(1–4), 503–509.

de Baar, H.J.W., 1994. von Liebig's law of the minimum and plankton ecology. Prog. Oceanogr. 33, 347–386.

DeAngelis, D., Gross, L. (Eds.) 1992. Individual-Based Models and Approaches in Ecology. Chapman and Hill, New York.

DeAngelis, D.L., Rose, K.A., 1992. Which individual-based model is most approriate for a given problem? In: DeAngelis, D.L., Gross, L.J. (Eds.), Individual-Based Models and Approaches in Ecology. Chapman and Hill, New York, pp. 67–87.

Dodson, S., Tollrian, R., Lampert, W., 1997. Daphnia swimming behavior during vertical migration. J. Plankton Res. 19, 969–978.

Doney, S.C., 1999. Major challenges confronting marine biogeochemical modeling. Global Biogeochem. Cycles 13, 705–714.

Droop, 1974. The nutrient status of algal cells in continuous cultures. J. Mar. Biol. Assoc. 54, 825–855.

Ebenhöh, W., Baretta-Bekker, J.G., Baretta, J.W., 1997. The primary production module in the marine ecosystem model ERSEM II, with emphasis on the light forcing. J. Sea Res. 38, 173–193.

Ebenhöh, W., Kohlmeier, C., Radford, P.J., 1995. The benthic biological submodel in the European regional seas ecosystem model. Netherlands J. Sea Res. 33(3–4), 423–452.

Eilola, K., Gustafsson, B., Kuznetsov, I., Meier, H., Neumann, T., Savchuk, O., 2011. Evaluation of biogeochemical cycles in an ensemble of three state-of-the-art numerical models of the baltic sea. J. Mar. Syst. 88(2), 267–284.

Emory, W.J., Lee, W.G., Maagard, L., 1984. Geographic and seasonal distribution of Brunt Väisälä frequency and Rossby radii in the North Pacific and North Atlantic. J. Phys. Oceanogr. 14, 293–317.

Eppley, R.W., 1972. Temperature and phytoplankton growth. Fish. Bull. 70, 1063–1085.

Evans, G.T., Parslow, J.S., 1985. A model of the annual plankton cycle. Biol. Oceanogr. 3, 327–347.

Fasham, M.J.R., Ducklow, H.W., McKelvie, S.M., 1990. A nitrogen-based model of plankton dynamics in the oceanic mixed layer. J. Mar. Res. 48, 591–639.

Fennel, W., 1988. Analytical theory of the steady state coastal ocean and equatorial ocean. J. Phys. Oceanogr. 18, 834–850.

Fennel, W., 1989. Inertial waves and inertial oscillations in channels. Cont. Shelf Res. 9, 403–426.

Fennel, K., 1999a. Convection and the timing of phytoplankton spring blooms in the western Baltic Sea. Estuar. Coast. Shelf Sci. 49, 113–128.

Fennel, K., 1999b. Interannual and regional variability of biological variables in a coupled 3-d model of the western Baltic. Hydrobiologica 393, 25–33.

Fennel, W., 1999c. Theory of the Benguela upwelling system. J. Phys. Oceanogr. 29, 177–190.

Fennel, W., 2001. Modeling of copepods with links to circulation models. J. Plankton Res. 23, 1217–1232.

Fennel, W., 2008. Towards bridging biogeochemical and fish production models. J. Mar. Syst. 71, 171–194.

Fennel, W., 2010. A nutrient to fish model for the example of the Baltic Sea. J. Mar. Syst. 81, 184–195.

Fennel, W., Johannessen, O., 1998. Wind forced oceanic responses near ice edges revisited. J. Mar. Sys. 14, 57–79.

Fennel, W., Neumann, T., 1996. The mesoscale variability of nutrients and plankton as seen in a coupled model. German J. Hydrography 48, 49–71.

Fennel, W., Neumann, T., 2003. Variability of copepods as seen in a coupled physical biological model of the Baltic Sea. In: ICES Marine Science Symposia 219.

Fennel, W., Junker, T., Schmidt, M., Mohrholz, V., 2012. Response of the Benguela upwelling systems to spatial variations in the wind stress. Cont. Shelf Res. 45, 65–77.

Fennel, W., Lass, H.U., Seifert, T., 1986. Some aspects of vertical and horizontal excursion of phytoplankton. Ophellia 4(Suppl.), 55–62.

Fennel, W., Radtke, H., Schmidt, M., Neumann, T., 2010. Transient upwelling in the central Baltic Sea. Cont. Shelf Res. 30, 2015–2026.

Fennel, W., Seifert, T., Kayser, B., 1991. Rossby radii and phase speeds in the Baltic Sea. Cont. Shelf Res. 11, 23–36.

Fennel, K., Spitz, Y., Letelier, R.M., Abott, M., Karl, D.M., 2002. A deterministic model for $N_2$ fixation at stn ALOHA in the subtropical North Pacific Ocean. Deep Sea Res. II 49, 149–174.

Fernö, A., Jørgensen, T., Løkkeborg, S., Winger, W., 2011. Variable swimming speeds in individual Atlantic cod (*Gadus morhua* L.) determined by high-resolution acoustic tracking. Mar. Biol. Res. 7(3), 310–113.

Fiksen, O., Utne, A., Aksnes, D., ane J.V. Helvik, K.E., Sundby, S., 1998. Modelling the influence of light, turbulence and ontogenetic on ingestion rates in larval cod and herring. Fish. Oceanogr. 7(3–4), 355–363.

Fonselius, S., 1981. Oxygen and hydrogen sulphide conditions in the Baltic Sea. Mar. Pollut. Bull. 12, 187–194.

Franks, P.J.S., Wroblewski, J.S., Flierl, G.R., 1986. Behavior of a simple plankton model with food-level acclimation by herbivores. Mar. Biol. 91, 121–129.

Fransz, H.G., Mommaerts, J.P., Radach, G., 1991. Ecological modelling of the North Sea. Netherlands J. Sea Res. 28, 67–140.

Geider, R.J., MacIntyre, H.L., Kana, T.M., 1998. A dynamic regulatoy model of phytoplankton acclimation to light, nutrients, and temperature. Limnol. Oceanogr. 43, 679–694.

Gerritsen, H., Boon, J.G., van der Kaaij, T., 2001. Integrated modelling of suspended matter in the North Sea. Estuar. Coast. Shelf Sci. 53, 581–594.

Graf, G., Rosenberg, R., 1997. Bioresuspension and biodeposition: a review. J. Mar. Syst. 11, 269–278.

Gran, H.H., Braarud, T., 1935. A quantitative study of phythoplankton in the Bay of Fundy and the gulf of Maine. J. Biol. Board Canada 1, 279–467.

Gregoire, M., Lacroix, G., 2001. Study of the oxygen budget of the Black Sea waters using a 3D coupled hydrodynamical–biogeochemical model. J. Mar. Syst. 31, 175–202.

Griffies, S., 2004. Fundamentals of Ocean Climate Models. Princeton University Press, Princeton, NJ, USA.

Griffies, S., Pacanowski, R., Schmidt, M., Balaji, V., 2001. Tracer conservation with an explicit free surface method for $z$-coordinate ocean models. Mon. Weather Rev. 129, 1081–1098.

Grimm, V., 1999. Ten years of individual-based modelling in ecology: what have we learned and what could we learn in the future? Ecol. Modell. 115, 129–148.

Grimm, V., Railsback, S.F., 2005. Individual-Based Modeling and Ecology. Princeton University Press, Princeton.

Gulland, J.A., 1974. The Management of Marine Fisheries. Scientechnica, Bristol.

Gupta, S., Lonsdale, D.J., Wang, D.P., 1994. The recruitment pattern of an estuarine copepod: a biological–physical model. J. Mar. Res. 52, 687–710.

Haidvogel, D.B., Arango, H.G., Hedstrom, K., Beckmann, A., Malanotte-Rizzoli, P., Shchepetkin, A.F., 2000. Model evaluation experiments in the North Atlantic Basin: simulations in nonlinear terrain-following coordinates. Dyn. Atmos. Oceans 32, 239–281.

Haidvogel, D.B., Beckmann, A., 1998. Numerical models of the coastal ocean. In: Brink, K.H., Robinson, A.R. (Eds.), The Global Coastal Ocean: Processes and Methods. The Sea 10. John Wiley and Sons, New York, pp. 457–482 (Chapter 17).

Haidvogel, D.B., Wilkins, J.L., Young, R.E., 1991. A semi-spectral primitive equation ocean circulation model using vertical sigma and orthogonal curvilinear horizontal coordinates. J. Comput. Phys. 94, 151–158.

Hammer, C., Dorrien, C., Ernst, P., Gröhsler, T., Köster, F., Mackenzieand, B., et al., 2008. Fish stock development under hdyrographic and hydrochemical aspects, the history of the Baltic Sea fisheries and its management. In: Feistel, R., Nausch, G., Wasmund, N. (Eds.), State and Evolution of the Baltic Sea. John Wiley and Sons, Inc., Hoboken, NJ, pp. 543–581.

Hansson, S., Hjerne, O., Harvey, C.J., Kitchell, J.F., Cox, S.P., Essington, T.E., 2007. Managing baltic sea fisheries under contrasting production and predation regimes: ecosystem model analyses. Ambio 60, 259–265.

HELCOM, 1994. Baltic Sea Environment Proceedings No. 39. Baltic Marine Environment Protection Commission, Airborne Pollution Load to the Baltic Sea 1986–1990.

HELCOM, 1996. Baltic Sea Environment Proceedings No. 64B. Baltic Marine Environment Protection Commission, Third Periodic Assessment of the state of the marine Environment of the Baltic Sea 1989–1993.

Hernroth, L., 1985. Recommendations on Methods for Marine Biological Studies in the Baltic Sea: Mesozooplankton Assessment, Baltic Marine Biologists WG 14, Publication No. 10.

Hinrichsen, H.-H., Johns, M.S., Gronkjaer, E.A.P., Voss, R., 2001. Testing the larval drift hypothesis in the Baltic Ssea: retention versus dispersion caused by wind driven circulation. ICES J. Mar. Sci. 58, 973–984.

Hirche, H.J., Meyer, U., Niehoff, B., 1997. Egg production of *Calanus finmarchius*: effect of temperature, food and seasons. Mar. Biol. 127, 609–620.

Hjerne, O., Hansson, S., 2002. The role of fish and fisheries in Baltic Sea nutrient dynamics. Limnol. Oceanogr. 47, 1023–1032.

Hjort, J., 1914. Fluctuations of the great fisheries of northern Europe in the light of biological research. Rapp. Proc. Verb. Reun. Cons. Perm. Int. Explor. Mer. 20, 1–228.

Holt, J.T., James, I.D., 1999. A simulation of the southern North Sea in comparison with measurements from the North Sea project part 2 suspended particulate matter. Cont. Shelf Res. 19, 1617–1642.

Horbowy, J., 1989. A multispecies model of fish stocks in the Baltic Sea. Dana 7, 23–43.

Horbowy, J., 1996. The dynamics of Baltic fish stocks on the basis of a multispecies stock-production model. Can. J. Fish. Aquat. Sci. 53, 2115–2125.

Humborg, C., Fennel, K., Pastuszak, M., Fennel, W., 2000. A box model approach for a long-term assessment of estuarine eutrophication, Szczecin Lagoon, southern Baltic. J. Mar. Syst. 25, 387–403.

Huse, I., Ona, E., 1996. Tilt angle distribution and swimming speed of overwintering Norwegian spring-spawning herring. ICES J. Mar. Sci. 53, 863–873.

IOC, SCOR, IAPSO, 2010. The international thermodynamic equation of seawater – 2010: calculation and use of thermodynamic properties, Manuals and Guides No. 56, UNESCO. Tech. Rep., IOC, SCOR and IAPSO.

Ivlev, V.S., 1945. The biological productivity of water. Usp. Sovrem. Biol. 19, 98–120.

Janssen, F., Schrum, C., Backhaus, J., 1999. A climatological dataset of temperature and salinity for the North Sea and the Baltic Sea. German J. Hydrography 9, 1–245.

Jansson, B.O., 1976. Modeling of Baltic ecosystems. Ambio 4, 157–169.

Jansson, B.O., 1978. Advance in oceanography. In: Charnock, H., Deacon G. (Eds.), The Baltic – A System Analysis of a Semi-Enclosed Sea, Plenum Publishing Corporation, New York, London, pp. 131–183.

Jørgensen, S.E., 1994. Fundamentals of Ecological Modelling. Elsevier, Amsterdam, London, New York.

Kahru, M., 1997. Using satellites to monitor large-scale environmental changes: a case study of cyanobactereial blooms in the Baltic Sea. In: Kahru, M., Brown, C.W. (Eds.), Monitoring Algal Blooms: New Techniques for Detecting Large-Scale Environmental Changes. Springer-Verlag and Landes Bioscience, Berlin, pp. 43–61.

Kahru, M., Nommann, S., 1990. The phytoplankton spring bloom in the Baltic Sea in 1985, 1986: multitude of spatio-temporal scales. Cont. Shelf Res. 10(4), 329–354.

Kaiser, W., Renk, H., Schulz, S., 1981. Die Primärproduktion der Ostsee. Geod. Geoph. Veröff. R.4 33, 27–48.

Kaiser, W., Schulz, S., 1976. Zur Ursache der zeitlichen und räumlichen Differenz des Beginns der Phytoplantonblüte in der Ostsee. Fischerei-Forschung 14, 77–81.

Kaiser, W., Schulz, S., 1978. On the causes for the differences in space and time of the commencement of phytoplankton bloom in the Baltic. Kieler Meeresforsch. Sonderh. 4, 161–170.

Karl, D.M., Letelier, R., Tupas, L., Dore, J., Christian, J., Hebel, D., 1997. The role of nitrogen fixation in biogeochemical cycling in the subtopical North Pacific Ocean. Nature 388, 533–538.

Kierstead, H., Slobodkin, L.B., 1953. The size of water masses containing plankton blooms. J. Mar. Res. 12, 141–147.

Klein-Breteler, W.C.M., Schogt, N., van der Meer, J., 1994. The duration of copepod life stages estimated from stage frequancy data. J. Plankton Res. 16, 1039–1057.

Klimontovich, Y.L., 1982. Kinetic Theory of Non Ideal Gases and Non Ideal Plasmas. Pergamon Press, Oxford.

Kullenberg, G., 1981. Physical oceanography. In: Voipio, A. (Ed.), The Baltic Sea. Elsevier, Amsterdam, pp. 135–181.

Kundu, P.K., Chao, S.Y., McCreary, J.P., 1983. Transient coastal currents and inertia-gravity waves. Deep Sea Res. 30, 1054–1082.

Lääne, A., Pitkänen, H., Arheimer, B., Behrendt, H., Jarosinski, W., Lucane, S., et al., 2002. Evaluation of the implementation of the 1988 ministerial declaration regarding nutrient load reductions in the Baltic Sea catchment area. Tech. Rep. 524, Finish Environment Institute.

Lacroix, G., Gregoire, M., 2002. Revisted ecosystem model MODECOGeL of the Ligurian Sea: seasonal and interannual variability due to atmospheric forcing. J. Mar. Syst. 37, 229–258.

Larsson, U., Elmgren, R., Wulff, F., 1985. Eutrophication and the Baltic Sea: causes and consequences. Ambio 14, 9–14.

Leipe, T., Harff, J., Meyer, M., Hille, S., Pollehne, F., Schneider, R., et al., 2008. Sedimentary records of environmental changes and antropogenic impacts during the past decades. In: Feistel, R., Nausch, G., Wasmund, N. (Eds.), State and Evolution of the Baltic Sea. John Wiley and Sons, Inc., Hoboken, NJ, pp. 395–439.

Levitus, S., 1982. Climatological Atlas of the World Ocean. NOAA, Prof. Pap. 13.

Lisitzin, E. (Ed.), 1974. Sea Level Changes. Elsevier, Amsterdam, 286 pp.

Livingstone, D.M., 1993. Lake oxygenation: application of a one-box model with ice cover. Int. Rev. Gesamten Hydrobiol. 78, 465–480.

Lynch, D.R., Gentleman, W.C., McGillycuddy, D.J., Davis, C.S., 1998. Biological–physical simulations of *Calanus finmarchius* population dynamics in the Gulf of Maine. Mar. Ecol. Prog. Ser. 169, 189–210.

Matthäus, W., 1995. Natural variability and human impact reflected in long term changes in the Baltic deep water conditions – a brief review. Dtsch. Hydrograph. Z. 47, 47–65.

Mayzaud, P., Poulet, S.A., 1978. The importance of the time factor in the response of zooplankton to varying concentrations of naturally occurring particulate matter. Limnol. Oceanogr. 23, 1144–1154.

McCreary, J.P., Kundu, P., Chao, S., 1987. On the dynamics of the California current system. J. Mar. Res. 45, 1–32.

McGillicuddy, D., Lynch, D.R., Wiebe, P., Runge, J., Durbin, E.G., Gentleman, W.C., et al., 2001. Evaluating the synopicity of the US GLOBEC Georges Bank broad-scale sampling pattern with observational system simulation experiment. Deep Sea Res. II 48, 483–499.

McGillicuddy, D., Robinson, A., Siegel, D., Jannasch, H., Johnson, R., Dickey, T., et al., 1998. Influence of mesoscale eddies on new production in the Sargasso sea. Nature 394, 263–266.

Megrey, B.A., Rose, K.A., Klumb, R.A., Hay, D.E., Werner, F.E., Eslinger, D.L., et al., 2007. A bioenergetics-based population model of Pacific herring (*Clupea harengus pallasi*) coupled to a lower trophic level nutrient–phytoplankton–zooplankton model: description, calibration, and sensitivity analysis. Ecol. Modell. 202, 144–164.

Meier, H.E.M., Andersson, H.C., Eilola, K., Gustafsson, B.G., Kuznetsov, I., Müller-Karulis, B., et al., 2011. Hypoxia in future climates: a model ensemble study for the Baltic Sea. Geophys. Res. Lett. 38(24), L24608.

Ménesguen, A., Cugier, P., Leblond, I., 2006. A new numerical technique for tracking chemical species in a multi-source, coastal ecosystem, applied to nitrogen causing Ulva blooms in the Bay of Brest (France). Limnol. Oceanogr. 51(1–2), 591–601.

Michaelis, L., Menten, M.I., 1913. Die Kinetik der Invirtinwirkung. Biochem. Z. 49, 333–369.

Miller, C.B., Lynch, D.R., Carlotti, F.C., Gentleman, W., Lewis, C.V.W., 1998. Coupling of an individual-based population model of Calanus finmarchicus to a circulation model of the Georges Bank region. Fish. Oceanogr. 7, 219–234.

Miller, C.B., Tande, K.S., 1993. Stage duration estimation for Calanus populations, a modelling study. Mar. Ecol. Prog. Ser. 102, 15–34.

Miller, C.B., Wheeler, P.A., 2012. Biological Oceanography, second ed. Wiley-Blackwell, Hoboken, NY.

Misund, O.A., Aglen, A., 1992. Swimming behaviour offish schools in the north sea during acoustic surveying and pelagic trawl sampling. ICES J. Mar. Sci. 49, 325–334.

Moll, A., Radach, G., 2003. Review of three-dimensional ecological modelling related to the North Sea shelf system, part 1: models and their results. Prog. Oceanogr. 57, 175–217.

Monod, J., 1949. The growth of bacterial cultures. Ann. Rev. Microbiol. 3, 371–394.

Murray, J.D., 1993. Mathematical Biology. Springer-Verlag, Berlin, Heidelberg, New York.

Nausch, G., Nehring, D., Nagel, K., 2008. Nutrient Concentrations, Trends and Their Relation to Eutrophication. In: Feistel, R., Nausch, G., Wasmund, N. (Eds.), State and Evolution of the Baltic Sea. Wiley & Sons, Inc., Hoboken, 337–366.

Nehring, D., 1981. Hydrographisch-chemische untersuchungen der ostsee von 1969–1978. Geod. Geoph. Veröff. R.4 35, 39–220.

Nehring, D., Matthäus, W., 1991. Current trends in hydrographic and chemical parameters and eutrophication in the Baltic Sea. Int. Rev. Gesamten Hydrobiol. 76, 276–316.

Neumann, T., 2000. Towards a 3D-ecosystem model of the Baltic Sea. J. Mar. Syst. 25(3–4), 405–419.

Neumann, T., 2007. The fate of river-borne nitrogen in the baltic sea – an example for the river oder. Estuar. Coast. Shelf Sci. 73(1Ǔ-2), 1–7.

Neumann, T., 2010. Climate-change effects on the baltic sea ecosystem: a model study. J. Mar. Syst. 81(3), 213–224.

Neumann, T., Fennel, W., 2006. A method to present seasonal vertical migration of zooplankton in 3D-Eulerian models. Ocean Modell. 12, 188–204.

Neumann, T., Fennel, W., Kremp, C., 2002. Experimental simulations with an ecosystem model of the Baltic Sea: a nutrient load reduction experiment. Global Biogeochem. Cycles 16, 7-1–7-19.

Neumann, T., Schernewski, G., 2008. Eutrophication in the Baltic Sea and shifts in the nitrogen fixation analyzed with a 3D ecosystem model. J. Mar. Syst. 74(1–2), 592–602.

Nielsen, P., 1994. Coastal Bottom Boundary Layer and Sediment Transport. Advanced Series on Ocean Engineering, vol. 4. World Scientific, Ithaca, NY.

Okubo, A., 1967. Some remarks on the importance of the shear effect on horizontal diffusion. J. Oceanogr. Soc. Jpn. 24, 60–69.

Okubo, A., 1978. Advection–diffusion in the presence of surface convergence. In: Bowman, M.J., Esaias, W.E. (Eds.), Oceanic Fronts in Coastal Processes. Springer-Verlag, Berlin, New York, pp. 23–28.

Oschlies, A., Garzon, V., 1998. Eddy-induced enhancement of primary production in a model of the north atlantic ocean. Nature 412, 638–641.

Oschlies, A., Schulz, K.G., Riebesell, U., Schmittner, A., 2008. Simulated 21st century's increase in oceanic suboxia by $CO_2$-enhanced biotic carbon export. Global Biogeochem. Cycles 22(4), GB4008.

Pacanowski, R.C., Dixon, K., Rosati, A., 1990. The GFDL modular ocean model users guide version 1.0. GFDL Technical Report No. 2., Geophysical Fluid Dynamic Laboratory, NOAA, Princeton University, Princeton.

Pacanowski, R.C., Griffies, S.M., 2000. MOM 3.0 Manual. Tech. Rep., Geophysical Fluid Dynamics Laboratory.

Pacanowski, R.C., Philander, S.G.H., 1981. Parameterization of vertical mixing in numerical models of tropical oceans. J. Phys. Oceanogr. 11, 1443–1451.

Parson, T.R., Takahashi, M., Hargrave, B., 1984. Biological Oceanographic Processes. Pergamon Press, Oxford, NY.

Puls, W., Doerfer, R., Sündermann, J., 1994. Numerical simulation and satellite observations of suspended matter in the North Sea. J. Oceanogr. Eng. 19(1), 3–9.

Radach, G., Moll, A., 1990. State of the art in algal bloom modelling. In: Lancelot, C., Billen, G., Barth, H. (Eds.), Water Pollution Report 12. EUR 12190 EN, Brussels, pp. 115–149.

Radtke, H., Neumann, T., Voss, M., Fennel, W., 2012. Modeling pathways of riverine nitrogen and phosphorus in the Baltic Sea. J. Geophys. Res. Oceans 117, C09024.

Radtke, H., Neumann, T., Fennel, W., 2013. A Eulerian nutrient to fish model of the Baltic Sea – a feasibility-study. J. Mar. Syst. 125, 61–76.

Redfield, A.C., Ketchum, B.H., Richards, B.H., 1963. The influence of organisms on the composotion of sea water. In: Hill, L. (Ed.), The Sea, vol. 2. Interscience, New York, pp. 26–77.

Riley, G.H., 1942. The relationship of vertical turbulence and spring diatom flowering. J. Mar. Res. 5, 67–87.

Riley, G.H., 1946. Factors controlling phytoplankton populations on the Georges Bank. J. Mar. Res. 6, 54–73.

Riley, G.H., 1947. A theoretical analysis of the zooplankton population of Geroges Bank. J. Mar. Res. 6, 104–113.

Rothschild, B.J., 1986. Dynamics of Marine Fish Populations. Harvard University Press, Cambridge, USA.

Rothschild, B.J., Osborn, T.R., 1988. Small-scale turbulence and plankton contact rates. J. Plankton Res. 10, 465–474.

Ruardij, P., van Raaphorst, W., 1995. Benthic nutrient regeneration in the ERSEM ecosystem model of the North Sea. Netherlands J. Sea Res. 33(3–4), 453–483.

Sarmiento, J.L., Slater, R.D., Fasham, M.J.R., Ducklow, H.W., Toggweiler, J.R., Evans, G.T., 1993. A seasonal three-dimensional ecosystem model of nitrogen cycling in the North Atlantic euphotic zone. Global Biogeochem. Cycles 7, 417–450.

Savchuk, O.P., 2002. Nutrient biogeochemical cycles in the Gulf of Riga: scaling up field studies with a mathematical model. J. Mar. Syst. 32, 253–280.

Scheffer, M., Baveco, J.M., DeAnglis, D.L., Rose, K.A., von Nes, E., 1995. Super-individuals a simple solution for modelling large populations on an individual basis. Ecol. Modell. 80, 161–170.

Schulz, S., Kaiser, W., Breuel, G., 1978. The annual course of some biological and chemical parameters at two stations in the Arkona and Bornholm Sea in 1975 and 1976. Kieler Meeresforsch. Sonderh. 4, 154–160.

Seifert, T., Kayser, B., 1995. A High Resolution Spherical Grid Topography of the Baltic Sea. Marine Science Report, No. 9. Institute for Baltic Sea Research, (ISSN 0939-396X), 72–88.

Shum, K.T., Sundby, B., 1996. Organic matter processing in continental shelf sediments – the subtidal pump revisited. Mar. Chem. 53, 81–87.

Siegel, H., Bruhn, R., Gerth, M., Nausch, G., Matthäus, W., Neumann, T., et al., 1998. The exceptional Oder flood in summer 1997 – distribution patterns of the Oder discharge in the Pomeranian Bight. Dtsch. Hydrograph. Z. 50(2–3), 145–167.

Six, K.D., Maier-Reimer, E., 1996. Effects of plankton dynamics on seasonal carbon fluxes in an ocean general circulation model. Global Biogeochem. Cycles 10, 559–583.

Sjøberg, B., Mork, M., 1985. Wind-induced stratified ocean response in the ice edge region: an analytical approach. J. Geophys. Res. 90, 7273–7285.

Sjöberg, S., 1980. A mathematical and conceptual framework for models of the pelagic ecosystem of the baltic sea, contributions from the Askö Laboratory, Univ. Stockholm, No.27.

Sjöberg, S., Willmot, W., 1977. System Analysis of a Spring Phytoplankton Bloom in the Baltic. Contributions from the Askö Laboratory, Univ. Stockholm, No. 20.

Slater, R.D., Sarmiento, J.L., R. Fasham, M.J., 1993. Some parametric and structural simulations with a three-dimensional ecosystem model of nitogen cycling in the north atlantic euphotic zone. In: Evans, G.T., Fasham, M.J.R. (Eds.), Towards a model of ocean biogeochemical processes. Springer, Berlin, pp. 261–294.

Smagorinsky, J., 1993. Some historical remarks on the use of nonlinear viscosities. In: Galperin, O. (Ed.), Large Eddy Simulation of Complex Engineering and Geophysical Flows. Cambridge University Press, Cambride, pp. 3–36 (Chapter 1).

Smetacek, V., Passow, U., 1990. Spring bloom initiation and Sverdrup's critical depth model. Limnol. Oceanogr. 35, 228–234.

Smetacek, V., von Bröckel, K., Zeitschel, B., Zenk, W., 1978. Sedimentation of particulate matter during a phytoplankton spring bloom in relation to the hydrographical regime. Mar. Biol. 47, 211–226.

Song, Y., Haidvogel, D., 1994. A semi-implicit ocean circulation model using a generalized topography – following coordinate system. J. Comput. Phys. 115, 228–244.

Spencer, J.W., 1971. Fourier series representation olf the position of the sun. Search 2, 172.

Steele, J., 1962. Environmental control of photosynthesis in the sea. Limnol. Oceanogr. 7, 137–150.

Steele, J.H., 1974. The Structure of Marine Ecosystems. Harvard University Press, Cambridge.

Stigebrandt, A., 2001. Physical oceanography of the Baltic Sea. In: Wulff, F., Rahm, L., Larson, P. (Eds.), Systems Analysis of the Baltic Sea. Springer-Verlag, Berlin, pp. 19–74.

Stigebrandt, A., Wulff, F., 1987. A model for the dynamics of nutrients and oxygen in the Baltic proper. J. Mar. Res. 45, 729–759.

Sverdrup, H., 1953. On the conditions for the vernal blooming of phytoplankton. J. Cons. Int. Explor. Mer. 18, 287–295.

Tett, P., 1987. The ecophysiology of exeptional blooms. Rapp. Proc. Verb. Reun. Cons. Perm. Int. Explor. Mer. 187, 47–60.

Totterdell, I.J., 1993. An annotated bibliography of marine biological models. In: Evans, G.T., Fasham, M.J.R. (Eds.), Towards a Model of Ocean Biogeochemical Processes. Springer-Verlag, Berlin, pp. 317–339.

Travers, M., Shin, Y.J., Jennings, S., Cury, P., 2007. Towards end-to-end models for investigating the effects of climate and fishing in marine ecosystems. Prog. Oceanogr. 75, 751–770.

Turnpenny, A.W.H., 1983. Swimming performance of juvenile sprat, *Sprattus sprattus* L., and herring, *Clupea harengus* L., at different salinities. J. Fish Biol. 23, 321–325.

Utne, K.R., Hjøllo, S.S., Huse, G., Skogen, M., 2012. Estimating the consumption of *Calanus finmarchicus* by planktivorous fish in the Norwegian Sea using a fully coupled 3D model system. Mar. Biol. Res. 8, 527–547.

Vinogradov, M.E., Menshutkin, V.V., Shushkina, E.A., 1972. On the mathematical simulation of a pelagic ecosystem in tropical waters of the ocean. Mar. Biol. 16, 261–268.

Voss, M., Emeis, K.-C., Hille, S., Neumann, T., Dippner, J.W., 2005. Nitrogen cycle of the Baltic Sea from an isotopic perspective. Global Biogeochem. Cycles 19(3), GB3001.

Wasmund, N., 1997. Occurrence of cyanobacterial blooms in the Baltic Sea in relation to environmental conditions. Int. Rev. Gesamten Hydrobiol. 82(2), 169–184.

Wolff, U., Strass, V., 1993. Seasonal and meridional variability of the remotely sensed fraction of euphotic zone chlorophyll predicted by a Lagrangian plankton mode. In: Barale, V., Schlittenhardt, P. (Eds.), Ocean Colour: Theory and Applications in a Decade of CZCS Expierence. Kluwer Academic Publishers, Doordrecht, Boston, London, pp. 319–329.

Woods, J., 2002. Primitive equation models of plankton ecology. In: Pinardi, N., Woods, J. (Eds.), Ocean Forcasting. Springer-Verlag, Berlin, Heidelberg, New York, pp. 377–465.

Woods, J., Barkmann, W., 1994. Simulating plankton ecosystems by the Lagrangian Ensemble method. Philos. Trans. R. Soc. B 343, 27–31.

Wroblewski, J.S., 1982. Interaction of currents and vertical migration in maintaining *Calanus marshallae* in the Oregon upwelling zone – a simulation. Deep Sea Res. 29, 665–686.

Wroblewski, J.S., O'Brien, J.J., 1976. A spatial model of phytoplankton patchiness. Mar. Biol. 35, 161–175.

Wroblewski, J.S., Sarmiento, J.L., Flierl, G.R., 1988. An ocean basin scale model of plankton dynamics in the North Atlantic. 1. Solutions for the climatological conditions in May. Global Biogeochem. Cycles 2, 199–218.

Wulff, F., Stigebrandt, A., 1989. A time-dependent budget model of the Baltic Sea. Global Biogeochem. Cycles 3, 63–78.

# Index

# Color Plates

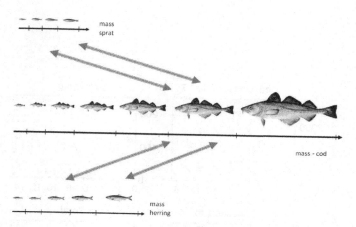

**FIGURE 4.21** Illustration of the mass-class structured model, showing the different mass axes for the different fishes. The arrows indicate the size-class structured predator–prey interaction.

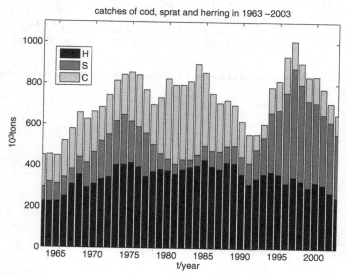

**FIGURE 4.22** Annual variations of the catches of sprat, herring and cod (data source: ICES).

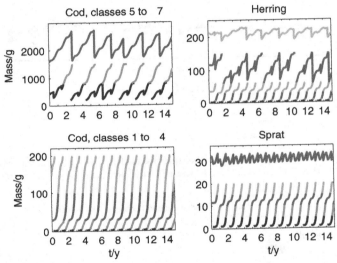

**FIGURE 4.23** The propagation of the mean individual masses through the mass classes of cod, herring and sprat.

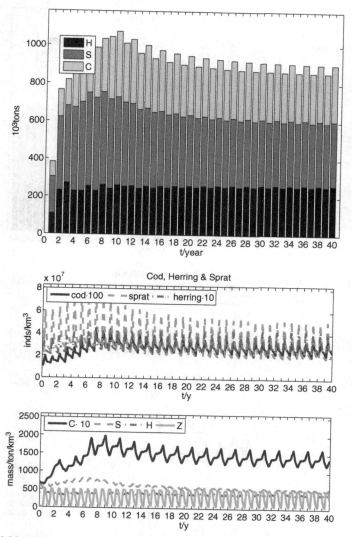

**FIGURE 4.24** Variation of catches and state variables: total abundance and biomass of sprat, herring, cod and biomass of zooplankton for constant rates (fishing mortality, reproduction and loads). (a) Catches for scenario 1 and (b) state variables for scenario 1

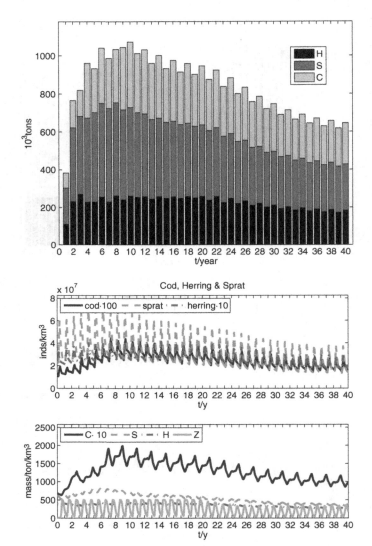

**FIGURE 4.25** Variation of catches and state variables in response to reduced nutrient discharge after 18 years: total abundance and biomass of sprat, herring, cod and biomass of zooplankton. (a) Catches for scenario 2 and (b) state variables for scenario 2.

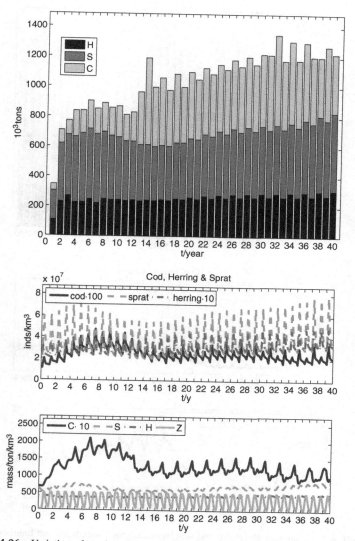

**FIGURE 4.26** Variation of catches and state variables: total abundance and biomass of sprat, herring and cod and the biomass of zooplankton for fishery scenario 3, beginning with low and then increasing fishing mortality. (a) Catches for scenario 3 and (b) state variables for scenario 3.

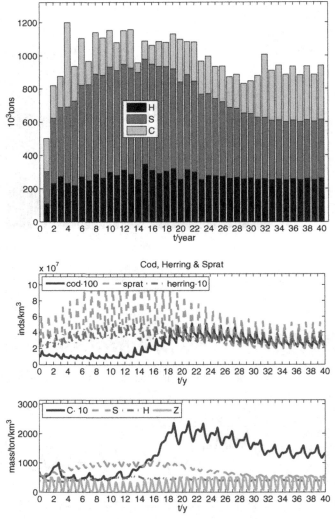

**FIGURE 4.27** Variation of catches and state variables: total abundance and biomass of sprat, herring and cod and the biomass of zooplankton for fishery scenario 4, beginning with high and then decreasing fishing mortality. (a) Catches for fishery scenario 4 and (b) State variables for fishery scenario 4.

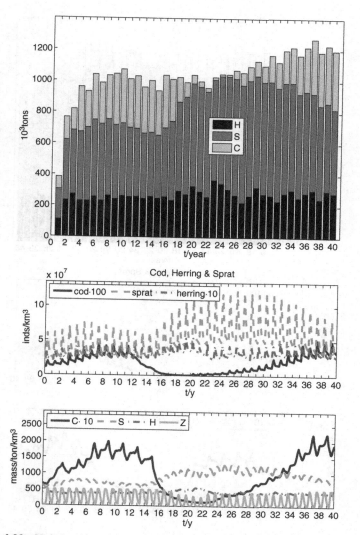

**FIGURE 4.28** Variation of catches and state variables: total abundance and biomass of sprat, herring and cod and the biomass of zooplankton for a low-reproduction scenario (scenario 5); in the time span from 12 to 20 years, there is no cod reproduction. (a) Catches for scenario 5 and (b) State variables for scenario 5.

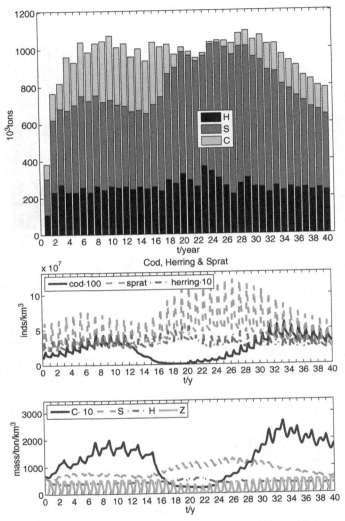

**FIGURE 4.29** Variation of catches and state variables: total abundance and biomass of sprat, herring and cod and the biomass of zooplankton for a mixed reproduction and fishery scenario (scenario 6); there is a reproduction gap from year 12 to year 20, but cod fishing is moderated after year 18.

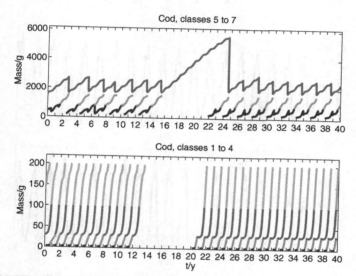

**FIGURE 4.30** The propagation of the mean individual masses through the mass classes of cod for scenario 6. Due to the period of no reproduction, the lower mass classes no longer exist, while the mass accumulates in the largest class.

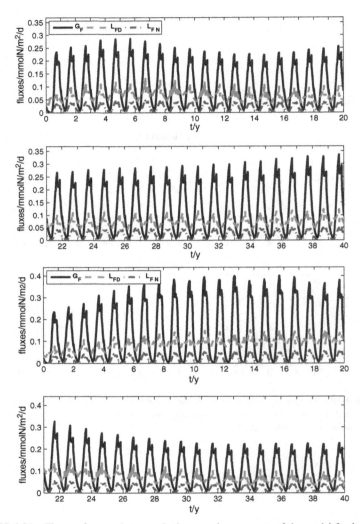

**FIGURE 4.31** Fluxes of matter between the lower and upper parts of the model food web for scenarios 3 (upper panels) and 4 (lower panels).

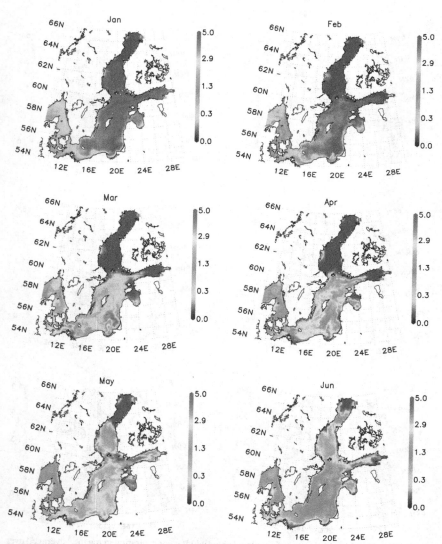

**FIGURE 6.10** Monthly means of simulated chlorophyll *a* concentration in mg m$^{-3}$ averaged over the upper 10 m for January to June of the model year 1984.

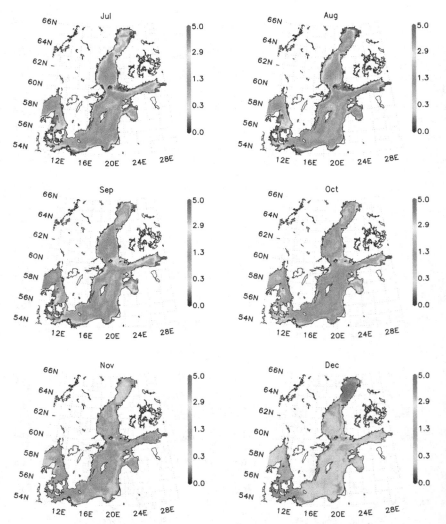

**FIGURE 6.11** Monthly means of simulated chlorophyll *a* concentration in mg m$^{-3}$ averaged over the upper 10 m for July to December of the model year 1984.

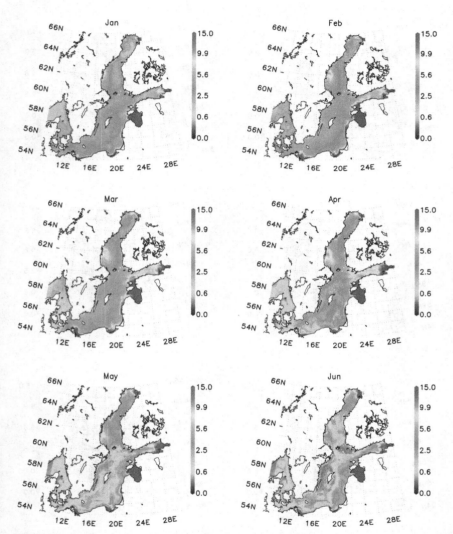

**FIGURE 6.12** Monthly means of simulated DIN (ammonium and nitrate) concentration in mmol m$^{-3}$ averaged over the upper 10 m for January to June of the model year 1984.

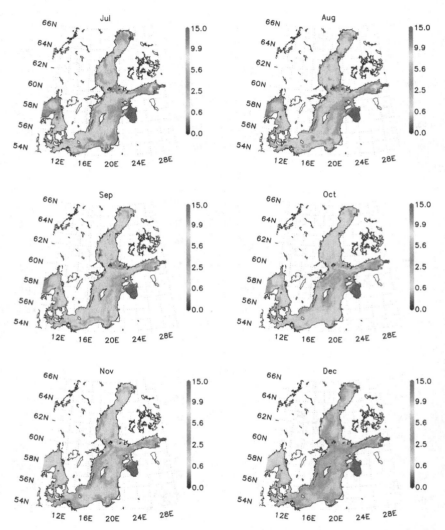

**FIGURE 6.13** Monthly means of simulated DIN (ammonium and nitrate) concentration in mmol m$^{-3}$ averaged over the upper 10 m for July to December of the model year 1984.

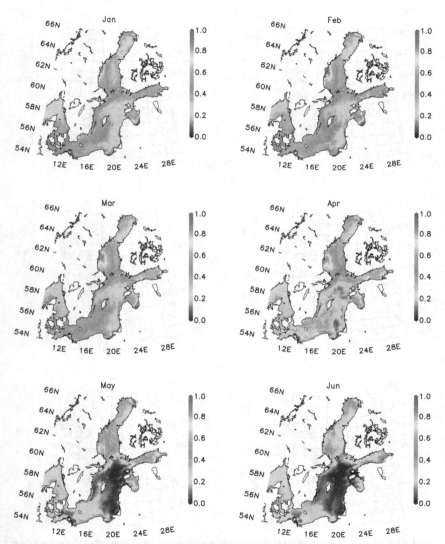

**FIGURE 6.14** Monthly means of simulated phosphate concentration in mmol m$^{-3}$ averaged over the upper 10 m for January to June of the model year 1984.

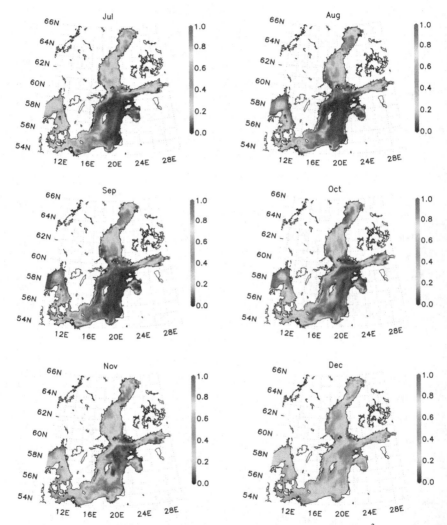

**FIGURE 6.15** Monthly means of simulated phosphate concentration in mmol m$^{-3}$ averaged over the upper 10 m for July to December of the model year 1984.

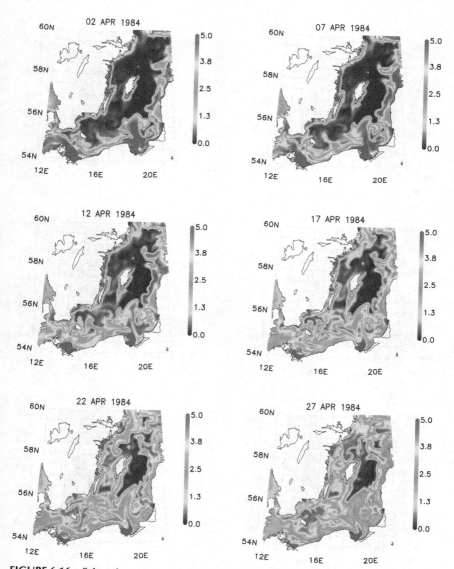

**FIGURE 6.16** Selected snapshots of 5-day means of the simulated model chlorophyll in mg m$^{-3}$ averaged over the upper 10 m.

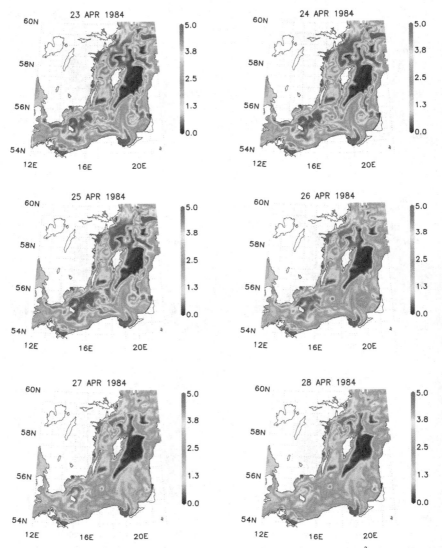

**FIGURE 6.17** Daily snapshots of surface chlorophyll *a* concentration in mg m$^{-3}$ averaged over the upper 10 m.

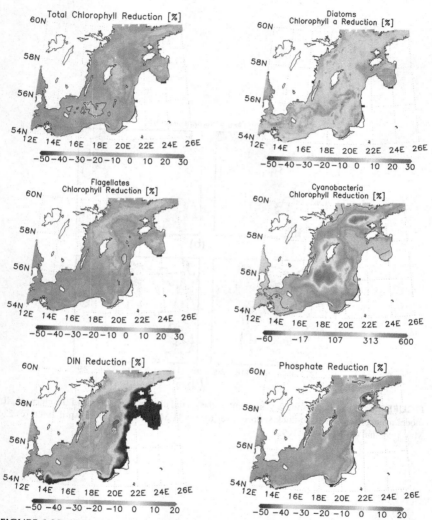

**FIGURE 6.33** Reduction effect on selected state variables after 10 years of integration in terms of annual means for the year 1990 (redrawn from Neumann et al., 2002).

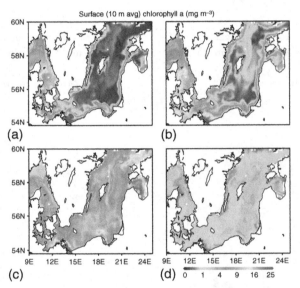

**FIGURE 7.3** Snapshots of the development of the spatial distribution patterns of the chlorophyll model averaged over a 10-m-thick surface layer for (a) 31 March 1999, (b) 22 April 1999, (c) 22 May 1999 and (d) 10 August 1999.

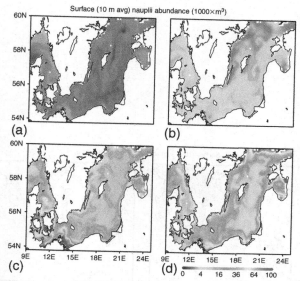

**FIGURE 7.4** Snapshots of the development of the spatial distribution patterns of the nauplii model averaged over a 10-m-thick surface layer for (a) 22 May 1999, (b) 21 June 1999, (c) 21 July 1999 and (d) 22 August 1999.

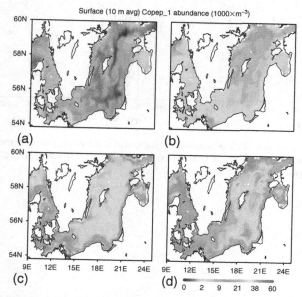

**FIGURE 7.5** Snapshots of the development of the spatial distribution patterns of the copepodites 1 model averaged over a 10-m-thick surface layer for (a) 22 May 1999, (b) 21 June 1999, (c) 21 July 1999 and (d) 22 August 1999.

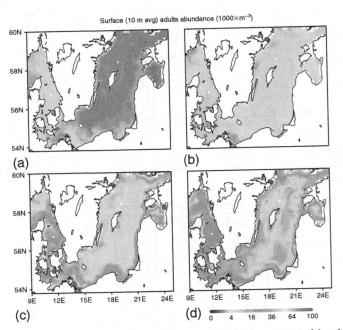

**FIGURE 7.6** Snapshots of the development of the spatial distribution patterns of the adults model averaged over a 10-m-thick surface layer for (a) 22 May 1999, (b) 21 June 1999, (c) 21 July 1999 and (d) 22 August 1999.

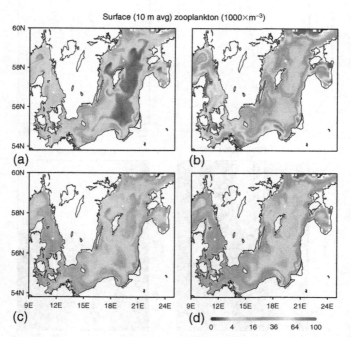

**FIGURE 7.7** Snapshots of the distribution patterns of (a) eggs, (b) nauplii, (c) copepodites 1 and (d) adults for 10 August.

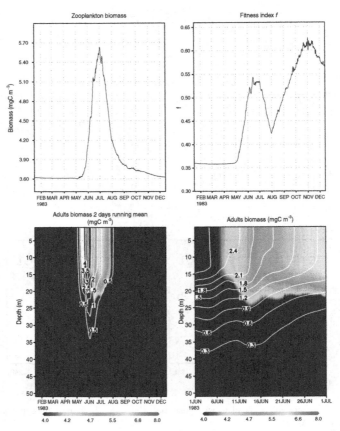

**FIGURE 7.9** Reference case without migration. Upper panel: Vertically averaged total zooplankton biomass concentration (left). Nutritional state (fitness) index for the mean individual adult (right). Lower panel: Variation of the vertical distribution of the adult biomass concentration: (left) annual; (right) during June. The variation of chlorophyll concentration is also indicated by isolines.

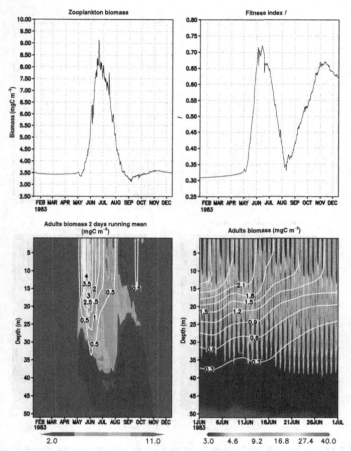

**FIGURE 7.10** As Fig. 7.9, but for experiment case 1 with vertical migration driven only by light-dependent avoidance behaviour.

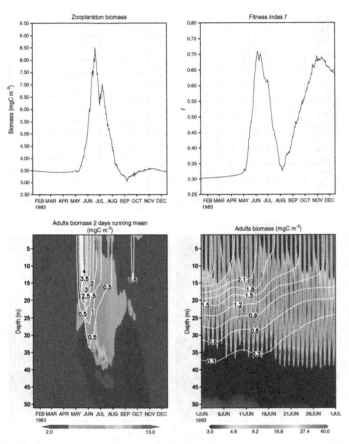

**FIGURE 7.11** As Fig. 7.9, but for experiment case 2 with vertical migration driven by both light-dependent avoidance behaviour and hunger.

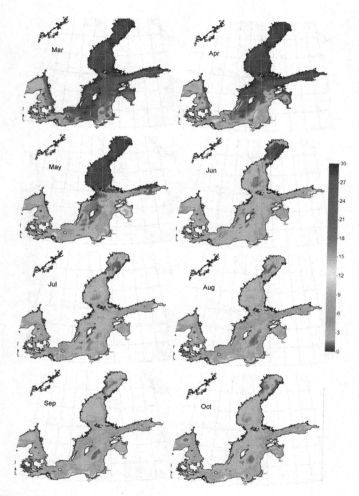

**FIGURE 7.14**  Distribution of the vertically integrated zooplankton model, averaged over the first 10 days of each month, from March to October in 1980 (t km$^{-2}$), after Radtke et al. (2013).

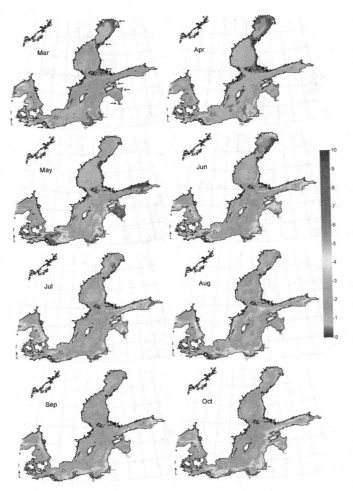

**FIGURE 7.15** Distribution of the total herring biomass model, averaged over the first 10 days of each month, from March to October in 1980 (t km$^{-2}$), after Radtke et al. (2013).

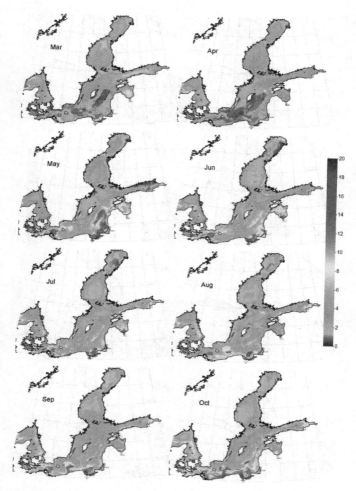

**FIGURE 7.16** Distribution of the total sprat biomass model, averaged over the first 10 days of each month, from March to October in 1980 (t km$^{-2}$), after Radtke et al. (2013).

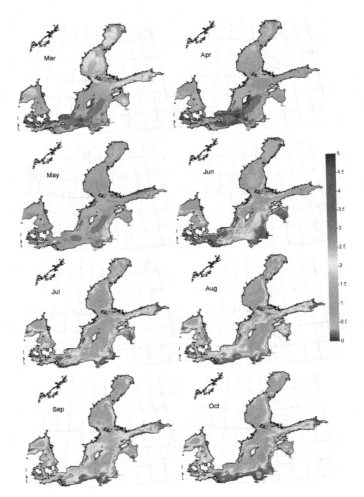

**FIGURE 7.17** Distribution of the total cod biomass model, averaged over the first 10 days of each month, from March to October in 1980 (t km$^{-2}$), after Radtke et al. (2013).

Printed in the United States
By Bookmasters